Algebraic Invariants
of Links

2nd Edition

K&E Series on Knots and Everything — Vol. 52

Algebraic Invariants
of Links

2nd Edition

Jonathan Hillman

The University of Sydney, Australia

 **World Scientific**

NEW JERSEY · LONDON · SINGAPORE · BEIJING · SHANGHAI · HONG KONG · TAIPEI · CHENNAI

Published by

World Scientific Publishing Co. Pte. Ltd.
5 Toh Tuck Link, Singapore 596224
USA office: 27 Warren Street, Suite 401-402, Hackensack, NJ 07601
UK office: 57 Shelton Street, Covent Garden, London WC2H 9HE

Library of Congress Cataloging-in-Publication Data
Hillman, Jonathan A. (Jonathan Arthur), 1947–
    Algebraic invariants of links / by Jonathan Hillman. -- 2nd ed.
    p. cm. -- (Series on knots and everything ; v. 52)
    Includes bibliographical references and index.
    ISBN-13: 978-981-4407-38-0 (hardcover : alk. paper)
    ISBN-10: 981-4407-38-0 (hardcover : alk. paper)
    1. Link theory. 2. Invariants. 3. Abelian groups. I. Title.
    QA612.2.H552 2012
    514'.224--dc23

                           2012012248

British Library Cataloguing-in-Publication Data
A catalogue record for this book is available from the British Library.

Printed in Singapore.

Contents

v

Preface

This book is intended as an introduction to links and a reference for the invariants of abelian coverings of link exteriors, and to outline more recent work, particularly that related to free coverings, nilpotent quotients and concordance. Knot theory has been well served with a variety of texts at various levels, but essential features of the multicomponent case such as link homotopy, I-equivalence, the fact that not all links are boundary links, longitudes, the role of the lower central series as a source of invariants and the homological complexity of the many-variable Laurent polynomial rings are all generally overlooked. Moreover, it has become apparent that for the study of concordance and link homotopy it is more convenient to work with disc links; the distinction is imperceptible in the knot theoretic case.

Invariants of these types play an essential role in the study of such difficult and important problems as the concordance classification of classical knots and the questions of link concordance arising from the Casson-Freedman analysis of topological surgery problems, and particularly in the applications of knot theory to other areas of topology. For instance, the extension of the Disc Embedding Lemma to groups of subexponential growth by Freedman and Teichner derived from computations using link homotopy and the lower central series. Milnor's interpretation of the multivariable Alexander polynomial as a Reidemeister-Franz torsion was refined by Turaev, to give "sign-determined" torsions and Alexander polynomials. These were used by Lescop to extend the Casson invariant to all closed orientable 3-manifolds, and by Meng and Taubes to identify the Seiberg-Witten invariant for 3-manifolds. The multivariable Alexander polynomial

also arises in McMullen's lower bound for the Thurston norm of a
torally bounded 3-manifold.

Links and the main equivalence relations relating them are defined in Chapter 1. (In particular, we include a proof of Giffen's
theorem relating F-isotopy and I-equivalence via shift-spinning.) In
Chapter 2 we review homology and cohomology with local coefficients, and Poincaré duality for covering spaces. The most useful
manifestations of duality are the Blanchfield pairings for abelian
coverings (considered in this chapter) and for free coverings of homology boundary links (considered in Chapter 9). Most of Chapter
3 is on the determinantal invariants of modules over a commutative noetherian ring (including the Reidemeister-Franz torsion for
chain complexes), but it also considers some special features of low-
dimensional rings and Witt groups of hermitean pairings on torsion
modules. These results are applied to the homology of abelian covers of link exteriors in the following five chapters. Chapter 4 is on
the maximal abelian cover. Some results well-known for knots are
extended to the many component case, and the connections between
various properties of boundary links are examined. Relations with
the invariants of sublinks, the total linking number cover, fibred links
and finite abelian branched covers are considered in Chapter 5.

In the middle of the book (Chapters 6-8) the above ideas are
applied in some special cases. Chapters 6 and 7 consider in more detail invariants of knots and of 2-component links, respectively. Here
there are some simplifications, both in the algebra and the topology.
In particular, surgery is used to describe the Blanchfield pairing of
a classical knot (in Chapter 6) and to give Bailey's theorem on presentation matrices of the modules of 2-component links (in Chapter
7). Symmetries of links and link types, as reflected in the Alexander
invariants, are studied in Chapter 8.

The later chapters (9-12) describe some invariants of nonabelian
coverings and their application to questions of concordance and link
homotopy. The links of greatest interest here are those concordant
to sublinks of homology boundary links ($cSHB$ links). The exteriors

of homology boundary links have covers with nontrivial free covering group. As free groups have cohomological dimension 1, the ideas
used in studying knot modules extend readily to the homology modules and duality pairings of such covers. This is done in Chapter 9,
which may be considered as an introduction to the work of Sato, Du
Val and Farber on high dimensional boundary links. We also give a
new proof of Gutiérrez' unlinking theorem for n-links, which holds
for all $n \geq 3$ and extends, modulo s-cobordism, to the case $n = 2$.

Although $cSHB$ links do not always have such free covers, their
groups have nilpotent quotients isomorphic to those of a free group.
More generally, the quotients of a link group by the terms of its
lower central series are concordance invariants of the link. (The
only other such invariants known are the Witt classes of duality
pairings on covering spaces.) Chapter 10 considers the connections
between the nilpotent quotients, Lie algebra, cohomology algebra
and minimal model of a group and more particularly the relations
between Massey products and Milnor invariants for a link group.
Although we establish the basic properties of the Milnor invariants
here, we refer to Cochran's book for further details on geometric
interpretations, computation and construction of examples.

The final two chapters are intended as an introduction to the
work of Levine (on algebraic closure and completions), Le Dimet
(on high dimensional disc links) and Habegger and Lin (on string
links). As this work is still evolving, and the directions of further
development may depend on the outcome of unproven conjectures,
some arguments in these chapters are only sketched, if given at all.
One of the difficulties in constructing invariants for links from the
duality pairings of covering spaces is that, in contrast to the knot
theoretic case, link groups do not in general share a common quotient with reasonable homological properties. The groups of all μ-
component 1-links with all Milnor invariants 0 and the groups of all
μ-component n-links for any $n \geq 2$ share the same tower of nilpotent
quotients. The projective limit of this tower is the nilpotent completion of the free group on μ generators, and is uncountable. This is

related to other notions of completion in Chapter 11. Another problem is that the set of concordance classes of links does not have a natural group structure. However "stacking" with respect to the last coordinate endows the set of concordance classes of n-disc links with such a structure. Chapter 12 considers disc links and their relation to spherical links.

The emphasis is on establishing algebraic invariants and their properties, and constructions for realizing such invariants have been omitted, for the most part. The reader is assumed to know some algebraic and geometric topology, and some commutative algebra (to the level of a first graduate course in each). We occasionally use spectral sequence arguments. Commutative and homological algebra are used systematically, and we avoid as far as possible accidental features, such as the existence of Wirtinger presentations. While the primary focus is on links in S^3, links in other homology spheres and higher dimensions and disc links in discs are also considered.

I would like to thank M.Morishita, D.S.Silver and V.G.Turaev for their detailed comments on earlier drafts of this book. The text was prepared using the AMS-LaTeX generic monograph package.

Jonathan Hillman

Added in February 2012. The topic considered in this book that has expanded most rapidly in the past decade is that of *Twisted Polynomial Invariants.* The final section of the former Chapter 5 has become a new Chapter 6, on this topic. In addition, Chapter 2 has been rewritten, and there is a new Chapter 10, on *Singularities of Plane Curves.* Material has been added to the chapters on *Knot Modules* and on *Nilpotent Quotients.* The errors noticed to date have been corrected, and equations have been displayed more often.

I would like to thank C.Livingston and S.Naik, and D.Silver and S.Williams for our recent collaborations, and J.C.Cha, T.D.Cochran and S.A.Melikhov for their observations.

Part 1

Abelian Covers

CHAPTER 1

Links

In this chapter we shall define knots and links and the standard equivalence relations used in classifying them. We shall also outline the most important geometric aspects. The later chapters shall concentrate largely on the algebraic invariants of covering spaces.

1.1. Basic notions

The standard orientation of $\mathbb{R}^n$ induces an orientation on the unit n-disc $D^n = \{(x_1, \ldots, x_n) \in \mathbb{R}^n \mid \Sigma x_i^2 \leq 1\}$ and hence on its boundary $S^{n-1} = \partial D^n$, by the convention "outward normal first". We shall assume that standard discs and spheres have such orientations. Qualifications shall usually be omitted when there is no risk of ambiguity. In particular, we shall often abbreviate $X(K)$, $M(K)$ and πK (defined below) as X, M and π, respectively. If μ is a positive integer and Y is a topological space $\mu Y = Y \times \{1, \ldots, \mu\}$, the disjoint union of μ copies of Y.

All manifolds and maps between them shall be assumed PL unless otherwise stated. The main exceptions arise when considering 4-dimensional issues.

A μ-*component* n-*link* is an embedding $L : \mu S^n \to S^{n+2}$ which extends to an embedding j of $\mu S^n \times D^2$ onto a closed neighbourhood N of L, such that $j(\mu S^n \times \{0\}) = L$ and ∂N is bicollared in S^{n+2}. (We may also use the terms *classical link* when $n = 1$, *higher dimensional link* when $n \geq 2$ and *high dimensional link* when $n \geq 3$.) With this definition and the above conventions on orientations, each link is oriented. It is determined up to (ambient) isotopy by its image $L(\mu S^n)$, considered as an oriented codimension 2 submanifold of S^{n+2}, and so we may let L also denote this submanifold. The ith

3

component of L is the *n-knot* (1-component *n*-link) $L_i = L|_{S^n \times \{i\}}$.
Most of our arguments extend to links in homology spheres.

Links are locally flat by definition. (However, PL embeddings of
higher dimensional manifolds in codimension 2 need not be locally
flat. The typical singularity is the cone over an $(n-1)$-knot; there
are no nontrivial 0-knots.) We may assume that the embedding
j of the product neighbourhood is orientation preserving, and it
is then unique up to isotopy *rel* $\mu S^n \times \{0\}$. The *exterior* of L is
the compact $(n+2)$-manifold $X(L) = S^{n+2} \setminus int\, N$ with boundary
$\partial X(L) \cong \mu S^n \times S^1$, and is well defined up to homeomorphism. It
inherits an orientation from S^{n+2}. Let $M(L) = X(L) \cup \mu D^{n+1} \times S^1$ be
the closed manifold obtained by surgery on L in S^{n+2}, with framing
0 on each component if $n = 1$. (Since $\pi_n(O(2)) = 0$ if $n > 1$, the
framing is then essentially unique.)

The *link group* is $\pi L = \pi_1(X(L))$. A *meridianal curve* for the
i^{th} component of L is an oriented curve in $\partial X(L_i) \subseteq \partial X(L)$ which
bounds a 2-disc in $S^{n+2} \setminus X(L_i)$ having algebraic intersection $+1$ with
L_i. The image of such a curve in πL is well defined up to conjugation,
and any element of πL in this conjugacy class is called an ith *merid-*
ian. A *basing* for a link L is a homomorphism $f : F(\mu) \to \pi L$ determined by a choice of one meridian for each component of L. The ho-
mology classes of the meridians form a basis for $H_1(X(L); \mathbb{Z}) \cong \mathbb{Z}^\mu$,
while $H_{n+1}(X(L); \mathbb{Z}) \cong \mathbb{Z}^{\mu-1}$ and $H_q(X(L); \mathbb{Z}) = 0$ for $1 < q < n+1$,
by Alexander duality.

A *Seifert hypersurface* for L is a locally flat, oriented codimension
1 submanifold V of S^{n+2} with (oriented) boundary L. By a standard
argument these always exist. (Using obstruction theory it may be
shown that the projection of $\partial X \cong \mu S^n \times S^1$ onto S^1 extends to
a map $q : X \to S^1$ [**Ke65**]. By transversality (TOP if $n = 2$!)
we may assume that $q^{-1}(1)$ is a bicollared, proper codimension 1
submanifold of X. The union $q^{-1}(1) \cup j(S^n \times [0,1])$ is then a Seifert
hypersurface for L.) In general there is no canonical choice of Seifert
surface. However, there is one important special case. A link L is
fibred if there is such a map $q : X \to S^1$ which is the projection
of a fibre bundle. The exterior is then the mapping torus of a self

homeomorphism θ of the fibre F of q. The isotopy class of θ is called the geometric monodromy of the bundle. Such a map q extends to a fibre bundle projection $\hat{q} : M(L) \to S^1$, with fibre $\widehat{F} = F \cup \mu D^{n+1}$, called the *closed fibre* of L. Higher dimensional links with more than one component are never fibred. (See Theorem 5.12.)

An n-link L is *trivial* if it bounds a collection of μ disjoint locally flat 2-discs in S^n. It is *split* if it is isotopic to one which is the union of nonempty sublinks L_1 and L_2 whose images lie in disjoint discs in S^{n+2}, in which case we write $L = L_1 \amalg L_2$, and it is a *boundary* link if it bounds a collection of μ disjoint orientable hypersurfaces in S^{n+2}. Clearly a trivial link is split, and a split link is a boundary link; neither implication can be reversed if $\mu > 1$. Knots are boundary links, and many arguments about knots that depend on Seifert hypersurfaces extend readily to boundary links.

1.2. The link group

If m_i is a meridian for L_i, represented by a simple closed curve on ∂X then $X \cup_{\{m_i\}} \bigcup D^2$ is a deformation retract of $S^{n+2} \setminus \mu\{*\}$ and so is 1-connected. (This is the only point at which we need the ambient homology sphere to be 1-connected.) Hence $\pi = \pi L$ is the normal closure of a set of meridians. (If S is a subset of a group G the *normal closure* $\langle\langle S \rangle\rangle_G$, or just $\langle\langle S \rangle\rangle$, is the smallest normal subgroup of G containing S, and G has *weight* m if $G = \langle\langle S \rangle\rangle_G$ for some subset S with m elements.) By Hopf's theorem, $H_2(\pi; \mathbb{Z})$ is the cokernel of the Hurewicz homomorphism from $\pi_2(X)$ to $H_2(X; \mathbb{Z})$.

If π is the group of a μ-component n-link L in S^{n+2} then

(1) π *finitely presentable;*

(2) π *is of weight μ;*

(3) $H_1(\pi; \mathbb{Z}) = \pi/\pi' \cong \mathbb{Z}^\mu$; and

(4) *(if $n > 1$)* $H_2(\pi; \mathbb{Z}) = 0$.

Conversely, any group satisfying these conditions is the group of an n-link, for every $n \geq 3$. If (4) is replaced by the stronger condition that π has deficiency μ then π is the group of a 2-link, but this stronger condition is not necessary [**Ke65'**]. If subcomplexes of aspherical 2-complexes are aspherical then a higher-dimensional link group π has geometric dimension at most 2 if and only $\operatorname{def}(\pi) = \mu$ (in which case it is a 2-link group).

The group of a classical link has geometric dimension at most 2. Moreover it has a Wirtinger presentation of deficiency 1 and satisfies (1-3), but satisfies (4) if and only if the link splits completely as a union of knots in disjoint balls. This is related to the presence of longitudes, nontrivial elements which commute with meridians. By the Loop Theorem, every 1-link L has a connected Seifert surface whose fundamental group injects into πL. The image is a non-abelian free subgroup of πL unless the Seifert surface is a disc or an annulus. In fact the unknot and the *Hopf link Ho* (2_1^2 in the tables of [**Rol**]) are the only 1-links with solvable link group.

Let L be a μ-component 1-link. An ith *longitudinal curve* for L is a closed curve in $\partial X(L_i)$ which intersects an ith meridianal curve transversely in one point and which is null homologous in $X(L_i)$. The ith meridian and i^{th} longitude of L are the images of such curves in πL, and are well defined up to simultaneous conjugation. If $*$ is a basepoint for $X(L)$ then representatives for the conjugacy classes of the meridians and longitudes may be determined on choosing paths joining each component of $\partial X(L)$ to the basepoint. The *linking number* $\ell_{ij} = \operatorname{lk}(L_i, L_j)$ is the image of the ith longitude in $H_1(X(L_j); \mathbb{Z}) \cong \mathbb{Z}$; in particular, $\ell_{ii} = 0$. It is not hard to show that $\ell_{ij} = \ell_{ji}$.

When chosen as above, the ith longitude and i^{th} meridian commute, since they both come from $\pi_1(\partial X(L_i)) \cong \mathbb{Z}^2$. In classical knot theory ($\mu = 1$) the longitudes play no role in connection with abelian invariants, as they always lie in the second commutator subgroup $(\pi K)''$. In higher dimensions there is no analogue of the longitude in the link group; there are longitudinal n-spheres, but these represent classes in $\pi_n(X(L))$ and so are generally inaccessible to computation. Let $F(r)$ denote the free group on r letters.

THEOREM 1.1. *A 1-link L is trivial if and only if πL is free.*

PROOF. The condition is clearly necessary. If πL is free then the ith longitude and i^{th} meridian must lie in a common cyclic group, for each $1 \leq i \leq \mu$, since a free group has no non-cyclic abelian subgroups. On considering the images in $H_1(X(L_i); \mathbb{Z}) \cong \mathbb{Z}$ we conclude that the ith longitude must be null homotopic in $X(L)$. Hence using the Loop Theorem inductively we see that the longitudes bound disjoint discs in S^3.   □

In Chapter 11 we shall show that if $n \geq 3$ an n-link L is trivial if and only if πL is freely generated by meridians and the homotopy groups $\pi_j(X(L))$ are all 0, for $2 \leq j \leq [\frac{n+1}{2}]$. These conditions are also necessary when $n = 2$, and if moreover $\mu = 1$ then L is topologically unknotted, by TOP surgery, since $F(1) = \mathbb{Z}$ is "good" [**FQ**]. However, it is not yet known whether such a knot is (PL) trivial, nor whether these conditions characterize triviality of 2-links with $\mu > 1$. (We show instead that such a 2-link is s-concordant to a trivial link. See §5 below re s-concordance.) The condition on meridians cannot be dropped if $n > 1$ and $\mu > 1$ ([**Po71**] – see §7 of Chapter 8 below).

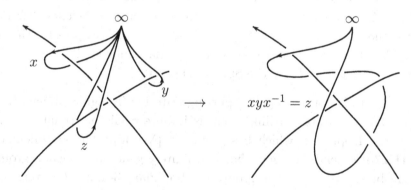

$$xyx^{-1} = z$$

Figure 1.

Any 1-link is ambient isotopic to a link L with image lying strictly above the hyperplane $\mathbb{R}^2 \times \{0\}$ in $\mathbb{R}^3 = S^3 \setminus \{\infty\}$ and for which composition with the projection to $\mathbb{R}^2$ is a local embedding with finitely many double points. Given such a link, the *Wirtinger presentation* is obtained as follows. For each component of the link minus the lower member of each double point pair assign a generator. (This corresponds to a loop coming down on a vertical line from ∞, going once around this component, and returning to ∞.) For the double point corresponding to the arc x crossing over the point separating arcs y and z, there is a relation $xyx^{-1} = z$, where the arcs are oriented as in Figure 1. Thus πL has a presentation

$$\langle x_{i,j} \mid u_{i,j} x_{i,j} u_{i,j}^{-1} = x_{i,j+1}, 1 \leq j \leq j(i), 1 \leq i \leq \mu \rangle,$$

where $u_{i,j} = x_{p,q}^{\pm 1}$ for some p, q and $x_{i,j(i)+1} = x_{i,1}$. It is not hard to see that one of these relations is redundant, and so πL has a presentation of deficiency 1. For an unsplittable link this is best possible.

THEOREM 1.2. *Let L be a 1-link. Then the following are equivalent:*

(1) *L is splittable;*
(2) *πL is a nontrivial free product;*
(3) *$\mathrm{def}(\pi L) > 1$.*

PROOF. Clearly (1) implies (2) and (2) implies (3). If C is the finite 2-complex determined by a presentation of deficiency ≥ 2 for πL then $\beta_2(\pi L) \leq \beta_2(C) \leq \mu - 2 < \beta_2(X(L)) = \mu - 1$. Hence $\pi_2(X(L)) \neq 0$ and so there is an essential embedded S^2 in $X(L)$, which must split L, by the Sphere Theorem. $\square$

There is not yet a good splitting criterion in higher dimensions.

The centre of a 1-link group is infinite cyclic or trivial, except for the Hopf link, which has group $\mathbb{Z}^2$ [Mu65]. The argument of [HK78] extends to show that any finitely generated abelian group can be the centre of the group of a boundary 3-link. However, the group of a 2-link with more than one component has no abelian normal subgroup of rank > 0. (See page 42 of [Hil]. In all known examples the centre is trivial.)

If G is a group and $x, y \in G$ let $[x, y] = xyx^{-1}y^{-1}$ be the commutator, and let $G' = [G, G]$ be the commutator subgroup. Define the lower central series $\{G_q\}_{q \geq 1}$ for G inductively by $G_1 = G$, $G_2 = G' = [G, G_1]$ and $G_{q+1} = [G, G_q]$. Let $G_\omega = \cap_{q \geq 1} G_q$. A group homomorphism $f : G \to H$ induces homomorphisms $f_q : G/G_q \to H/H_q$, for all $1 \leq q \leq \omega$. It is *homologically 2-connected* if $H_1(f; \mathbb{Z})$ is an isomorphism and $H_2(f; \mathbb{Z})$ is an epimorphism. These notions are related in the following result of Stallings.

THEOREM 1.3. **[St65]** *Let $f : G \to H$ be a homologically 2-connected group homomorphism. Then $f_q : G/G_q \to H/H_q$ is an isomorphism, for all $q \geq 1$. If f is an epimorphism then $f_\omega : G/G_\omega \to H/H_\omega$ is also an isomorphism.*

PROOF. The LHS spectral sequence for G as an extension of G/G_q by G_q gives an exact sequence

$$H_2(G; \mathbb{Z}) \to H_2(G/G_q; \mathbb{Z}) \to G_q/G_{q+1} \to 0,$$

for all $q \geq 1$. Since G/G_{q+1} is a central extension of G/G_q by G_q/G_{q+1} the result follows by the Five-Lemma and induction. $\square$

The group of a link L may be given a *pre-abelian* presentation

$$\langle x_i, y_{ij} \mid [v_{ij}, x_i]y_{ij}, [w_i, x_i], 2 \leq j \leq j(i), 1 \leq i \leq \mu \rangle,$$

where the words x_i and w_i represent i^{th} meridians and longitudes. The images of the x_is generate the nilpotent quotients; for links there is a more precise result due to Milnor.

THEOREM 1.4. **[Mi57]** *Let π be the group of a μ-component 1-link. Then π/π_q has a presentation*

$$\langle x_i, 1 \leq i \leq \mu \mid [w_{i,q}, x_i] = 1, \ 1 \leq i \leq \mu, \ F(\mu)_q \rangle,$$

where x_i and $w_{i,q}$ represent the images in π/π_q of the ith meridian and longitude, respectively. There are words $y_i \in F(\mu)$ such that $\Pi y_i[w_{i,q}, x_i]y_i^{-1} \in F(\mu)_q$.

PROOF. (From **[Tu76]**.) Fix a basepoint $* \in X$ and choose arcs α_i from $*$ to $\partial X(L_i)$ which meet only at $*$. Let N be a closed regular neighbourhood of $\cup \alpha_i$ in X and let $D_i = N \cap \partial X(L_i)$. Then $N \cong D^3$, $D_i \cong D^2$ and $\overline{\partial X(L_i) \setminus D_i}$ is a punctured torus. Let $W = \overline{X \setminus N}$ and $G = \pi_1(W, *)$. Since $H_1(W; \mathbb{Z}) \cong \mathbb{Z}^\mu$ and $H_2(W; \mathbb{Z}) = 0$ the inclusion of meridians induces isomorphisms from $F(\mu)/F(\mu)_q$ to G/G_q for all $q \geq 1$, by Theorem 1.3. Since $X = W \cup (\cup_{i=1}^{i=\mu} D_i) \cup N$ we see that $\pi \cong G/\langle\langle \partial D_i \mid 1 \leq i \leq \mu \rangle\rangle$. Clearly ∂D_i represents the commutator of curves in W whose images in π are an ith meridian-longitude pair.

The final assertion follows from the fact that $W \cap N = \overline{\partial N \setminus \partial X}$ is a punctured sphere with boundary $\cup_{i=1}^{i=\mu} \partial D_i$.   □

If L is the closure of a pure braid we may take $w_{i,q} = w_i$ for all q, for πL then has a presentation $\langle x_i, 1 \leq i \leq \mu \mid [w_i, x_i], 1 \leq i \leq \mu \rangle$. (See Theorem 2.2 of [**Bir**].)

1.3. Homology boundary links

Classical boundary links were characterized by Smythe, and his result was extended to higher dimensions by Gutiérrez.

THEOREM 1.5. [**Sm66, Gu72**] *A μ-component link L is a boundary link if and only if there is an epimorphism $f : \pi L \to F(\mu)$ which carries a set of meridians to a free basis.*

PROOF. Suppose that L has a set of disjoint Seifert hypersurfaces U_j, with disjoint product neighbourhoods $N_j \cong U_j \times (-1, 1)$ in X. Let $p : X \to \vee^\mu S^1$ be the map which sends $X \setminus \cup N_j$ to the basepoint and which sends $(n, t) \in N_j$ to $e^{\pi i(t+1)}$ in the jth copy of S^1, for $1 \leq j \leq \mu$. Then $f = \pi_1(p)$ sends a set of meridians for L to the standard basis of $\pi_1(\vee^\mu S^1) \cong F(\mu)$.

Conversely, such a homomorphism $f : \pi L \to F(\mu)$ may be realized by a map $F : X \to \vee^\mu S^1$, since $\vee^\mu S^1$ is aspherical. We may also assume that $F|_{\partial X}$ is standard, since f sends meridians to generators, and that F is transverse to $\mu e^{-\pi i}$, the set of midpoints of the circles. The inverse image $F^{-1}(\mu e^{-\pi i})$ is then a family of disjoint hypersurfaces spanning L.   □

An equivalent characterization that is particularly useful in questions of concordance and surgery is that a μ-component n-link L is a boundary link if and only if there is a degree 1 map of pairs from $(X(L), \partial X(L))$ to the exterior of the trivial link which restricts to a homeomorphism on the boundary [**CS80**]. A boundary n-link L is *simple* if there is such a degree 1 map which is $[\frac{n+1}{2}]$-connected. (Thus L is simple if πL is freely generated by meridians and $\pi_j(X(L)) = 0$ for $1 < j < [\frac{n+1}{2}]$, and so every such degree 1 map is $[\frac{n+1}{2}]$-connected.)

If the condition on meridians is dropped L is said to be a *homology boundary link*. Smythe showed also that a classical link L is a homology boundary link if and only if there are μ disjoint oriented codimension 1 submanifolds $U_i \subset X(L)$ with $\partial U_i \subset \partial X(L)$ and such that the image of ∂U_i in $H_n(\partial X(L); \mathbb{Z})$ is homologous to the image of the ith longitudinal n-sphere. This characterization extends to all higher dimensions. In Chapter 2 such *singular Seifert hypersurfaces* are used to construct covering spaces of $X(L)$.

If L is a homology boundary link the epimorphism from $\pi = \pi L$ to $F(\mu)$ satisfies the hypotheses of Stallings' Theorem, and so $\pi/\pi_q \cong F(\mu)/F(\mu)_q$ for all $q \geq 1$. Moreover $\pi/\pi_\omega \cong F(\mu)$, since free groups are residually nilpotent.

If L is a higher dimensional link $H_2(\pi L; \mathbb{Z}) = H_2(F(\mu); \mathbb{Z}) = 0$ and hence a basing f induces isomorphisms on all the nilpotent quotients $F(\mu)/F(\mu)_q \cong \pi L/(\pi L)_q$, and a monomorphism $F(\mu) \to \pi L/(\pi L)_\omega$, by Stallings' Theorem, since in any case $H_1(f; \mathbb{Z})$ is an isomorphism. (In particular, if $\mu \geq 2$ then πL contains a non-abelian free subgroup.) The latter map is an isomorphism if and only if L is a homology boundary link.

An *SHB link* is a sublink of a homology boundary link. Although sublinks of boundary links are clearly boundary links, *SHB* links need not be homology boundary links. (See Chapter 8 below.)

1.4. $Z/2Z$-boundary links

A μ-component n-link is a $Z/2Z$-*boundary link* if there is an embedding $P : U = \amalg_{i=1}^{i=\mu} U_i \to S^{n+2}$ of μ disjoint $(n+1)$-manifolds U_i such that $L = P|_{\partial U}$. (We do not require that the hypersurfaces are orientable.) The simplest nontrivial example is spanned by two simply linked Möbius bands. (See the link 9_{61}^2 of the tables of [**Rol**].)

THEOREM 1.6. *A link L is a $Z/2Z$-boundary link if and only if there is an epimorphism from πL to $*^\mu(Z/2Z)$ which carries some i^{th} meridian to the generator of the ith factor, for all $1 \leq i \leq \mu$.*

PROOF. Let L be a $Z/2Z$-boundary n-link with spanning surfaces U_i, and let ν_i be the normal bundle of U_i in X. Crushing

the complement of a disjoint family of open regular neighbourhoods
of the U_i to a point collapses X onto the wedge of Thom spaces
$\vee T(\nu_i)$. The bundles ν_i are induced from the canonical line bundle
η_N over $\mathbb{RP}^N$ (for N large) by classifying maps $n_i : U_i \to \mathbb{RP}^N$, and
these maps induce a map $T(n) : \vee T(\nu_i) \to \vee^\mu T(\eta_N)$. Now $T(\eta_N)$
is homeomorphic to $\mathbb{RP}^{N+1}$ by a homeomorphism carrying the zero
section to the hyperplane at infinity. Hence we obtain a map from X
to $\vee^\mu \mathbb{RP}^\infty = K(*^\mu(Z/2Z), 1)$, which determines a homomorphism
$f : \pi L \to *^\mu(Z/2Z)$. The map from X to $\vee^\mu \mathbb{RP}^{N+1}$ carries a loop
which meets U_i transversely in one point and is disjoint from U_j for
$j \neq i$ to the Thom space of the restriction of η_N over a point, in
other words to a curve which meets $\mathbb{RP}^N$ in one point. Thus this
curve is essential in $\mathbb{RP}^{N+1}$, and so in $\mathbb{RP}^\infty$. Hence the image of the
corresponding meridian generates the ith factor of $*^\mu(Z/2Z)$.

Conversely, such a homomorphism $f : \pi L \to *^\mu(Z/2Z)$ may be
realized by a map $F : X \to \vee^\mu \mathbb{RP}^\infty$. Since $X(L)$ has the homotopy
type of an $(n+1)$-dimensional complex, we may assume that F maps
X to $\vee^\mu \mathbb{RP}^{n+1}$. We may also assume that $F|_{\partial X}$ is standard, since f
sends meridians to generators, and that F is transverse to $\amalg^\mu \mathbb{RP}^n$,
the disjoint union of the hyperplanes at infinity. Then $F^{-1}(\amalg^\mu \mathbb{RP}^n)$
is a family of disjoint hypersurfaces spanning L. $\square$

The normal bundles for orientable hypersurfaces are trivial, and
the universal trivial line bundle $\mathbb{R}$ (with base space a point) has
Thom space $T(\mathbb{R}) = S^1 = K(\mathbb{Z}, 1)$. In the characterization of boundary links this plays the part which $T(\eta) = \mathbb{RP}^\infty$ plays here. Finite dimensional approximations $\mathbb{RP}^N$ have been used to emphasize
the distinction between the base space ($\mathbb{RP}^N$) and the Thom space
($\mathbb{RP}^{N+1}$) of the universal line bundle.

A similar application of transversality to high dimensional lens
spaces shows that L has μ disjoint spanning complexes, the ith being a $Z/p_i Z$-manifold with no singularities on the boundary, if and
only if there is an epimorphism from π to $*_{i=1}^{i=\mu}(Z/p_i Z)$ which carries
meridians to generators of the factors.

Smythe's characterization of homology boundary links suggests
several possible definitions for the unoriented analogue. The most

useful seems to be as follows. A link L is a $Z/2Z$-*homology boundary link* if and only if there are μ disjoint codimension 1 submanifolds $U_i \subset X(L)$ with $\partial U_i \subset \partial X(L)$ and such that the images of ∂U_i and the ith longitudinal n-sphere are homologous in $H_n(\partial X(L); \mathbb{F}_2)$. There is an analogous characterization, which we shall not prove.

THEOREM 1.7. *A link L is a $Z/2Z$-homology boundary link if and only if there is an epimorphism from πL to $*^\mu(Z/2Z)$ such that composition with abelianization carries some i^{th} meridian to the generator of the ith summand of $(Z/2Z)^\mu$, for all $1 \le i \le \mu$.* $\qquad\square$

1.5. Isotopy, concordance and I-equivalence

A *link type* is an ambient isotopy class of links. A locally flat isotopy is an ambient isotopy, but even an isotopy of 1-links need not be locally flat. For instance, any knot is isotopic to the unknot, but no such isotopy of a nontrivial knot can be ambient. However, a theorem of Rolfsen [**Ro72**] shows that the situation for links is no more complicated.

Two μ-component n-links L and L' are *locally isotopic* if there is an embedding $j : D^{n+2} \to S^{n+2}$ such that $D = L^{-1}(j(D^{n+2}))$ is an n-disc in one component of μS^n and $L|_{(\mu S^n)\setminus D} = L'|_{(\mu S^n)\setminus D}$.

THEOREM. [**Ro72**] *Two n-links L and L' are isotopic if and only if L' may be obtained from L by a finite sequence of local isotopies and an ambient isotopy.* $\qquad\square$

In other words, L and L' are isotopic if and only if L' may be obtained from L by successively suppressing or inserting small knots in one component at a time.

An I-*equivalence* between two embeddings $f, g : A \to B$ is an embedding $F : A \times [0,1] \to B \times [0,1]$ such that $F|_{A\times\{0\}} = f$, $F|_{A\times\{1\}} = g$ and $F^{-1}(B \times \{0,1\}) = A \times \{0,1\}$. Here we do *not* assume the embeddings are PL. Clearly isotopy implies I-equivalence. The next result is clear.

THEOREM 1.8. *Let $\mathcal{L}$ be an I-equivalence between μ-component n-links L and L'. Then the inclusions of $X(L)$ and $X(L')$ into $X(\mathcal{L})$ induce isomorphisms on homology.* $\qquad\square$

A *concordance* between two μ-component n-links L and L' is a locally flat PL I-equivalence $\mathcal{L}$ between L and L'. Let $C_n(\mu)$ denote the set of concordance classes of such links, and let $C_n = C_n(1)$. The concordance is an *s-concordance* if its exterior is an *s*-cobordism (*rel* ∂) from $X(L)$ to $X(L')$. In high dimensions this is equivalent to ambient isotopy, by the *s*-cobordism theorem, but this is not known when $n = 2$. (*s*-Concordant 1-links are isotopic, by standard 3-manifold topology.) A link L is *null concordant* (or *slice*) if it is concordant to a trivial link. Thus L is a slice link if and only if it extends to a locally flat embedding $C : \mu D^{n+1} \to D^{n+3}$ such that $C^{-1}(S^{n+2}) = \mu S^n$. It is an attractive conjecture that every even-dimensional link is a slice link. This has been verified under additional hypotheses on the link group. In particular, even-dimensional SHB links are slice links [**Co84, De81**].

A μ-component n-link L is *doubly null concordant* or *doubly slice* if there is a trivial μ-component $(n+1)$-link U which is transverse to the equatorial $S^{n+2} \subset S^{n+3}$ and such that U_i meets S^{n+2} in L_i, for $1 \le i \le \mu$. Doubly slice links are clearly boundary links, as they are spanned by the intersections of S^{n+2} with μ disjoint $(n+2)$-discs spanning U.

THEOREM. [**Ro85**] *Two n-links L and L' are PL I-equivalent if and only if L' may be obtained from L by a finite sequence of local isotopies and a concordance.* $\qquad\qquad\Box$

A concordance between boundary links L and L' is a *boundary concordance* if it extends to an embedding of disjoint orientable $(n+2)$-manifolds which meet $S^{n+2} \times \{0\}$ and $S^3 \times \{1\}$ transversely in systems of disjoint spanning surfaces for L and L', respectively. There is a parallel notion of $Z/2Z$-boundary concordance.

The process of replacing L_i by a knot K contained in a regular neighbourhood N of L_i (disjoint from the other components) such that K is homologous to L_i in N is called an *elementary F-isotopy* on the ith component of L. (The elementary F-isotopy is *strict* if the maximal abelian covering space of $N \setminus K$ is acyclic.) Two μ-component n-links L and L' are *(strictly) F-isotopic* if they may be related by a sequence of (strict) elementary F-isotopies.

Giffen found a beautiful elementary construction which related F-isotopy and I-equivalence. As his "shift-spinning" construction has never been published, we present it here.

THEOREM 1.9. [Gi76] F-isotopic 1-links are I-equivalent.

PROOF. Let K be a knot in the interior of $S^1 \times D^2$ which is homologous to the core $S^1 \times \{0\}$. Let Δ be a 2-disc properly embedded in $S^1 \times D^2$, with $\partial\Delta$ essential in $S^1 \times S^1$, and which is transverse to K. Assume that the number $w = |K \cap \Delta|$ is minimal. (This is the geometric winding number of K in $S^1 \times D^2$.) Suppose that $K \subset S^1 \times \rho D^2$, where $0 < \rho < 1$, and split $S^1 \times D^2$ along Δ to obtain a copy of $D^2 \times [0,1]$, with a 1-submanifold L.

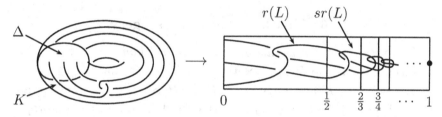

Figure 2.

Let $f : [0,1]^2 \to [\frac{1}{2}, 1]$ be a continuous function such that $f(x,t) = \frac{1}{2}$ if $0 \le x \le (1-t)\rho$ and $f(x,t) = 1$ if $(1 + (1-t)\rho)/2 \le x \le 1$. Let r and s be the self maps of $D^2 \times [0,1]$ given by $r(z,t) = (2^{-t}z, \frac{t}{2})$ and $s(z,t) = (f(|z|,t)z, 1/(2-t))$, for all $(z,t) \in D^2 \times [0,1]$, and let $\kappa = (\cup_{n \ge 0} s^n r(L)) \cup \{(0,1)\}$. Then κ is the union of finitely many arcs in $D^2 \times [0, \frac{1}{2})$ with a "periodic" Fox-Artin arc which tapers towards the core as the interval coordinate increases and converges to $(0,1)$, which is the one wild point, and $s(\kappa) \subset \kappa$. (See Figure 2.)

Now form the mapping torus of the pair $(s, s|_\kappa)$. The result is a wild annulus in $M(s) \cong S^1 \times D^2 \times [0,1]$ with boundary $K \amalg C$, where $C = S^1 \times \{(0,1)\}$ is the core of $S^1 \times D^2 \times \{1\}$. On embedding this solid torus appropriately in $S^3 \times [0,1]$, we obtain an I-equivalence from K to C. (See Figure 3.) This clearly implies the theorem. $\square$

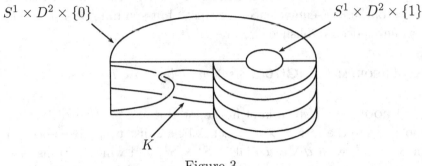

$S^1 \times D^2 \times \{0\}$ $\qquad\qquad\qquad\qquad\qquad$ $S^1 \times D^2 \times \{1\}$

K

Figure 3.

If K is as depicted in Figure 2 then $K \amalg \partial\Delta$ is I-equivalent to the Hopf link $C \amalg \partial\Delta$, but there is no PL I-equivalence - see §3 of Chapter 8.

1.6. Link homotopy and surgery

Two μ-component n-links L and L' are *link homotopic* if they are connected by a map $H : \mu S^n \times [0, 1] \to S^{n+2}$ such that $H|_{\mu S^n \times \{0\}} = L$, $H|_{\mu S^n \times \{1\}} = L'$ and $H(S^n \times \{(i, t)\}) \cap H(S^n \times \{(j, t)\}) = \emptyset$ for all $t \in [0, 1]$ and all $1 \le i < j \le \mu$. Thus a link homotopy is a homotopy of the maps L and L' such that at no time do the images of distinct spheres intersect (although self intersections are allowed). This is interesting only in the classical case as every higher dimensional link is link homotopic to a trivial link [**BT99, Ba01**]. (However, the link homotopy classification of higher-dimensional "link maps" is nontrivial. See also [**Kai**].)

THEOREM 1.10. [**Gi79, Go79**] *Concordant 1-links are link homotopic.*

PROOF. Let $\mathcal{L}$ be a concordance from L to L'. After an isotopy if necessary, we may assume that $\mathcal{L}$ has an embedded handle decomposition in which the levels at which the handles are added increase with the degree. Since the domain of $\mathcal{L}$ is a product, we may assume that the 0-handles cancel with the 1-handles added below level $\frac{1}{2}$ (and that none are added at this level). It can then be shown that the link at this level is link homotopic to L. Viewed from the other end, the duals of the remaining 1-handles cancel the duals of the 2-handles, and so this link is also link homotopic to L'. See [Ha92] for details. □

It is well known that a 1-knot K may be unknotted by "replacing certain of the undercrossings by overcrossings"; this idea is made precise and extended to links in the following lemma.

Let L be a 1-link and $D \subset S^3$ an oriented 2-disc which meets one component L_i transversely in two points, with opposite orientations, and is otherwise disjoint from L. Let N be an open regular neighbourhood of ∂D in $X(L)$, Then there is an orientation preserving homeomorphism $D^2 \times S^1 \cong S^3 \setminus N$. Fix such a homeomorphism f and define a self homeomorphism h of $S^3 \setminus N$ by $hf(z, s) = f(sz, s)$, for all $s \in S^1$ and $z \in D^2$. The links L and $h(L)$ are then said to be obtained from each other by an *elementary surgery*.

LEMMA 1.11. *Let L and L' be 1-links. Then the following are equivalent:*

(1) *L and L' are link homotopic;*
(2) *there is a sequence $L(0) = L, \ldots, L(n) = L'$ of links such that $L(i)$ is obtained from $L(i-1)$ by an elementary surgery, for all $1 \leq i \leq n$.*

PROOF. Up to isotopy, any link homotopy may be achieved by a sequence of elementary homotopies, involving the crossing of two arcs in a small ball B. Clearly such an elementary homotopy is equivalent to an elementary surgery. □

The correct choices of "twisting" homeomorphisms h are important here.

The disc D used in such an elementary surgery may be isotoped to avoid a finite set of disjoint discs, and so the surgeries of (2) can be performed simultaneously. Thus the conditions of the lemma imply

ADDENDUM. *If L and L' are link homotopic then there is an embedding T of $mS^1 \times D^2$ in $X(L)$ with core $T|_{mS^1 \times \{0\}}$ a trivial link and a self homeomorphism h of $X(T) = S^3 \setminus int\,T(mS^1 \times D^2)$ such that*

(1) $\text{lk}(T_i, L_j) = 0$, *for each* $1 \leq i \leq m$ *and each component* L_j
    of L;

(2) $h(T_i(z, s)) = T_i(s^{e(i)}z, s)$, *for some* $e(i) = \pm 1$ *and for all*
    $(z, s) \in \partial(S^1 \times D^2)$ *and* $1 \leq i \leq m$;

(3) $h \circ L = L'$.   □

We shall say that two links L and L' related by such surgeries are
surgery equivalent. The requirement that the core be trivial ensures
that the 3-manifold resulting from the surgeries is again S^3; the
linking number condition implies that the surgery tori lift to abelian
covers of L. Surgery equivalent links need not be link homotopic, as
the cores of the surgery tori may link distinct components of L.

1.7. Ribbon links

A μ-*component* n-*ribbon* is a map $R : \mu D^{n+1} \to S^{n+2}$ which
is locally an embedding and whose only singularities are transverse
double points, the double point sets being a disjoint union of discs,
and such that $R|_{\mu S^n}$ is an embedding. A μ-component n-link L is a
ribbon link if there is a ribbon R such that $L = R|_{\mu S^n}$.

If D is a component of the singular set of R then either D is
disjoint from $\partial(\mu D^{n+1})$ or $\partial D = D \cap \partial(\mu D^{n+1})$: we call such a
component a *slit* or a *throughcut*, respectively. We may assume that
each component of the graph with vertices the components of the
complement of the throughcuts and edges the throughcuts has at
most one vertex of degree > 2, and that the slits are in components
corresponding to terminal vertices [**Yj69**]. In our examples below
(here and in Chapter 8) each vertex has degree ≤ 2.

An n-link L is a *homotopy ribbon* link if it bounds a properly
embedded $(n + 1)$-disc in D^{n+3} whose exterior W has a handlebody
decomposition consisting of 0-, 1- and 2-handles. The dual decomposition of W relative to $\partial W = M(L)$ has only $(n + 1)$- and $(n + 2)$-
handles, and so the inclusion of M into W is n-connected. (The
definition of "homotopically ribbon" for 1-knots given in Problem
4.22 of [**Ki97**] requires only that this latter condition be satisfied.)
Every ribbon link is homotopy ribbon and hence slice [**Ht79**]. It is

unknown whether every classical slice knot is ribbon, but in higher dimensions there are slice knots which are not even homotopy ribbon.

THEOREM 1.12. *Let L be a μ-component ribbon n-link. Then L is a sublink of a ν-component ribbon n-link $\widehat{L}$ such that $M(\widehat{L}) \cong \natural^\nu(S^1 \times S^{n+1})$. In particular, $\widehat{L}$ is a homology boundary link and L is an SHB link.*

PROOF. Let R be a ribbon for L, with slits $\{S_i \mid 1 \leq i \leq \sigma\}$. Choose disjoint regular neighbourhoods N_i for each slit in the interior of the corresponding $(n+1)$-disc. Let $\nu = \mu + \sigma$ and let $\widehat{L} = L \cup R|_{\partial N}$, where $N = \cup N_i$. Let $W = D^{n+3} \cup \nu(D^{n+1} \times D^2)$ be the trace of surgery on $\widehat{L}$ (with framing 0 on each component if $n = 1$). Then $M(\widehat{L}) = \partial W$.

Now $\widehat{L}$ may be replaced by a ribbon link with one less singularity, by adding a pushoff of $\widehat{L}|_{\partial N_i}$ to the component of L bounding the $(n+1)$-disc containing N_i. Moreover, if $n = 1$ each component of the new link still has framing 0. Continuing thus, $\widehat{L}$ may be replaced by a ribbon link $\widetilde{L}$ for which the only singularities are those corresponding to the components ∂N_i. These may be slipped off the ends of the other components of the new ribbon and so $\widetilde{L}$ is trivial. Adding pushoffs of link components to one another corresponds to sliding $(n+1)$-handles of W across one another, which leaves unchanged the topological type of W. Hence $M(\widehat{L}) \cong M(\widetilde{L}) \cong \natural^\nu(S^1 \times S^{n+1})$, and $\pi\widehat{L}$ maps onto $F(\nu)$. □

If $n = 1$ the homomorphism $\pi\widehat{L} \to F(\nu)$ is an isomorphism if and only if $\widehat{L}$ is trivial, in which case L is also trivial. If $n > 1$ this homomorphism is an isomorphism, but need not carry any set of meridians to a basis. This is so if and only if $\widehat{L}$ is a boundary link. If $n > 2$ it is then trivial and so L is also trivial.

If $M(K) \cong S^1 \times S^2$ then K is the unknot [**Ga87**]. Is there a nontrivial boundary 1-link L such that $M(L) \cong \natural^\mu(S^1 \times S^2)$?

There is the following partial converse.

THEOREM 1.13. *If L is a ν-component n-link such that $M(L) \cong \natural^\nu(S^1 \times S^{n+1})$ then L is a homology boundary link and is null concordant. Hence also any sublink of L is nullconcordant.*

PROOF. That L is a homology boundary link is clear. Let $U(L)$ be the trace of the surgeries on L, so $\partial U(L) = S^{n+2} \amalg \natural^\nu (S^1 \times S^{n+1})$. The $(n+3)$-manifold $D(L) = U(L) \cup \natural^\nu(D^2 \times S^{n+1})$ is contractible and has boundary S^{n+2}, and so is homeomorphic to D^{n+3}. The link L clearly bounds ν disjoint $(n+1)$-discs in $D(L)$.   □

This argument rests on the TOP 4-dimensional Poincaré conjecture when $n = 1$. This dependance can be partially sidestepped. A relatively simple argument using the TOP Schoenflies Theorem shows that if the result of 0-framed surgery on the first ρ components of L is $\natural^\rho(S^1 \times S^2)$, for each $\rho \leq \nu$, then L is TOP null concordant [**Ru80**]. Is every slice link an SHB link?

THEOREM 1.14. *A finitely presentable group G is the group of a μ-component sublink of a ν-component n-link L with group $\pi L \cong F(\nu)$ (for some ν and any $n \geq 2$) if and only if it has deficiency μ and weight μ.*

PROOF. The conditions are clearly necessary. Suppose that G has a presentation $\langle x_i, 1 \leq i \leq \nu \mid r_j, 1 \leq j \leq \nu - \mu \rangle$ and that the images of $s_1, \ldots, s_\mu \in F(\nu)$ in G generate G normally. The words r_j and s_k may be represented by disjoint embeddings ρ_j and σ_k of $S^1 \times D^{n+1}$ in $\natural^\nu(S^1 \times S^{n+1})$. If surgery is performed on all the ρ_j and σ_k the resulting manifold is a homotopy $(n+2)$-sphere, and $Y = \natural^\nu(S^1 \times S^{n+2}) \backslash \cup_{j=1}^{j=\nu-\mu} \rho_j(S^1 \times D^{n+1}) \backslash \cup_{k=1}^{k=\mu} \sigma_k(S^1 \times D^{n+1})$ is the complement of a ν-component n-link in this homotopy sphere, with link group $F(\nu)$. Therefore if surgery is performed on the ρ_j only, the space $Y \cup (\nu - \mu)(D^2 \times S^n)$ is the complement of a μ-component sublink with group G.   □

When $n = 2$ the resulting link is merely TOP locally flat.

Let $G(i, j)$ be the group with presentation $\langle x, y, z \mid x[z^i, x][z^j, y] \rangle$. Then the generators y and z determine a homomorphism from $F(2)$ to $G(i, j)$ which induces isomorphisms on all nilpotent quotients, and $G(i, j)_\omega = 1$, but $G(i, j)$ is not free unless $ij = 0$ [**Ba69**]. As $G(i, j)$ is the normal closure of the images of y and z, and the presentation $\langle x, y, z \mid x[z^i, x][z^j, y], y, z \rangle$ of the trivial group is AC-equivalent to the empty presentation, this group can be realized by a PL locally flat link in S^4. The higher dimensional links constructed from this presentation as in Theorem 1.14 are sublinks of 3-component homology boundary links but are not homology boundary links. See Chapter 8 for examples of ribbon 1-links which are not homology boundary links (although they are SHB links, by Theorem 1.12).

An immediate consequence of Theorems 1.12 and 1.14 is that if $n > 1$ the group of a μ-component ribbon n-link has a presentation of deficiency μ. Thus the 2-twist spin of the trefoil knot is slice [**Ke65**], but not ribbon [**Yj64**]. (See also Theorems 4.3 and 4.6 below.)

We may use a ribbon map R extending a 1-link L to construct a concordance $\mathcal{C}$ from L to a trivial link U, such that the only singularities of the composite $f = pr_2 \circ \mathcal{C} : \mu D^2 \to S^3 \times [0,1] \to [0,1]$ are saddle points corresponding to the throughcuts. Capping off the components of U in D^4 and doubling gives a μ-component 2-link DR. The *ribbon group* of R is $H(R) = \pi DR$. Each throughcut T determines a conjugacy class $g(T) \subset \pi L$ represented by the oriented boundary of a small disc neighbourhood in R of the corresponding slit. (The standard orientation on D^2 induces an orientation on this neighbourhood via the local homeomorphism R.) Let TC be the normal subgroup determined by the throughcuts of R.

THEOREM 1.15. *Let L be a ribbon 1-link with group $\pi = \pi L$ and R a ribbon map extending L. Then $H(R) = \pi L/TC$ and has a Wirtinger presentation of deficiency μ. The longitudes of L are in the normal subgroup TC, which is contained in π_ω. Hence the projection of π onto π/π_ω factors through $H(R)$.*

PROOF. It is clear from the description of the construction in the above paragraph that the inclusion of $X(L)$ into $X(\mathcal{C})$ induces an isomorphism from $\pi L/TC$ to $\pi_1(X(\mathcal{C}))$. Hence $\pi DR \cong \pi L/TC$, by Van Kampen's Theorem.

Each longitude is represented up to conjugacy by a curve on and near the boundary of the corresponding disc, which is clearly homotopic to a product of conjugates of loops about the slits on the disc. Thus the longitudes of L are in TC.

We may show by induction on q that $TC \leq \pi_q$ for all $q \geq 1$. This is clear for $q = 1$. If $TC \leq \pi_n$ the image of each conjugacy class $g(T)$

in π/π_{n+1} is a central element $g_{n+1}(T)$. If T and T' are adjacent representatives of $g(T)$ and $g(T')$ differ only by commutators involving loops around slits in the segment of the ribbon between T and T', and so $g_{n+1}(T) = g_{n+1}(T')$. Moving along the ribbon, we find that $g_{n+1}(T) = 1$, and so $g(T) \subset \pi_{n+1}$, for all T. Thus $TC \leq \pi_\omega$.

We may choose a generic projection of the ribbon with no triple points. The Wirtinger generators of the link group corresponding to the subarcs of the link which "lie under" a segment of the ribbon may be deleted, and the two associated relations replaced by one stating that either adjacent generator is conjugate to the other by a loop around the overlying segment.

Any loop about a segment of the ribbon dies in $H(R)$, for the only obstructions to deforming it onto a loop around the throughcut at an end of the segment are elements in the conjugacy classes of the throughcuts between the loop and that end. Hence the remaining generators corresponding to subarcs of the boundary of a given component of the complement of the throughcuts coalesce in $H(R)$. Conversely the presentation obtained from the Wirtinger presentation by making such deletions and identifications is as claimed, and presents a group in which the image of each $g(T)$ is trivial, for the image of $g(T)$ is trivial if and only if the pair of generators corresponding to arcs meeting the projection of T are identified. Thus the group is exactly $H(R)$. $\square$

Conversely, any such presentation can be realized by some ribbon map $R : \mu D^2 \to S^3$. A similar argument shows that a group G is the group of a μ-component ribbon n-link for any $n \geq 2$ if and only if G has a Wirtinger presentation of deficiency μ and $G/G' \cong Z^\mu$. The generators correspond to meridianal loops transverse to the components of the complements of the throughcuts, and there is one relation for each throughcut. Thus although the group of an unsplittable 1-link has no presentation of deficiency > 1, the groups of ribbon links have quotients with deficiency μ. (See [**Si80**] for some connections between Wirtinger presentations and homology.)

Much of this theorem can be deduced from Theorem 1.12, by arguing as in Theorem 1.14 to adjoin $\nu - \mu$ relations to $F(\nu)$. In general, π/π_ω, $H(R)$ and $\pi/\langle\langle longitudes\rangle\rangle$ are distinct groups, even when $\mu = 1$. (Consider the square knot $3_1\sharp - 3_1$.) If one ribbon R_1 is obtained from another R_2 by knotting the ribbon or inserting full twists then $H(R_1) = H(R_2)$, as such operations do not change the pattern of the singularities.

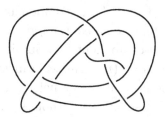

Figure 4.

Let R be the ribbon disc of Figure 4 and let $K = \partial R$. Then $\Pi(R)$ has the presentation $\langle a, b, c, d \mid aca^{-1} = d, dad^{-1} = b, dcd^{-1} = b\rangle$, and so $H(R) \cong Z$. It can be shown that there is a homomorphism from πK to $\mathrm{SL}(2, \mathbb{F}_7)$ with non-abelian image. The corresponding ribbon 2-knot is trivial, and so K is a nontrivial slice of a trivial 2-knot [**Yn70**]. (It is in fact the Kinoshita-Terasaka 11-crossing knot with Alexander polynomial 1 [**KT57**]. See Figure 3(a) of [**Wa94**].)

Figure 5.

Similarly, $\langle a, w, x, y, z \mid axa^{-1} = y, wyw^{-1} = z, zwz^{-1} = x\rangle$ leads to a 2-component homology boundary link which is a slice of a 2-link with group $F(2)$. (See Figure 5.) This 1-link is not a boundary link ([**Cr71**] - see also §7 of Chapter 8 below). Hence the 2-link with group $F(2)$ of which it is a slice is not one either, illustrating the result of Poenaru [**Po71**].

The above results may be usefully extended by the notion of fusion. A fusion band for an n-link L is a pair $\beta = (b, u)$, where $b : [0, 1] \to S^{n+2}$ is an embedded arc with endpoints on L and u is a unit normal vector field along b such that $u|_{\{0,1\}}$ is normal to L, and such that the orientations are compatible. These data determine a band $B : [0, 1] \times D^n \to S^{n+2}$ which may be used to form the connected sum of two of the components of L. The resulting $(\mu - 1)$-component link is called the *fusion* of L (along β). The *strong fusion* is the μ-component link obtained by adjoining to the fusion the boundary of an $(n + 1)$-disc transverse to b.

When $n > 1$ the normal vector field u is unique up to isotopy, but in the classical case any two choices differ by an element of $\pi_1(SO(2)) \cong Z$, and so it determines the twisting of the band B.

Ribbon links are fusions of trivial links. The argument of Theorem 1.12 can be extended to show that a fusion of a boundary link is an SHB link [**Co87**]. Moreover any SHB link is concordant to a fused boundary link [**CL91**]. If a strong fusion of a link is an homology boundary link then so was the original link [**Ka93**].

Concordance of 1-links is generated as an equivalence relation by fusions $L \to L +_\beta \partial R$, where $R : D^2 \to X(L)$ is a ribbon map with image disjoint from L and where $+_\beta$ denotes fusion along a band β from some component of L to ∂R [**Tr69**].

1.8. Link-symmetric groups

Let $r_n : S^n \to S^n$ be the map which changes the sign of the last coordinate. Then every (PL) homeomorphism of S^n is isotopic to id_{S^n} or r_n, depending on whether it preserves or reverses the orientation. An n-knot K is *invertible*, $+amphicheiral$ or $-amphicheiral$ if it is ambient isotopic to $K\rho = K \circ r_n$, $rK = r_{n+2} \circ K$ or $-K = rK\rho$, respectively. If a knot has two of these properties then it has all three. Conway has suggested the alternative terminology *reversible, obversible, inversible*, as $-K$ represents the inverse of the class of K in the knot concordance group [**Co70**].

These notions have been extended to links as follows. The extended symmetric group on μ symbols is the semidirect product

$(Z/2Z)^\mu \rtimes S_\mu$, where S_μ acts on the normal subgroup $(Z/2Z)^\mu$ by permutation of the symbols. Then the *link-symmetric group of degree* μ is $LS(\mu) = (Z/2Z) \times ((Z/2Z)^\mu \rtimes S_\mu)$. A μ-component n-link L *admits* $\gamma = (\epsilon_0, \ldots, \epsilon_\mu, \sigma) \in LS(\mu)$ if L is ambient isotopic to $\gamma L = r_{n+2}^{\epsilon_0} \circ L \circ (\mathrm{II} r_n^{\epsilon_i}) \circ \tilde{\sigma}$, (where $\tilde{\sigma}$ permutes the components, and where $Z/2Z$ is identified with $\{\pm 1\}$). A link L is *invertible* if it admits $(1, -1, \ldots, -1, id)$, ϵ-*amphicheiral* if it admits $(-1, \epsilon, \ldots, \epsilon, id)$, and *interchangeable* if it admits γ with image $\sigma \in S_\mu$ not the identity permutation.

The group of symmetries of a link L is the subgroup $\Sigma(L) \leq LS(\mu)$ consisting of the elements admitted by L. This group depends only on the ambient isotopy type of L. Changing the orientation of one component or the order of the components replaces $\Sigma(L)$ by a conjugate subgroup. (See [**Wh69**].)

1.9. Link composition

Let L be a μ-component n-link, and choose homeomorphisms ϕ_i from $S^n \times D^2$ onto disjoint regular neighbourhoods of the components L_i, for $1 \leq i \leq \mu$. If $n = 1$ assume that the circles $\phi_i(S^1 \times \{d\})$ corresponding to different values of $d \in D^2$ have mutual linking number 0. Let $K(i)$ be a ν_i-component link in $S^n \times D^2$ and let $K(i)^+$ be the $(\nu_i + 1)$-component n-link in S^{n+2} obtained by adjoining $S^n \times \{1\} \subset \partial(S^n \times D^2)$. Then the *composite* of L with $\mathcal{K} = \{K(i)\}_{1 \leq i \leq \mu}$ is $L \circ \mathcal{K} = \bigcup_{1 \leq i \leq \mu} \phi_i \circ K(i)$. (This link has $\nu = \Sigma \nu_i$ components.) As $X(L \circ \mathcal{K}) \cong X(L) \cup \bigcup_{1 \leq i \leq \mu} X(K(i)^+)$, this construction is well adapted to applications of the Van Kampen and Mayer-Vietoris Theorems. If $K(i) = S^n \times \{0\}$ for all i then $L \circ \mathcal{K} = L$. We shall assume henceforth that $K(i) = S^n \times \{0\}$ for $i \neq j$.

If $\mu = 1 = \nu$ then $L \circ \mathcal{K}$ is a *satellite* of L; in particular, if $K = K(1)$ has geometric winding number 1 in $S^n \times D^2$ (i.e., intersects some disc $\{s\} \times D^2$ transversely in one point) this gives the *sum* $K \sharp L$ of the knots K and L.

If $\nu_j = 1$ and $K(j)$ is homologous to $S^1 \times \{0\}$ in $S^1 \times D^2$ then $L \circ \mathcal{K}$ is obtained from L by an elementary F-isotopy on the jth

component. If L' is obtained from L by an elementary F-isotopy then $X(L)$ is a retract of $X(L')$, since $\partial X(h)$ is a retract of $X(h)$ for any 2-component link h with linking number 1.

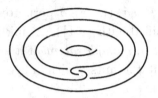

Figure 6. $\theta \circ Wh_2$

Let $Wh : 2S^1 \to S^3$ be the *Whitehead link* (5_1^2 in the tables of [**Rol**]), and let $\theta : X(Wh_1) \to S^1 \times D^2$ be a homeomorphism such that $\theta(\phi_1(u,v)) = (v,u)$, for all $u, v \in S^1$. If $K(j) = \theta \circ Wh_2$ then $L \circ \mathcal{K}$ is obtained from L by *Whitehead doubling* the jth component. (See Figure 6.) When $\mu = 1$ this is an untwisted double of the knot L. Since each component of the Whitehead link bounds a punctured torus in the complement of the other component, Whitehead doubling every component of a link gives a boundary link.

Figure 7. $\theta \circ Bo_{2,3}$

Similarly, if $Bo : 3S^1 \to S^3$ is the *Borromean ring* link (6_2^3 in the tables of [**Rol**]) let $\theta : X(Bo_1) \to S^1 \times D^2$ be a homeomorphism such that $\theta(\phi_1(u,v)) = (v,u)$, for all $u, v \in S^1$. If $Bo_{2,3}$ is the union of the second and third components of Bo and $K(j) = \theta \circ Bo_{2,3}$ then $L \circ \mathcal{K}$ is obtained from L by *Bing doubling* the jth component. (See Figure 7.) In the latter two cases there are further mild ambiguities, related to the definition of the Whitehead link, etc.

CHAPTER 2

Homology and Duality in Covers

The primary algebraic invariants of knots and links are the homology groups of covering spaces of the exterior, considered as modules over the group ring of the covering group, together with the bilinear pairings determined by Poincaré duality. In high dimensions simple knots (i.e., knots with highly connected Seifert surfaces) are completely classified by such invariants, and the knot concordance group is isomorphic to an algebraically defined Witt group of equivalence classes of pairings.

We review these homology and cohomology invariants, and the spectral sequences and duality pairings relating them. We shall also sketch the construction of the maximal free cover of the exterior of a homology boundary link by splitting along singular Seifert surfaces, In Chapters 4–10 our main interest is in invariants of abelian covers, but in Chapter 11 we shall consider the modules and pairings associated to such maximal free covers. (Analogous invariants of solvable covers have been used to study classical knot concordance, but we shall only hint at this, in Chapter 12.)

2.1. Homology and cohomology with local coefficients

Let X be a connected finite cell complex with universal cover $\tilde{p} : \tilde{X} \to X$. If H is a normal subgroup of $G = \pi_1(X)$ we may lift the cellular decomposition of X to an equivariant cellular decomposition of the corresponding covering space X_H. The cellular chain complex C_* of X_H with coefficients in a commutative ring R is then a complex of left $R[G/H]$-modules, with respect to the action of the covering group G/H. Moreover each module C_q has a finite free basis, obtained by choosing a lift of each q-cell of X.

The ith *equivariant homology* of X with coefficients $R[G/H]$ is the left module $H_i(X; R[G/H]) = H_i(C_*)$, which is clearly isomorphic to $H_i(X_H; R)$ as an R-module, with the action of the covering group determining its $R[G/H]$-module structure. The ith *equivariant cohomology* of X with coefficients $R[G/H]$ is the right module $H^i(X; R[G/H]) = H^i(C^*)$, where $C^* = Hom_{R[G/H]}(C_*, R[G/H])$ is the associated cochain complex of right $R[G/H]$-modules. (This may be regarded as the ith cohomology with compact supports of X_H.) More generally, if A and B are right and left $\mathbb{Z}[G/H]$-modules (respectively) we may define $H_j(X; A) = H_j(A \otimes_{\mathbb{Z}[G/H]} C_*)$ and $H^{n-j}(X; B) = H^{n-j}(Hom_{\mathbb{Z}[G/H]}(C_*, B))$. (These groups only have natural module structures if A and B are bimodules.) There is a Universal Coefficient Spectral Sequence (UCSS)

$$E_2^{pq} = Ext_{R[G/H]}^q(H_p(X; R[G/H]), R[G/H]) \Rightarrow H^{p+q}(X; R[G/H]),$$

whose r^{th} differential d_r has bidegree $(1 - r, r)$. If J is a normal subgroup of G which contains H there is also a Cartan-Leray spectral sequence relating the homology of X_H and X_J,

$$E_{pq}^2 = Tor_p^{R[G/H]}(H_q(X; R[G/H]), R[G/J]) \Rightarrow H_{p+q}(X; R[G/J]),$$

whose r^{th} differential d^r has bidegree $(-r, r - 1)$. There are similar spectral sequences for the equivariant homology and cohomology of pairs of cell complexes. (See [**McC**] for more details on such spectral sequences.)

2.2. Covers of link exteriors

As the abelianization of a link group is free abelian, link exteriors have canonical free abelian covers. In particular, if L is a μ-component n-link the *maximal abelian cover* $p' : X(L)' \to X(L)$ has covering group $\mathbb{Z}^\mu$. Similarly, the exteriors of homology boundary links have canonical free covers. (These notions coalesce for knots.) Infinite cyclic covers of arbitrary link exteriors and the maximal abelian and maximal free covers of the exteriors of homology boundary links may be constructed by *splitting along Seifert surfaces*. Applying the Mayer-Vietoris sequence then leads to presentations of the equivariant homology modules, which are often quite efficient, in that few generators or relations are needed.

We shall sketch this construction for the *maximal free cover* of a homology boundary link. A homomorphism $f : \pi = \pi L \to F(\mu) = F(x_1, \ldots, x_\mu)$ may be realized by a map $F : X = X(L) \to \vee^\mu S^1 = K(F(\mu), 1)$, which may be assumed transverse to $\{P_1, \ldots, P_\mu\}$, where P_i is a point in the ith copy of S^1, other than the wedge point. Let $U_i = F^{-1}(P_i)$ and let $Y = X \setminus W$, where W is an open regular neighbourhood of $U = \cup U_i$. There are two embeddings $i_\pm$ of U in ∂Y and $X^\omega = Y \times F(\mu)/ \sim$, where $(i_+(u_j), hx_j) \sim (i_-(u_j), h)$ for all $u_j \in U_j$ and $h \in F(\mu)$. (Here x_j is a generator of $F(\mu)$ corresponding to a loop in X which meets U_j transversally in one point and avoids the other components of U.) Projection of $Y \times F(\mu)$ onto its first factor induces a covering projection $p_\omega : X^\omega \to X$ with covering group $F(\mu)$. This is just the pull-back of the corresponding construction of the universal cover of $\vee^\mu S^1$.

There is then a Mayer-Vietoris sequence

$$\cdots \to \mathbb{Z}[F(\mu)] \otimes H_q(U) \xrightarrow{d_q} \mathbb{Z}[F(\mu)] \otimes H_q(Y) \to H_q(X; \mathbb{Z}[F(\mu)])$$

where $d_q(\gamma \otimes v_j) = \gamma x_j \otimes i_{j+*}(v_j) - \gamma \otimes i_{j-*}(v_j)$ for $\gamma \in \mathbb{Z}[F(\mu)]$ and $v_j \in H_q(U_j; \mathbb{Z})$. If L is a boundary link we may assume that the U_j are Seifert surfaces; if also L is a 1-link then the longitudes must be null-homologous in X^ω, and so lie in $(\pi_\omega)'$, since they bound the U_j, which lift to X^ω.

A similar construction gives the covering space X_N, and hence a Mayer-Vietoris sequence for $H_*(X; \mathbb{Z}[\pi/N])$, corresponding to any normal subgroup $N \trianglelefteq \pi$ containing π_ω. (In particular, this applies to subgroups π_n in the lower central series.)

The maximal abelian cover of any 1-link can be constructed in a similar way, on using Seifert surfaces which intersect in a controlled way [Cp82]. (Abelian covers for composite links $L \circ \mathcal{K}$ may be assembled from copies of abelian covers for L and the $K(i)^+$.)

2.3. Some terminology and notation

Let R be a ring with an involution $\bar{\ }$. The key example for us is $\Lambda_\mu = \mathbb{Z}[\mathbb{Z}^\mu] = \mathbb{Z}[t_1, t_1^{-1}, \ldots, t_\mu, t_\mu^{-1}]$. If N is a left R-module let $\overline{N}$ denote the *conjugate* right R-module, with the same underlying abelian group but with module action given by $n.r = \bar{r}n$ for all $r \in R$ and $n \in N$. (Similarly, right modules have conjugate left modules.)

If R is a factorial domain (such as Λ_μ) and N is a finitely generated R-module let TN be its torsion submodule and let zN be the submodule of N generated by elements whose annihilator ideal is not contained in a proper principal ideal. (This is the maximal pseudonull submodule of N, cf. Chapter 3.) Let $\hat{t}N = TN/zN$.

Let M_1, M_2 and N be R-modules. A map $c : M_1 \times M_2 \to N$ is a *sesquilinear pairing* if it is R-linear in the first variable and R-antilinear in the second variable, i.e., if $c(rm_1, m_2) = rc(m_1, m_2) = c(m_1, \bar{r}m_2)$ for all $m_1 \in M_1$, $m_2 \in M_2$ and $r \in R$. The *transposed* pairing is defined by $c^\dagger(m_2, m_1) = \overline{c(m_1, m_2)}$. The *adjoint* of c is the homomorphism $Ad(c) : \overline{M_1} \to Hom_R(M_2, N)$ given by $Ad(c)(m_1)(m_2) = c(m_1, m_2)$. The pairing is *left primitive* if $Ad(c)$ is injective, *right primitive* if $Ad(c^\dagger)$ is injective and *non-singular* if $Ad(c)$ is an isomorphism.

Suppose now that R is an integral domain with field of fractions R_0. Let $\varepsilon = \pm 1$. An ε-*hermitean pairing* on an R-torsion module M is a sesquilinear pairing $b : M \times M \to R_0/R$ such that $b = \varepsilon b^\dagger$. The *conull space* of b is $\mathrm{Cok}(Ad(b))$. A ε-*linking pairing* over R is a (left and right) primitive ε-hermitean pairing b on a finitely generated R-torsion module M.

2.4. Poincaré duality and the Blanchfield pairings

If M is a compact orientable m-manifold with boundary ∂M and $\pi = \pi_1(M)$ the pair $(M, \partial M)$ is homotopy equivalent to a pair of finite cell complexes (X, Y) which satisfies equivariant Poincaré duality, in the following sense. Let C_* be the cellular chain complex of $\widetilde{X}$ and P_* be the relative chain complex of the pair $(\widetilde{X}, p^{-1}(Y))$. The group ring $\mathbb{Z}[\pi]$ has a canonical involution $\bar{\ }$, given by $\bar{g} =$

g^{-1}. Then $H_m(X, Y; \mathbb{Z}) = H_m(\mathbb{Z} \otimes_{\mathbb{Z}[\pi]} P_*) \cong \mathbb{Z}$, and cap product with a generator $[X, Y]$ determines a chain homotopy equivalence from $\overline{C^*}$ to P_{m-*}, which induces isomorphisms from $\overline{H^j(C^*)}$ to $H_{m-j}(P_*)$. More generally, it gives rise to ($\mathbb{Z}$-linear) isomorphisms from $H^j(M; B)$ to $H_{m-j}(M, \partial M; \overline{B})$ (and similarly from $H^j(M, \partial M; B)$ to $H_{m-j}(M; \overline{B})$) where B is any left $\mathbb{Z}[\pi]$-module of coefficients. (See Chapter 2 of [**Wal**] for further details.)

In particular, if H is a normal subgroup of π and $B = \mathbb{Z}[\pi/H]$ we obtain isomorphisms between the homology and the compactly-supported cohomology of the covering space X_H. Together with the UCSS these lead to duality pairings of various kinds on (subquotients of) the homology modules. Poincaré duality gives isomorphisms $D_q : H_{m-q}(M, \partial M; \mathbb{Z}[\pi/H]) \to H^q(M; \mathbb{Z}[\pi/H])$ (since $\overline{\mathbb{Z}[\pi/H]} \cong \mathbb{Z}[\pi/H]$ as bimodules), and evaluation of cohomology classes $D_q(v)$ on homology classes defines a $\mathbb{Z}[\pi/H]$-valued pairing of $H_q(M; \mathbb{Z}[\pi/H])$ with $H_{m-q}(M, \partial M; \mathbb{Z}[\pi/H])$. These pairings are sesquilinear, i.e., $\mathbb{Z}[\pi]$-linear in the first variable and conjugate linear in the second variable.

These pairings may be interpreted geometrically in terms of intersections of dual cells in the covering space X_H. Fix a triangulation of X. Let C_* be the cellular chain complex of X_H with respect to the induced triangulation, and let P_* be the cellular chain complex of $(X_H, \partial X_H)$ with respect to the dual triangulation. Intersection numbers of transverse cells of complementary dimensions determine pairings $I_q : C_q \times P_{m-q} \to \mathbb{Z}$ such that $I_q(\partial u, v) = (-1)^{q+1} I_{q+1}(u, \partial v)$ if $u \in C_{q+1}$ and $v \in P_{m-q}$. (We may write I_{X_H} instead of I_q when the dimensions are clear but it is important to distinguish the space.) Therefore intersections induce well-defined pairings on homology. The covering group acts isometrically on I_p and so the pairing $\tilde{I}_q : C_q \times P_{m-q} \to \mathbb{Z}[\pi/H]$ given by $\tilde{I}_q(x, y) = \Sigma_{g \in \pi/H} I_q(x, gy)g$ is sesquilinear. Hence we may extend these pairings to pairings on homology with other $\mathbb{Z}[\pi/H]$-module coefficients by taking suitable tensor products. These pairings agree with the ones defined by means of the isomorphisms D_q and so are independent of the triangulation used. (See [**Mc06**]. This relies upon Poincaré duality for singular homology!)

For knot and link complements the most important manifestations of duality are the *Blanchfield pairings* associated to free abelian covers [**Bl57**]. Let $p_H : X_H \to X$ be a connected cover of a compact oriented PL m-manifold, with covering group $\mathbb{Z}^\mu$. The natural involution on $\Lambda_\mu = \mathbb{Z}[\mathbb{Z}^\mu]$ extends to its field of fractions $(\Lambda_\mu)_0 = \mathbb{Q}(t_1, \ldots, t_\mu)$ and to the quotient module $S(\Lambda_\mu) = (\Lambda_\mu)_0/\Lambda_\mu$ in an obvious way. The Blanchfield pairings associated to the cover are pairings of (quotients of) the Λ_μ-torsion submodules of $H_q(X; \Lambda_\mu)$ and $H_{m-q-1}(X, \partial X; \Lambda_\mu)$, with values in $S(\Lambda_\mu)$, for all $q < m$. We shall first describe them algebraically, and then give a more geometric interpretation.

Let $e^i N = Ext^i_{\Lambda_\mu}(N, \Lambda_\mu)$, for all $i \geq 0$. (The value of μ shall always be clear from the context.) There is a natural isomorphism from $e^1 TN$ to $Hom_{\Lambda_\mu}(TN, S(\Lambda_\mu))$, since $\mathbb{Q}(t_1, \ldots, t_\mu)$ is injective and TN is a torsion module.

Let θ_q be the composite of the homomorphism from $H^{q+1}(X; \Lambda_\mu)$ to $e^1 H_q(X; \Lambda_\mu)$ given by the UCSS, restriction to $e^1 TH_q(X; \Lambda_\mu)$, and the natural isomorphism to $Hom_{\Lambda_\mu}(TH_q(X; \Lambda_\mu), S(\Lambda_\mu))$, and let $T = TH_{m-q-1}(X, \partial X; \Lambda_\mu)$. Then $\theta_q D_{q+1}|_T = \rho D_q^{S(\Lambda)} \beta|_T^{-1}$, where $\rho : H^q(X; S(\Lambda_\mu)) \to Hom_\Lambda(TH_q(X; \Lambda_\mu), S(\Lambda_\mu))$ is the evaluation homomorphism, $D_q^{S(\Lambda)}$ is the duality isomorphism with coefficients $S(\Lambda_\mu)$, $T = TH_{m-q-1}(X, \partial X; \Lambda_\mu)$ and β is the Bockstein homomorphism associated to the coefficient sequence

$$0 \to \Lambda_\mu \to (\Lambda_\mu)_0 \to S(\Lambda_\mu) \to 0.$$

Then the Blanchfield pairing has value $V(u, v) = \theta_q(D_{q+1}(v))(u)$, for $u \in TH_q(X; \Lambda_\mu)$ and $v \in TH_{m-q-1}(X, \partial X; \Lambda_\mu)$. If x is a q-cycle of X_H and y is a relative $(m-1-q)$-cycle of $(X_H, \partial X_H)$ such that $\alpha x = \partial u$ and $\beta y = \partial v$ for some $(q+1)$-chain u and $(m-q)$-chain v and some nonzero $\alpha, \beta \in \Lambda_\mu$, then

$$V(x, y) = [\alpha^{-1} \tilde{I}_q(u, y)] = [\bar{\beta}^{-1} \tilde{I}_q(x, v)] = [\bar{\beta}^{-1} \Sigma_{g \in \pi/H} I_q(x, gv)g]$$

in $S(\Lambda_\mu)$. If $m = 2q + 1$ this determines a sesquilinear pairing $\langle\,,\,\rangle$ of $TH_q(X;\Lambda_\mu)$ with itself by $\langle x, y\rangle = V(x, j_*y)$, where $j : (X, \emptyset) \to (X, \partial X)$ is the inclusion. Moreover, the geometric properties of intersections of cells in manifolds imply that $\langle x, y\rangle = (-1)^{q+1}\overline{\langle y, x\rangle}$, since we may compute one side via the transposed pairing. (See [**Mc06**].) However, in general there may be $x \neq 0$ such that $\langle x, y\rangle = 0$ for all y.

We shall modify this definition below to obtain a linking pairing.

2.5. The total linking number cover

Let L be a μ-component n-link and let $\tau : \pi = \pi L \to \mathbb{Z}$ be the homomorphism which sends each meridian to $1 \in \mathbb{Z}$. The *total linking number cover* of X is the cover $p_\tau : X^\tau \to X$ associated to $\mathrm{Ker}(\tau)$. A loop in X lifts to a loop in X^τ if and only if the sum of its linking numbers with the various components of L in S^{n+2} is 0, whence the name "total linking number cover".

Let $e : \mathbb{R} \to S^1$ be the exponential map, given by $e(r) = e^{2\pi i r} \in S^1$ for all $r \in \mathbb{R}$. Then e is the universal cover of S^1. Since $S^1 = K(Z, 1)$ there is a unique homotopy class of maps $\tilde{\tau} : X \to S^1$ inducing τ, and p_τ is the pullback $\tilde{\tau}^*e$. Let $\Lambda = \mathbb{Z}[t, t^{-1}] = \Lambda_1$. The Wang sequence for $\tilde{\tau}$ is also the long exact sequence associated to the exact sequence of coefficient modules

$$0 \to \Lambda \to \Lambda \to \mathbb{Z} \to 0.$$

THEOREM 2.1. *Let L be a μ-component n-link, and let $X = X(L)$. If $\mu > 0$ then*

(1) *the Λ-modules $H_q(X;\Lambda)$ and $H^q(X;\Lambda)$ are finitely generated, and are 0 if $q > n + 1$;*

(2) *$\mathbb{Z} \otimes_\Lambda H_1(X;\Lambda) \cong \mathbb{Z}^{\mu-1}$ and $\mathbb{Z} \otimes_\Lambda TH_1(X;\Lambda)$ is finite, if $n > 1$;*

(3) *$\mathbb{Z} \otimes_\Lambda H_q(X;\Lambda) = 0$ and so $H_q(X;\Lambda)$ is a torsion module, for $1 < q \leq n$;*

(4) *$H_{n+1}(X;\Lambda) \cong \Lambda^{\mu-1}$ and $H_1(X;\Lambda)$ has rank $\mu - 1$, if $n > 1$;*

(5) *$H_2(X;\Lambda)$ maps onto $H_2(\pi;\Lambda)$;*

(6) *there are short exact sequences*

$$0 \to \mathbb{Z} = e^1 \mathbb{Z} \to H^1(X; \Lambda) \to e^0 H_1(X; \Lambda) \to 0,$$

$$0 \to e^2 H_{q-2}(X; \Lambda) \to H^q(X; \Lambda) \to e^1 H_{q-1}(X; \Lambda) \to 0$$

(for $1 < q \le n$), and an exact sequence

$$0 \to J_{n,1} \to H^{n+1}(X; \Lambda) \to \Lambda^{\mu-1} \to e^2 H_n(X; \Lambda) \to 0,$$

where $J_{n,1}$ is an extension of $e^1 H_n(X; \Lambda)$ by $e^2 H_{n-1}(X; \Lambda)$.

PROOF. Since $\mu > 0$ the $(n + 2)$-manifold X has nonempty boundary, and so it is homotopy equivalent to a finite complex of dimension $\le n+1$. Hence the singular chain complex of X^τ is chain homotopy equivalent to a finitely generated free Λ-complex which is trivial in dimensions greater than $n + 1$. Since Λ is noetherian the homology and cohomology modules of such a complex are finitely generated.

Alexander duality in S^{n+2} gives $H_1(X; \mathbb{Z}) \cong \mathbb{Z}^\mu$, $H_q(X; \mathbb{Z}) = 0$ for $1 < q \le n$ and $H_{n+1}(X; \mathbb{Z}) \cong \mathbb{Z}^{\mu-1}$. It then follows from the Wang sequence that $\mathbb{Z} \otimes_\Lambda H_1(X; \Lambda) \cong \mathbb{Z}^{\mu-1}$ and $\mathbb{Z} \otimes_\Lambda H_q(X; \Lambda) = 0$, for $1 < q \le n$. Since surjective endomorphisms of noetherian modules are isomorphisms, $t - 1$ acts invertibly on $H_q(X; \Lambda)$, for $1 < q \le n$. If $n > 1$ multiplication by $t - 1$ is injective on $H_1(X; \Lambda)$ and $\mathbb{Z} \otimes_\Lambda H_{n+1}(X; \Lambda) \cong \mathbb{Z}^{\mu-1}$. Since $TH_1(X; \Lambda)$ is a noetherian torsion Λ-module, $\mathbb{Z} \otimes_\Lambda TH_1(X; \Lambda)$ must then be finite. The module $H_{n+1}(X; \Lambda)$ is free since it is the kernel of a homomorphism between free Λ-modules, and so $H_{n+1}(X; \Lambda) \cong \Lambda^{\mu-1}$.

Since ∂X^τ is a union of copies of $S^n \times R$ the natural homomorphism $H_q(X; \Lambda) \to H_q(X, \partial X; \Lambda)$ is an isomorphism if $q \neq 0, 1, n$ or $n + 1$, a monomorphism if $q = 1$ or $n + 1$ and an epimorphism if $q = 0$ or n. The kernels and cokernels are torsion modules, since $H_0(\partial X; \Lambda) \cong H_n(\partial X; \Lambda) \cong \mathbb{Z}^\mu$. Therefore $H_{n+1}(X, \partial X; \Lambda)$ has rank $\mu - 1$. The rest of (4) now follows by Poincaré duality and the UCSS, while (5) is a case of Hopf's Theorem. Since $H_q(X; \Lambda)$ is a torsion module for $1 < q \le n$ the UCSS collapses to give (6). $\square$

If M is a torsion Λ-module then $e^1 M \cong Hom_\Lambda(M, \mathbb{Q}(t)/\Lambda)$, and zM is the maximal finite submodule. If, moreover, M is finite there is also a natural isomorphism $e^2 M \cong Hom_{\mathbb{Z}}(M, \mathbb{Q}/\mathbb{Z})$ (see [**Le77**]). The Poincaré duality isomorphisms in conjunction with the above exact sequences give rise to pairings:

(Blanchfield) $\langle \, , \, \rangle : \hat{t}H_q(X; \Lambda) \times \hat{t}H_{n+1-q}(X; \Lambda) \to \mathbb{Q}(t)/\Lambda$

(Farber-Levine) $[\, , \,] : zH_q(X; \Lambda) \times zH_{n-q}(X; \Lambda) \to \mathbb{Q}/\mathbb{Z}$

The Blanchfield pairings are sesquilinear while the Farber-Levine pairings are $\mathbb{Z}$-linear and isometric ($[t\alpha, t\beta] = [\alpha, \beta]$ for any $\alpha \in zH_q(X; \Lambda)$ and $\beta \in zH_{n-q}(X; \Lambda)$). In each case the pairings are perfect, i.e., the adjoint homomorphisms are isomorphisms. If $n = 2q + 1$ then $\langle y, x \rangle = (-1)^q \overline{\langle x, y \rangle}$, for all $x, y \in \hat{t}H_q(X; \Lambda)$, while if $n = 2q$ it can be shown that $[x, y] = (-1)^{q-1}[y, x]$, for all $x, y \in zH_q(X; \Lambda)$. If $n = 3$ there is a Rochlin constraint on the pairing on $\hat{t}H_2(X; \Lambda)$. These pairings play central roles in the classification of simple $(2q+1)$-knots (with $q \geq 1$) and simple $2q$-knots (with $q \geq 4$), respectively [**Le70, Ke75, Fa83**].

Conversely, given $n > 1$ and any system of modules H_q for $1 \leq q \leq [n/2]$ satisfying the conditions of the Lemma and such that $TH_{[n/2]}$ or $zH_{n/2}$ admits such a self-dual pairing (for n odd or even, respectively) and the additional condition $p.d._\Lambda H_1 \leq 1$ (which is not necessary) there is a μ-component n-link L such that $H_q(X; \Lambda) \cong H_q$ for $1 \leq q \leq [n/2]$. There is a parallel result in which the link group π is also specified (subject to a constraint stronger than $\text{def}(\pi) = \mu$ if $n = 2$) and H_2 maps onto $H_2(\pi; \Lambda)$. (See [**Ne88**]. These results extend the work of [**Le77**] on knot modules and duality pairings.)

2.6. The maximal abelian cover

In the remaining sections we shall consider only classical links, as little is known about the homology modules or duality pairings on the maximal abelian covers of higher dimensional links. (Since Λ_μ has global dimension $\mu + 1$ it is rather difficult to extract useful results about such pairings if $\mu > 1$. However, we shall give an analogue of Theorem 2.1 at the end of Chapter 4.)

Let L be a μ-component 1-link and let $p' : X' \to X$ be the maximal abelian cover of the exterior $X = X(L)$. The meridians of L determine an isomorphism of the covering group $\pi/\pi' = Aut(X'/X)$

with $\mathbb{Z}^{\mu}$, and so we may identify $\mathbb{Z}[\pi/\pi']$ with Λ_{μ}. When $\mu = 1$ the natural map from $H_1(X; \Lambda)$ to $H_1(X, \partial X; \Lambda)$ is an isomorphism, $H_1(X; \Lambda)$ is a Λ-torsion module and $zH_1(X; \Lambda) = 0$. The UCSS and duality imply that $e^1 H_1(X; \Lambda) \cong \overline{H_1(X; \Lambda)}$. (See §7 below for an analogue for boundary links.) The outer maps of the sequence

$$H_1(\partial X; \Lambda_{\mu}) \to H_1(X; \Lambda_{\mu}) \to H_1(X, \partial X; \Lambda_{\mu}) \to H_0(\partial X; \Lambda_{\mu})$$

are nonzero, in general. However, the extreme terms vanish after localization with respect to any multiplicative system containing $\Pi_{i=1}^{i=\mu}(t_i - 1)$. The minimal option is the multiplicative system S generated by these monomials; another interesting choice is the system Σ generated by all nonzero 1-variable polynomials. For knots multiplication by $t - 1$ is invertible on $H_1(X; \Lambda)$, and so no information is lost on localization with respect to S.

Let $V : TH_1(X; \Lambda_{\mu}) \times TH_1(X, \partial X; \Lambda_{\mu}) \to S(\Lambda_{\mu})$ be the sesquilinear pairing given by the construction of §4. Since $zS(\Lambda_{\mu}) = 0$ we have $V(x, y) = 0$ if $x \in zH_1(X; \Lambda_{\mu})$ or $y \in zH_1(X, \partial X; \Lambda_{\mu})$. It may easily be verified by localizing at principal prime ideals that $zH_p(X; \Lambda_{\mu})$ is the kernel of $\theta_{2-p} D_{3-p}|_{TH_p(X;\Lambda_{\mu})}$. Moreover, since $\mathbb{Z}_S = 0$ localization simplifies the UCSS, and we obtain isomorphisms

$$TH_1(X, \partial X; \Lambda_{\mu})_S \cong TH^2(X; \Lambda_{\mu})_S \cong e^1 H_1(X; \Lambda_{\mu})_S.$$

The *localized Blanchfield pairings* of L are the induced pairings $b_S(L)$ and $b_{\Sigma}(L)$ of the localizations of $\hat{t}H_1(X; \Lambda_{\mu})$ with itself into $S(\Lambda_{\mu})$. These are $(+1)$-linking pairings.

Localization with respect to Σ annihilates knot modules. It follows easily from this and from Rolfsen's theorem on isotopies that $b_{\Sigma}(L)$ is invariant under isotopy. (Note also that as the localization of I_{μ} with respect to any such multiplicative system is free, the localized link module sequence as defined in Chapter 4 splits.)

2.7. Boundary 1-links

Let L be a μ-component boundary 1-link and $X = X(L)$. Let $V \simeq \vee^{\mu} S^1$ be the subcomplex of $X = X(L)$ formed by the unions of arcs from a basepoint to a meridian on each boundary component. Then $(X, \partial X)$ retracts onto $(V, V \cap \partial X)$. Since the longitudes of L lift to null-homologous loops in X' the image of $H_1(\partial X; \Lambda_\mu)$ in $H_1(X; \Lambda_\mu)$ is trivial. On comparing the long exact sequences of $(X, \partial X)$ and $(V, V \cap \partial X)$ we see that $TH_1(X; \Lambda_\mu) \cong TH_1(X, \partial X; \Lambda_\mu)$. Moreover, $zH_1(X; \Lambda_\mu) = 0$ and so $\hat{t}H_1(X; \Lambda_\mu) = TH_1(X; \Lambda_\mu)$, by Lemma 2.2 below (or by Corollary 4.15.1). Then

$$e^1 TH_1(X; \Lambda_\mu)_S = e^1 H_1(X; \Lambda_\mu)_S \cong \overline{TH_1(X; \Lambda_\mu)_S},$$

and so the adjoint map of $b_S(L)$ is an isomorphism. (Note also that $TH_1(X; \Lambda_\mu)$ is a direct summand of $H_1(X; \Lambda_\mu)$. The UCSS then implies that $e^1 H_1(X; \Lambda_\mu) \cong e^1 TH_1(X; \Lambda_\mu) \oplus e^3 \mathbb{Z}$. Hence the Blanchfield pairing defined on the *unlocalized* module $TH_1(X; \Lambda_\mu)$ is already non-singular if $\mu \neq 3$.)

Let U_j $(1 \leq j \leq \mu)$ be disjoint orientable surfaces spanning L. These are naturally oriented, by the conventions of Chapter 1. As in §2 there is a Mayer-Vietoris sequence

$$\cdots \Lambda_\mu \otimes H_1(U) \xrightarrow{d_1} \Lambda_\mu \otimes H_1(Y) \to H_1(X; \Lambda_\mu) \to \Lambda_\mu \otimes H_0(U)$$

where $U = \cup U_j$, Y is the complement of an open regular neighbourhood of U in X, and $d_1|_{H_*(U_j) \otimes \Lambda} = (i_{j+})_* \otimes t_j - (i_{j-})_* \otimes 1$.

LEMMA 2.2. $\hat{t}H_1(X; \Lambda_\mu) = TH_1(X; \Lambda_\mu) = \mathrm{Coker}(d_1)$.

PROOF. Clearly $TH_1(X; \Lambda_\mu) \leq \mathrm{Coker}(d_1)$, since $H_0(U) \otimes \Lambda_\mu$ is a free module. Let $\{\alpha_{jm} \mid 1 \leq m \leq m(j)\}$ be a basis for $H_1(U_j)$, for $1 \leq j \leq \mu$. Linking in S^3 establishes the Alexander duality between $H_1(U)$ and $H_1(Y) = H_1(S^3 \setminus U)$; let $\{\hat{\alpha}_{jm} \mid 1 \leq m \leq m(j), 1 \leq j \leq \mu\}$ be the dual basis, so that $\mathrm{lk}(\alpha_{jm}, \hat{\alpha}_{kn}) = 1$ if $j = k$ and $m = n$, and is 0 otherwise. Let $M = \Sigma m(j)$.

Let A be the matrix of $(i_+)_* = \oplus(i_{j+})_* : H_1(U) \to H_1(Y)$ with respect to these bases. Then d_1 is represented by the $M \times M$ matrix $\Delta = \Theta A - A^{tr}$ where Θ is the diagonal matrix $diag[t_1, \ldots, t_2, \ldots, t_\mu]$, where there are $m(i)$ repetitions of t_i, for $1 \leq i \leq \mu$. As $A - A^{tr}$ is the matrix in the corresponding Mayer-Vietoris sequence for $H_*(X; \mathbb{Z})$, it is invertible, and so $\det(\Delta) \neq 0$. Therefore d_1 is injective and $\mathrm{Cok}(d_1)$ is a torsion module. Hence there is an exact sequence

$$0 \to (\Lambda_\mu)^M \to (\Lambda_\mu)^M \xrightarrow{\phi} TH_1(X; \Lambda_\mu) \to 0,$$

so $zH_1(X; \Lambda_\mu) = 0$ and $\hat{t}H_1(X; \Lambda_\mu) = TH_1(X; \Lambda_\mu)$, by Theorem 3.8. □

Using the same symbols to denote 1-chains in U and Y representing the classes α_{jm}, $\hat{\alpha}_{kn}$ respectively, consider the 2-chain $[-1,1] \times \alpha_{jm}$ in X'. Then $\partial([-1,1] \times \alpha_{jm}) = t_j i_{j+}(\alpha_{jm}) - i_{j-}(\alpha_{jm})$. Therefore $\partial([-1,1] \times \Sigma q_{jm}\alpha_{jm})$ represents $\Sigma(\Delta q)_{kn}\hat{\alpha}_{kn}$. Let $\delta = \det(\Delta)$. Then the matrix $\Delta^{-1}\delta$ has coefficients in Λ_μ, and $\delta\Sigma r_{kn}\hat{\alpha}_{kn}$ is the class of $\partial([-1,1] \times \Sigma(\Delta^{-1}\delta r)_{jm}\alpha_{jm})$. Hence the Blanchfield pairing on $TH_1(X; \Lambda_\mu)$ is given by

$$\langle\phi(r), \phi(s)\rangle = V(\phi(r), j_*\phi(s))$$
$$= [\delta^{-1}I_{X'}([-1,1] \times \Sigma(\Delta^{-1}\delta r)_{jm}\alpha_{jm}, \Sigma s_{kn}\hat{\alpha}_{kn})]$$
$$= [\delta^{-1}\Sigma\overline{s_{kn}}\Sigma(\Delta^{-1}\delta r)_{jm}I_{X'}([-1,1] \times \alpha_{jm}, \hat{\alpha}_{kn})]$$
$$= [\delta^{-1}\Sigma\overline{s_{kn}}\Sigma(\Delta^{-1}\delta r)_{jm}(1 - t_j)\delta_{jk}\delta_{mn}]$$
$$= [s^{tr}(I - \Theta)\Delta^{-1}r] \quad \text{in } S(\Lambda_\mu) = (\Lambda_\mu)_0/\Lambda_\mu,$$

for all $r, s \in (\Lambda_\mu)^M$ (considered as column vectors). (See [**Ke75**, **Tr77**] for the knot-theoretic case.)

A similar argument works for homology boundary links. Splitting along a system of singular Seifert surfaces and using Corollary 4.15.1 gives a short free resolution of $\hat{t}H_1(X; \Lambda_\mu) = TH_1(X; \Lambda_\mu)/P$, where P is the submodule generated by the images of the longitudes. However, it is better to localize and thus have an invariant applicable to all links, as there are ribbon links which are not homology boundary links, and so the latter class is not closed under concordance.

In general it is not clear whether $e^1H_1(X; \Lambda_\mu)_S \cong e^1TH_1(X; \Lambda_\mu)_S$. However restriction from $e^1H_1(X; \Lambda_\mu)$ to $e^1TH_1(X; \Lambda_\mu)$ is always a pseudo-isomorphism.

2.8. Concordance

The *sum* of two ε-linking pairings (M, b) and (M', b') over R is the pairing $(M, b) \oplus (M', b')$ with underlying module $M \oplus M'$ and with map sending $((m_1, m_1'), (m_2, m_2'))$ to $b(m_1, m_2) + b'(m_1', m_2')$. A pairing is *neutral* if M contains a submodule N such that $N =$

$N^\perp = \{m \in M \mid b(n,m) = 0 \; \forall n \in N\}$. It is *hyperbolic* if it is the direct sum of two such submodules. Two pairings are *Witt equivalent* if there are neutral pairings (N,c) and (N',c') such that $(M,b) \oplus (N,c) \cong (M',b') \oplus (N',c')$. The next result is clear.

THEOREM 2.3. *The set of Witt equivalence classes of ε-linking pairings over a ring R with an involution is an abelian group, with respect to the addition determined by the sum of pairings and with $(M,-b)$ representing the inverse of the class of (M,b).* $\qquad \square$

If R is a localization of Λ_μ let $W_\varepsilon(\mathbb{Q}(t_1, \ldots t_\mu), R, -)$ denote this *Witt group*. If L is a 1-link let $B_S(L)$ and $B_\Sigma(L)$ denote the Witt equivalence classes of $b_S(L)$ and $b_\Sigma(L)$ in $W_{+1}(\mathbb{Q}(t_1, \ldots t_\mu), \Lambda_\mu S, -)$ and $W_{+1}(\mathbb{Q}(t_1, \ldots t_\mu), \Lambda_\mu \Sigma, -)$, respectively.

THEOREM 2.4. *Let $L(0)$ and $L(1)$ be concordant links. Then $B_S(L(0)) = B_S(L(1))$.*

PROOF. Let $\mathcal{L}$ be a concordance from $L(0)$ to $L(1)$. Let $N(\mathcal{L})$ be an open regular neighbourhood for the image of $\mathcal{L}$, and let $Z = S^3 \times I \setminus N(\mathcal{L})$. Then $\partial Z = X_0 \cup \mu(S^1 \times S^1 \times I) \cup X_1$, where $X_i = Z \cap (S^3 \times \{i\})$ is the complement of $L(i)$ (for $i = 0,1$) and where the jth boundary component of X_0 is identified with the jth boundary component of X_1 via an orientation reversing map. The inclusions of X_0 and X_1 into Z each induce isomorphisms on homology. The Mayer-Vietoris sequence of the triple $(\partial Z', X_0', X_1')$ gives an isomorphism $H_1(\partial Z; \Lambda_\mu S) \cong H_1(X_0; \Lambda_\mu S) \oplus H_1(X_1; \Lambda_\mu S)$. Clearly the Blanchfield pairing on $\hat{t}H_1(\partial Z; \Lambda_\mu S)$ is the direct sum of the Blanchfield pairing on $\hat{t}H_1(X_0; \Lambda_\mu S)$ with the negative of the Blanchfield pairing on $\hat{t}H_1(X_1; \Lambda_\mu S)$. Thus to show that $B_S(L(0)) = B_S(L(1))$ it shall suffice to show that $\hat{t}H_1(\partial Z; \Lambda_\mu)$ contains a submodule which is its own annihilator with respect to the pairing.

The differentials in the chain complex for the maximal abelian cover (Z', X_0') reduce to those of (Z, X_0), and so $H_2(Z, X_0; \Lambda_\mu)$ is a Λ_μ-torsion module, since $H_2(Z, X_0; \mathbb{Z}) = 0$. Therefore the image of $H_2(Z; \Lambda_\mu)$ is contained in the torsion submodule of $H_2(Z, \partial Z; \Lambda_\mu)$, since $(Z, \emptyset) \subseteq (Z, X_0) \subseteq (Z, \partial Z)$. It follows that the sequence

$$TH_2(Z, \partial Z; \Lambda_\mu) \xrightarrow{\;\partial\;} TH_1(\partial Z; \Lambda_\mu) \to TH_1(Z; \Lambda_\mu)$$

is exact. Let P be the image of $TH_2(Z, \partial Z; \Lambda_\mu)$ in $\hat{t}H_1(\partial Z; \Lambda_\mu)$. We claim that $P^\perp$ is its own annihilator.

For let Q, R be relative 2-cycles on $(Z', \partial Z')$ representing torsion classes in $H_2(Z, \partial Z; \Lambda_\mu)$ and let q, r be the boundaries of Q, R, respectively, which are 1-cycles on $\partial Z'$ representing classes in P. Then $\alpha q = \partial u$ for some nonzero $\alpha \in \Lambda_\mu$ and some 2-chain u on $\partial Z'$. Let $\hat{q}, \hat{u}$ denote q, u considered as chains on Z'. Then

$$V_{\partial Z'}(q, r) = [\bar{\alpha}^{-1}\Sigma_{\gamma \in Z^\mu} I_{\partial Z'}(u, \gamma r)\gamma] = [\bar{\alpha}^{-1}\Sigma_{\gamma \in Z^\mu} I_{Z'}(u, \gamma R)\gamma],$$

and so

$$V_{\partial Z'}(q, r) = V_{Z'}(\hat{q}, R) = V_{Z'}(\partial Q, R) = 0.$$

Thus $P \le P^\perp$. Now let w be a 1-cycle on $\partial Z'$ representing a torsion class in $TH_1(\partial Z; \Lambda_\mu)$ (so that $\beta w = \partial W$ for some nonzero $\beta \in \Lambda_\mu$ and some 2-chain W on $\partial Z'$) and suppose $V_{\partial Z'}(w, r) = 0$ for every 1-cycle r as above, representing a class in P. Then $V_{Z'}(\hat{w}, R) = V_{\partial Z'}(w, r) = 0$. On localizing at principal prime ideals and using Poincaré duality for $(Z', \partial Z')$, we see that $\hat{w}$ represents an element of $zH_1(Z')$. Therefore $P^\perp/P$ is pseudonull. Since $S(\Lambda_\mu)$ has no nontrivial pseudonull submodule it follows that $P^\perp = P^{\perp\perp}$. This proves the theorem. □

COROLLARY 2.4.1. *If L is null concordant then $B_S(L) = 0$.* □

The analogous results for $B_\Sigma(L)$ are also immediate consequences of Theorem 2.4: $B_\Sigma(L(0)) = B_\Sigma(L(1))$ if $L(0)$ and $L(1)$ are PL I-equivalent, and $B_\Sigma(L) = 0$ if L is PL I-equivalent to a trivial link.

The high dimensional knot concordance groups were first computed in terms of Seifert matrices [**Le69**], and the Witt class of the Blanchfield pairing is a complete invariant of concordance [**Ke75'**]. Although this is not so in the classical case, the known invariants there may be derived from duality pairings on covering spaces [**COT03**].

2.9. Additivity

Let L_- and L_+ be μ-component n-links. After an ambient isotopy of each link, if necessary, it may be supposed that $\text{Im}(L_-) \subset D_-^{n+2}$, $\text{Im}(L_+) \subset D_+^{n+2}$ (where $S^{n+2} = D_-^{n+2} \cup D_+^{n+2}$) and that for each $1 \leq i \leq \mu$ the ith component of L_- meets $\mathrm{Im}(L_+)$ only in an arc contained in the ith component of L_+, which receives opposing orientations from L_- and L_+. Then the closure of the union of arcs $\mathrm{Im}(L_-) \cup \mathrm{Im}(L_+) - (\mathrm{Im}(L_-) \cap \mathrm{Im}(L_+))$ is the image of a compatibly oriented link. If one of the links $L_\pm$ is split the ambient isotopy type of the new link is well defined. In particular, this connected sum induces an addition on C_n, the set of knot concordance classes, and the Blanchfield pairing gives rise to a homomorphism from C_{2q+1} to $W_\varepsilon(\mathbb{Q}(t), \Lambda_1, -)$, where $\varepsilon = (-1)^q$.

In general, however, the connected sum of two links is not well defined, even modulo concordance. For instance the Hopf link may be summed with its reflection to give either a 2-component trivial link or a link L with $\alpha(L) = 1$. It may also be summed with itself to give a link L with linking number 2 and $TA(L) = \Lambda_2/(t_1 t_2 + 1)$. The localization of the latter module with respect to Σ has length 1 as a $\Lambda_{2\Sigma}$-module, and so cannot support any pairing whose Witt class is divisible by 2, as might be expected if additivity always held.

Nevertheless some additivity results may be obtained by restricting the classes of links considered. (An alternative approach is to consider *based* links - see Chapter 14.) Let $L_-\sharp L_+$ denote any link formed in the above fashion from two such links.

In the first place, if $\mathcal{L}$ is a concordance then $\mathcal{L}|_{\mu_* \times I}$ embeds μ disjoint arcs and so by general position and the isotopy extension theorem it may be assumed that $\mathrm{Im}(\mathcal{L})$ is contained in $D_-^{n+2} \times I$ and meets $S^{n+1} \times I$ in $(\mu D^n) \times I$. Hence if L_0 and L are concordant to L_0' and L', respectively, then any link of the form $L_0 \sharp L$ is concordant to some link of the form $L_0' \sharp L'$. In particular, the set of concordance classes containing split links forms an abelian group, isomorphic to $(C_n)^\mu$, which acts on the set $C_n(\mu)$ of concordance classes of μ-component links. (In fact $C_n(\mu)$ is equivariantly isomorphic to $(C_n)^\mu \times C_n^I(\mu)$, where $C_n^I(\mu)$ is the set of PL I-equivalence classes of links. The map from $C_1^I(\mu)$ to $C_1(\mu)$ sends the class of L to the class of $L\sharp(\mathrm{II}(-L_i))$, where $\mathrm{II}(-L_i)$ is the split link whose i^{th} component is the reflected inverse of L_i.)

Secondly, if $L_\pm$ are each boundary links, and bound systems of disjoint surfaces in $D_\pm^{n+2}$ then $L_-\natural L_+$ is a boundary link. If $n = 1$ it is clear from the interpretation of the Blanchfield pairing in terms of Seifert surfaces (see §7 above) that $B_S(L_-\natural L_+) = B_S(L_-)+B_S(L_+)$. Hence the image of the set of boundary 1-links in $W_+(\mathbb{Q}(t_1,\ldots t_\mu), \Lambda_\mu S, -)$ is a subgroup. (Since $B_S(-L) = -B_S(L)$ it is clearly closed under inversion.) However, the ribbon link $L_1 \cup L_2$ of Figure 8.1 is a connected sum of two copies of the trivial link, so a connected sum of boundary links need not be a boundary link. (In this example the homology modules are all torsion free, and so the Blanchfield pairings still add.)

2.10. Signatures

In the odd dimensional knot theoretic cases the Witt groups $W_\varepsilon(\mathbb{Q}(t), \Lambda, -)$ are isomorphic to $\mathbb{Z}^\infty \oplus (Z/4Z)^\infty \oplus (Z/2Z)^\infty$, for $\varepsilon = (-1)^q$ [Le69]. The Z summands are detected by signatures.

LEMMA 2.5. *If b is a primitive ε-hermitean pairing on a finitely generated R-torsion module M and if N is a finitely generated submodule of M such that $N \le N^\perp$ then (M, b) is Witt-equivalent to $(N^\perp/N^{\perp\perp}, b_N)$, where b_N is the primitive ε-hermitean pairing induced on $N^\perp/N^{\perp\perp}$ by b.*

PROOF. The pairing b_N is defined by $b_N([n], [n']) = b(n, n')$ for all $n, n' \in N^\perp$ is clearly a primitive ε-hermitean pairing. Let $P = \{(p, [p]) \mid p \in N^\perp\}$ be the image of $N^\perp$ in $M \oplus (N^\perp/N^{\perp\perp})$ under the diagonal embedding. Then P is self-orthogonal with respect to $b \oplus (-b_N)$, for if $m \in M$ and $n \in N^\perp$ are such that $b \oplus (-b_N)((m, [n]), (p, [p])) = 0$ for all $p \in N^\perp$, then $b(m - n, p) = 0$ for all $p \in N^\perp$ and so $m - n \in N^{\perp\perp}$. In particular, m is in $N^\perp$ and $(m, [n]) = (m, [m])$ is in P. Therefore $(M, b) \oplus (N^\perp/N^{\perp\perp}, -b_N)$ is neutral. □

Let R be a discrete valuation ring with maximal ideal $\mathfrak{p}$ and field of fractions R_0, and with an involution $\bar{\ }$. If (M, b) is an ε-linking pairing over R and N is a submodule of M then clearly $N^{\perp\perp} \le N$, and as these submodules have the same length they are in fact equal. If M is annihilated by $\mathfrak{p}^n$ then b takes values in the submodule $\mathfrak{p}^{-n}R/R$ of R_0/R. Thus if $n > 1$ and $N = \mathfrak{p}^{n-1}M$ we have $N \le N^\perp$. Hence (M, b) is Witt-equivalent to $(N^\perp/N, b_N)$, where b_N is the ε-linking pairing induced on $N^\perp/N$ by b. Repeating this argument, we see that (M, b) is Witt equivalent to some (M', b') where $\mathfrak{p}M = 0$. As M' is then a vector space over the residue field $k_\mathfrak{p} = R/\mathfrak{p}$ and b' takes values in $\mathfrak{p}^{-1}R/R \cong k_\mathfrak{p}$, we may regard (m', b') as an ε-hermitean form on M'. Conversely, every such ε-hermitean form on a $k_\mathfrak{p}$-vector space may be viewed as a ε-linking pairing on a torsion $R_\mathfrak{p}$-module. Hence $W_\varepsilon(F, R_\mathfrak{p}, -) \cong W_\varepsilon(k_\mathfrak{p}, -)$. Any $(+1)$-hermitean form (V, b) over an involuted field K splits as an orthogonal direct sum of pairings on 1-dimensional subspaces.

If M is a finitely generated $\mathbb{R}[t]$-torsion module then $M \cong \oplus M_\mathfrak{p}$, where $M_\mathfrak{p} \cong R_\mathfrak{p} \otimes_R M$ is the $\mathfrak{p}$-primary submodule, and the summation is over all prime ideals $\mathfrak{p} = (p(t))$ of $\mathbb{R}[t]$. If moreover b is an ε-linking pairing on M then it is easily seen that the summands $M_\mathfrak{p}$ and $M_\mathfrak{q}$ are orthogonal if $\mathfrak{p} \ne \bar{\mathfrak{q}}$. Moreover if $\mathfrak{p} \ne \bar{\mathfrak{p}}$ then the restriction of the pairing to $M_\mathfrak{p} \oplus M_{\bar{\mathfrak{p}}}$ is neutral (since $(M_\mathfrak{p})^\perp = M_\mathfrak{p}$ in this submodule). Let $\mathbb{R}\Lambda = \mathbb{R}[t, t^{-1}]$. Then $W_\varepsilon(\mathbb{R}(t), \mathbb{R}\Lambda, -)$ is naturally isomorphic to $\oplus W_\varepsilon(\mathbb{R}(t)/(p(t)), -)$, where the sum is taken over all irreducible real polynomials such that $p(t) \doteq p(t^{-1})$. Apart from $t + 1$ and $t - 1$, which play no part in knot theory, any such polynomial must be a quadratic of the form $p_\theta(t) = t^2 - \cos(\theta)t + 1$, for some $0 \le \theta \le \pi$. The induced maps of $W_\varepsilon(\mathbb{Q}(t), \Lambda, -)$ onto $W_\varepsilon(\mathbb{R}(t)/(p_\theta(t)), -) \cong \mathbb{Z}$ are the *Milnor signatures* σ_θ^M. (See [**Sto**] for a detailed study of the Witt groups $W_\varepsilon(\mathbb{Q}(t), \Lambda, -)$.)

These signatures were first defined for a 1-knot K in terms of the Milnor pairing b_K^M on $H = H^1(X', \partial X'; \mathbb{R})$, given by $b_K^M(\alpha, \beta) = (\alpha \cup t^*\beta + \beta \cup t^*\alpha)[X', \partial X']$ for all $\alpha, \beta \in H$, where t is the generator of $Aut(X'/X)$ corresponding to a meridian and $[X', \partial X']$ is the generator of $H_2(X', \partial X'; \mathbb{Z})$ determined by the orientation of X. If θ is irreducible and $\theta \dot= \bar{\theta}$ then the θ-primary submodule of H is an orthogonal direct summand and $\sigma_\theta^M(K)$ is the signature of the restriction of b_K^M to this summand [**Ke79, Li84**]. With respect to a suitable basis b_K^M has matrix $A + A^{tr}$, where A is a Seifert matrix for

K [**Er69**]. In this guise b_K^M and its signature $\sigma(K)$ (the sum of all the $\sigma_\theta^M(K)$) were originally due to Trotter [**Tr62**]. The pairing and signature were extended to links by Murasugi [**Mu65'**]. The sum $\sigma(L) + \Sigma_{i<j}\ell_{ij}$ is invariant under changes of orientation [**Mu70'**].

More generally, if A is the Seifert matrix associated to the homology in degree q of a Seifert hypersurface for a $(2q+1)$-link L and ξ is a complex number of modulus 1 then $(1-\bar{\xi})A+(1-\xi)A^{tr}$ is a $(-1)^{q+1}$-hermitean matrix. The signatures $\sigma_A(\xi) \in \mathbb{Z}$ depend only on L, and so define a function $\sigma_L : S^1 \to \mathbb{Z}$. (This is due to [**Tr69**] for 1-links and [**Le69**] for odd-dimensional knots.) The function σ_L is piecewise continuous, with jumps only at the roots of $\Delta(t) = \det(tA - A^{tr})$. The jump in σ_L at $\theta \in S^1$ is just σ_θ^M [**Ma77, Li84**]. If ξ is a pth root of unity for some prime p then $\sigma_L(\xi)$ is a concordance invariant [**Tr69**]. Since roots of unity of prime order are dense in S^1 the function obtained from σ_L by replacing the value at discontinuities by the average of the limits from either side is also a concordance invariant. The function σ_L is also a skein invariant, for links with nonzero first Alexander polynomial [**Li90**].

Viro reinterpreted these signatures for roots of unity in terms of the G-Index Theorem. Let V_k be the k-fold branched cyclic cover of D^{2q+4}, branched over a proper codimension 2 submanifold spanning a $(2q + 1)$-link L. The covering group Z/kZ has a canonical generator t, and acts isometrically on $H_{q+2}(V_k; \mathbb{C})$ with respect to the intersection pairing. The signature of the restriction of this pairing to the ξ-eigenspace of the action of t is independent of the choice of spanning submanifold, and is a concordance invariant if k is a prime power, by a G-Index Theorem argument. Moreover if $q = 0$ and ξ is a kth root of unity this signature is $\sigma_L(\xi)$ [**Vi73**]. (See also [**KT76**] for the case $q = 0$ and $\xi = -1$.)

Smolinsky has extended the G-Index Theorem approach to define multisignatures associated to finite abelian covers of the exteriors of $(2q + 1)$-links [**Sm89**]. In the classical case similar multisignatures were defined by Cooper [**Cp82**]. These are again concordance invariants. Can the link signature invariants also be interpreted as homomorphisms from a Witt group to $\mathbb{Z}$?

There are also signature invariants associated to pairings on canonical subquotients of a knot module supporting a Blanchfield pairing. Let (M, b) be a $+1$-linking pairing over $\mathbb{R}\Lambda_{\mathfrak{p}}$, where $\mathfrak{p} = (p_\theta(t))$. is generated by an irreducible self-conjugate quadratic polynomial. For each $r \geq 1$ let $M(r) = \{m \in M \mid p_\theta(t)^r = 0\}$ and $M^r = M(r)/(M(r-1) + p_\theta(t)M(r+1))$. Let $\langle [m], [n] \rangle_r = b(p_\theta(t)^{r-1}m, n)$ for $m, n \in M$. Then M^r is a finite dimensional vector space over the field $\mathbb{R}\Lambda/\mathfrak{p} \cong \mathbb{C}$, and $\langle -, - \rangle_r$ is an hermitean pairing on M^r. This pairing is determined up to isomorphism by its signature $\sigma_{\mathfrak{p}}^r(M, b)$. These signatures are not Witt invariant, but are 0 if M is the direct sum of two self-annihilating submodules. In [**Ke81**] it is shown that these signatures provide invariants of DNC-equivalence for odd dimensional knots. (For $(4q-1)$-knots the Blanchfield pairing is skew hermitean, but multiplication by $t - t^{-1}$ gives a $+1$-linking pairing.)

We shall mention only one of the projections to a torsion summand, the $Z/2Z$-valued *Arf invariant*, which plays an important role in applications of links to questions about surfaces in 4-manifolds. Robertello defined this invariant for 1-knots in terms of a 4-manifold pair (W^4, D^2) with boundary (S^3, K) and showed that it was a concordance invariant and that it may be computed in terms of a quadratic enhancement of the Seifert form [**Ro65**]. Levine defined the Arf invariant for odd dimensional knots and showed that $Arf(K) = 0$ in $Z/2Z$ if and only if $\Delta_K(-1) \equiv \pm 1 \mod (8)$, where $\Delta_K(t)$ is the Alexander polynomial of K [**Le66**]. Since the figure eight knot 4_1 has Alexander polynomial $t^2 - 3t + 1$ its Arf invariant is nontrivial, and since 4_1 is $-$amphicheiral it represents an element of order at most 2 in C_1. Hence the Arf invariant homomorphism splits off a $Z/2Z$ summand of C_1.

The Arf invariant has been extended to proper links. A μ-component 1-link L is *proper* if the total linking number of each component L_i with all the other components is even, i.e., if $\Sigma_{j \neq i}\ell_{ij} \equiv 0 \mod (2)$ for $1 \leq i \leq \mu$. (Note that sublinks of proper links need not be proper.) Robertello showed that if there is a knot K and a planar surface in $S^3 \times [0, 1]$ with boundary $L \times \{0\} \cup K \times \{1\}$ then $Arf(K)$ depends only on L, provided that L is proper, and so we may set $Arf(L) = Arf(K)$. (Such knots and planar surfaces may be obtained by iterated fusions of distinct components of the link.) It is not yet known whether $Arf(L)$ admits a simple description in terms of other invariants of the link, although $Arf(L)$ may be derived from the second derivative of the Conway polynomial of L and its sublinks if all the linking numbers ℓ_{ij} are even [**Ho84**]. (Hoste's formula is well defined for all links, but it is not known whether it agrees with the Arf invariant for all proper links, nor what its significance may be for other links.) See also [**Gi93, KY00, Kw02**].

Duality in other covering spaces leads to further invariants. See also [**Sm89', CO93, CK99**] and [**Le00**].

CHAPTER 3

Determinantal Invariants

The main determinantal invariants of modules and chain complexes over a noetherian ring are the elementary ideals, their divisorial hulls, Reidemeister-Franz torsion and the Steinitz-Fox-Smythe invariant. We shall describe these, and also consider some special features of low-dimensional rings and of Witt groups of hermitean pairings on torsion modules. Our principal technique shall be to reduce to the case of a discrete valuation ring by localizing at height 1 prime ideals. (In our applications the ring shall usually be a quotient of a ring of Laurent polynomials over $\mathbb{Z}$ or over a field.)

3.1. Elementary ideals

Let R be a commutative noetherian ring. Let $R^\times$ denote the group of units (invertible elements) of R. If two elements $r, s \in R$ generate the same principal ideal, i.e., if $r = us$ for some $u \in R^\times$, we shall write $r \doteq s$. If I is an ideal of R then $\sqrt{I}$ is the intersection of the prime ideals that contain I. In particular, $rad\, R = \sqrt{0}$ is the ideal generated by all nilpotent elements of R. A nonzero ideal I of a noetherian domain R is *invertible* if it is projective as an R-module. The ring R has *global dimension* d if every R-module M has a projective resolution of length d.

Let M be a finitely generated R-module. The *R-torsion submodule* of M is $TM = \{m \in M \mid rm = 0 \text{ for some non-zerodivisor } r\}$, and M is an R-torsion module if $TM = M$. The *annihilator ideal* of M is $Ann(M) = \{r \in R \mid rm = 0 \,\forall m \in M\}$. If R is an integral domain the *rank* of M is the dimension of $M_0 = R_0 \otimes_R M$ as a vector space over the field of fractions R_0.

We may assume that there is an epimorphism $\phi : R^q \to M$ whose kernel is generated by p elements. If we view the elements of R^q as *row* vectors then a set of p generators for $\text{Ker}(\phi)$ determines a $p \times q$ presentation matrix Q for M. This presentation has *deficiency* $q - p$. If Q is injective the exact sequence

$$0 \to R^p \xrightarrow{\ Q\ } R^q \to M \to 0$$

is a *short free resolution* of M. For each $k \geq 0$ the kth *elementary ideal* of M is the ideal $E_k(M)$ generated by the $(q - k) \times (q - k)$ sub-determinants of the matrix representing Q if $k < q$ and by 1 if $k \geq q$. Two such finite presentations for M may be related by a sequence of elementary operations ("Tietze moves") and it is not hard to see that these ideals depend only on M [**CF**]. Clearly $E_k(M) \leq E_{k+1}(M)$ and $E_k(M_S) = E_k(M)_S$ for any multiplicative system S in R. More generally, if $f : R \to R'$ is a ring homomorphism then $f(Q)$ is a presentation matrix for $M' = R' \otimes_R M$ over R', and so $E_k(M')$ is the ideal generated by $f(E_k(M))$ in R'.

Let $\wedge_k M$ be the kth exterior power of M, and $\alpha_k M = Ann(\wedge_k M)$, for each $k \geq 0$. This notation is due to Auslander and Buchsbaum, who showed that if R is a local domain and $\alpha_k M$ is principal for all k then M is a direct sum of cyclic modules, and used this to give criteria for projectivity [**AB62**]. Since $\wedge_k(M_S) = (\wedge_k M)_S$ it follows that $\alpha_k(M_S) = (\alpha_k M)_S$, while clearly $\alpha_k M \leq \alpha_{k+1} M$.

We shall usually invoke *Cramer's rule* in the following form. Let A be an $a \times a$ R-matrix and let $d \neq 0$ divide each of the $(a-1) \times (a-1)$ subdeterminants of A. If u is a $a \times 1$ column vector or a $1 \times a$ row vector then $(\det(A)/d)u$ is an R-linear combination of the columns or rows of A, respectively.

THEOREM 3.1. *Let M be a finitely generated R-module of rank r. Then*
   (1) $E_0(M) \leq Ann(M) = \alpha_1 M$;
   (2) $\sqrt{E_k(M)} = \sqrt{\alpha_{k+1} M}$ *for each $k \geq 0$.*

PROOF. (1) We may assume $E_0(M) \neq 0$. Let D be a $q \times q$ submatrix of a presentation matrix Q for M, with $\delta = \det(D) \neq 0$. Then $\delta R^q \leq D(R^q) \leq Q(R^p)$, by Cramer's rule, and so $\delta\phi(u) =$

$\phi(\delta u)$ is in $\mathrm{Im}(\phi Q) = 0$ for all u in R^q. Hence δ is in $Ann(M)$ and so $E_0(M) \leq Ann(M)$.

(2) Let $\wp$ be a prime ideal of R. We must show that $E_k(M) \leq \wp$ if and only if $\alpha_{k+1}M \leq \wp$. We may localize with respect to $S = R \setminus \wp$ and thus assume that P is the unique maximal ideal of R. Let q be the dimension of $M/\wp M$ as a vector space over the field $R/\wp$. Then $\alpha_k(M/\wp M) = 0$ if $k \leq q$ and $\alpha_{q+1}(M/\wp M) = R/\wp$, so $\alpha_k M \leq \wp$ if and only if $k \leq q$. Moreover, M has a presentation with q generators, by Nakayama's Lemma. Since $M/\wp M$ has dimension q, all the entries of the presentation matrix are in $\wp$, and hence $E_k(M) \leq \wp$ if and only if $k < q$, that is, if and only if $\alpha_{k+1}M \leq \wp$. In other words, $\sqrt{E_k(M)} = \sqrt{\alpha_{k+1}M}$. $\square$

In fact $E_k(M) \leq \alpha_{k+1}M \leq (E_k(M) : E_{k+1}(M))$, for all $k \geq 0$ [**BE77**]. (This paper also gives sufficient conditions for these ideals to be equal.) Is $E_r(M) = E_0(TM)E_r(M/TM)$ always? (See Theorem 3.12 below.)

Assume for the remainder of this section that R is an integrally closed noetherian domain. (Most of our arguments apply also to any Krull domain.) If I is an ideal in R its *divisorial hull* $\tilde{I}$ is the intersection of the principal ideals of R which contain I. If S is a multiplicative system in R then $\widetilde{I_S} = \tilde{I}_S$ as ideals of the localization R_S, while if R is factorial and $I \neq 0$ then $\tilde{I}$ is the principal ideal generated by the highest common factor of the elements of I.

LEMMA 3.2. *The divisorial hull of I is $\cap I_\wp$, the intersection of all of its localizations at height 1 prime ideals $\wp$ of R.*

PROOF. Since R is an integrally closed noetherian domain it is the intersection of all of its localizations at height 1 primes, and these are each discrete valuation rings. Therefore if I is an ideal of R then $\tilde{I}_\wp = I_\wp$ so $\tilde{I} \leq \cap(\tilde{I})_\wp = \cap \tilde{I}_\wp = \cap I_\wp$. On the other hand, if $I \leq (a)$ then $\cap I_\wp \leq \cap(a)_\wp = (a) \cap R_\wp = (a)$, so $\cap I_\wp \leq \tilde{I}$. $\square$

If R is factorial and M is a finitely generated R-module let $\Delta_k(M)$ be any generator of the principal ideal $\widetilde{E_k}(M)$, for all $k \geq 0$, and let $\lambda_k(M) = \Delta_{k-1}(M)/\Delta_k(M)$, for all $k > r$. We shall also let $\lambda_k(M) = 0$ for $k \leq r$. If M is an R-torsion module its *order* is $\Delta_0(M)$.

We shall say that M *satisfies the Elementary Divisor Theorem* if it is isomorphic to a direct sum $\oplus_{i \geq 1}(R/(\theta_i))$, with θ_{i+1} dividing θ_i for all $i \geq 1$. Finitely generated modules over principal ideal domains (PIDs) satisfy the Elementary Divisor Theorem; this is one formulation of the structure theorem for such modules.

LEMMA 3.3. *If R is a discrete valuation ring and M is a finitely generated R-module of rank r then $\alpha_k M = 0$ if $k \leq r$ and $\alpha_{r+j}M = \alpha_j TM = (\lambda_{r+j}(M))$, for all $j \geq 1$.*

PROOF. Let $\wp$ be the maximal ideal of R. Then $M \cong R^r \oplus TM$ and $TM \cong \bigoplus_{i=1}^{i=n}(R/\wp^{e(i)})$, where $e(i) \geq e(i+1) > 0$ for $1 \leq i < n$, by the Elementary Divisor Theorem. Therefore $E_k(M) = 0$ if $k < r$ and $E_{r+j}(M) = E_j(TM) = \wp^{s_j}$, where $s_j = \Sigma_{i=j+1}^{i=n}e(i)$, for each $j \geq 0$. Moreover, $\wedge_k M \cong \bigoplus_{j=0}^{j=k}(\wedge_{k-j}TM)^{\binom{r}{j}}$. Hence $\alpha_k M = 0$ if $k \leq r$ and $\alpha_{r+j}M = \alpha_j TM$ for all $j \geq 0$, and so we may assume $r = 0$.

Let $I(k) = \{\iota \in \mathbb{N}^k \mid 1 \leq \iota(1) < \cdots < \iota(k) \leq n\}$. Then

$$\wedge_k M \cong \bigoplus_{\iota \in I(k)} (R/\wp^{e(\iota(k))}) \cong \bigoplus_{i=1}^{i=n}(R/\wp^{e(i)})^{f(i)},$$

where $f(i) = |\{\iota \in I(k-1) \mid \iota(k-1) < i\}|$, for $i \leq n$. Clearly $f(i) = 0$ if $i < k$ and $f(k) = 1$. Therefore $\alpha_k M = Ann(\wedge_k M) = \wp^{e(k)} = (\lambda_k(M))$. $\square$

It follows that for modules M over a Krull domain R the ideal $\alpha_{r+j}(M)$ and the ideal quotient $(E_{r+k-1}(M) : E_{r+k}(M))$ have the same divisorial hull.

THEOREM 3.4. *If R is factorial and M is a finitely generated R-module of rank r then $\alpha_k M = 0$ for all $k \leq r$ and $\widetilde{\alpha_k}M = (\lambda_k(M))$ for all $k > r$. Hence $\lambda_{k+1}(M)$ divides $\lambda_k(M)$ for all $k > r$ and $\Delta_{r+j}(M) \doteq \Delta_j(TM)$ for each $j \geq 0$.*

PROOF. By Lemma 3.2 it is enough to prove that the localizations of the ideals at height 1 primes are the same. This follows from

Lemma 3.3 on observing that every step (forming exterior powers, annihilators, etc.) is compatible with localization. ☐

COROLLARY 3.4.1. *If M is an R-torsion module with a square presentation matrix then $Ann(M) = (\lambda_1(M))$.*

PROOF. It follows easily from Cramer's rule that $(\lambda_1(M)) \leq Ann(M) = \alpha_1 M \leq \widetilde{\alpha_1} M = (\lambda_1(M))$. ☐

A finitely generated R-module M is *pseudonull* if $M_\wp = 0$ for every height 1 prime ideal $\wp$ of R. A homomorphism $f : M \to N$ is an *elementary pseudoisomorphism* if $\mathrm{Ker}(f)$ and $\mathrm{Cok}(f)$ are pseudonull. *Pseudoisomorphism* is the equivalence relation generated by elementary pseudoisomorphisms.

THEOREM 3.5. *Let M and N be finitely generated R-torsion modules. Then M and N are pseudoisomorphic if and only if $\widehat{E}_k(M) = \widehat{E}_k(N)$, for each $k \geq 0$. In particular, M is pseudonull if and only if $\widetilde{\alpha_1} M = R$.*

PROOF. Since $E_k(M)_\wp = E_k(M_\wp) = E_k((N)_\wp) = E_k(N)_\wp$ the necessity of these conditions is clear. Suppose that they hold, and let $\widetilde{M} = \oplus_\wp M_\wp$ and $\widetilde{N} = \oplus_\wp N_\wp$, where the sums are taken over the height 1 primes $\wp$ which divide $\widehat{E}_0(M) = \widehat{E}_0(N)$. Let $j_M : M \to \widetilde{M}$ and $j_N : N \to \widetilde{N}$ be the canonical homomorphisms. There is an isomorphism $\theta : \widetilde{N} \cong \widetilde{M}$, by the Elementary Divisor Theorem. Let $P = j_M(M) + \theta j_N(N)$. Then j_M and θj_N determine elementary pseudoisomorphisms from M and N to P. The final observation is similar. ☐

See Theorem VII.4.5 of [**Bou**] for an equivalent result. In general, if M and N are pseudoisomorphic then $\widetilde{\alpha_k} M = \widetilde{\alpha_k} N$ and $\widehat{E}_k(M) = \widehat{E}_k(N)$, for each $k \geq 0$. Are modules of rank > 0 determined up to pseudoisomorphism by their elementary ideals?

LEMMA 3.6. *Let A be a $b \times c$ R-matrix of rank d, and suppose that there is a $d \times d$ submatrix D such that $\det(D)$ divides every $d \times d$ subdeterminant of A. Then there are invertible square matrices B and C such that $BAC = \left(\begin{smallmatrix} D & 0 \\ 0 & 0 \end{smallmatrix} \right)$.*

PROOF. After permuting the rows and columns if necessary, we may assume that D is in the top left hand corner of A. We may then apply Cramer's rule to annihilate the partial rows and columns below and to the right of D. The bottom right hand corner block of the resulting matrix must be zero as the ranks of BAC and D are both d. $\square$

A finitely generated R-module is projective if and only if the localization $M_\wp$ is a free $R_\wp$-module, for each prime ideal $\wp$ of R. (Thus an ideal J is invertible if and only if every localization $J_\wp$ is principal.)

THEOREM 3.7. *Let M be a finitely generated R-module of rank r. Then $E_r(M)$ is invertible if and only if $P = M/TM$ is projective and $p.d._R TM \leq 1$. If so, then $M \cong P \oplus TM$ and $E_{r+j}(M) = E_j(TM)$ for each $j \geq 0$.*

PROOF. Suppose that $E_r(M)$ is invertible and let Q be a presentation matrix for M. Let $P = M/TM$. On localizing at each prime ideal $\wp$ we may apply Lemma 3.6, for any generating set of a principal ideal in a local domain must contain a generator, by Nakayama's Lemma. Thus $M_\wp \cong R_\wp^r \oplus TM_\wp$, and $TM_\wp$ has a short free resolution. Thus $P_\wp$ is free. Hence P is projective and so $M \cong P \oplus TM$. Moreover, if $\psi : R^c \to TM$ is any epimorphism then $\mathrm{Ker}(\psi)$ is locally free, by Schanuel's Lemma, and hence projective, so $p.d._R TM \leq 1$.

Conversely, if P is projective and $p.d._R TM \leq 1$ then $P_\wp$ is free and $TM_\wp$ has a square presentation matrix, since it is a torsion module. Hence $E_r(M)_\wp = E_r(M_\wp) = E_r(P_\wp \oplus TM_\wp) = E_0(TM_\wp)$ is principal and so $E_r(M)$ is invertible. The final assertion is clear. $\square$

COROLLARY 3.7.1. *M is projective if and only if $E_r(M) = R$.* $\square$

THEOREM 3.8. *Let M be a finitely generated R-module of rank r and such that $p.d._R M \leq 1$. Then M has no nontrivial pseudonull submodule and $Ann(TM) = (\lambda_{r+1}(M))$.*

PROOF. Let N be a pseudonull submodule of M and $\wp$ a prime ideal of R. Then $N_\wp$ is a pseudonull submodule of $M_\wp$. Since $N = 0$

if and only if $N_\wp = 0$ for all $\wp$, and since localization is exact, we may therefore assume that R is local, and hence that there is an epimorphism $\phi : R^q \to M$ with free kernel. We shall induct on the rank r of M.

If $r = 0$ then M has a square presentation matrix Q, so $E_0(M)$ is principal and $\widetilde{E_0}(M) = E_0(M)$. Since N is pseudonull $\widetilde{E_0}(M/N) = \widetilde{E_0}(M)$, by Theorem 3.5. Hence the inclusions $E_0(M) \leq E_0(M/N) \leq \widetilde{E_0}(M/N)$ are all equalities. Since M/N is a quotient of M it has a presentation matrix of the form $\begin{pmatrix} Q \\ Q_1 \end{pmatrix}$, which by the argument of Lemma 3.6 may be changed to $\begin{pmatrix} Q \\ 0 \end{pmatrix}$ by row operations. Hence the projection of M onto M/N is an isomorphism, so $N = 0$.

If $r \geq 1$ then we may assume that $m = \phi((1, 0, \ldots, 0))$ generates a free submodule of M. Let $M' = M/Rm$ and let $f : M \to M'$ be the canonical epimorphism. Then $p.d._R M' \leq 1$ and M' has rank $r - 1$, so by the inductive hypothesis M' has no nontrivial pseudonull submodule. But f maps N isomorphically into M', as Rm is torsion free. Therefore $N = 0$.

Since $(\widetilde{Ann}(TM))TM$ is pseudonull $Ann(TM) = \widetilde{Ann}(TM) = (\lambda_{r+1}(M))$, by Theorem 3.4. $\square$

If M is a torsion module then $E_0(M)$ principal implies that M has no nontrivial pseudonull submodule, by Theorems 3.7 and 3.8. If R has global dimension ≤ 2 and M is a finitely generated R-module with no nontrivial pseudonull submodule then $p.d._R M \leq 1$, by Theorem 3.22 below. However, in general these implications are strict. (For instance, if $R = \mathbb{Z}[X]$ and M is any maximal ideal then $E_0(M) = M$ is not principal, although M is torsion free, while if $M = \mathbb{Z} \oplus (Z/2Z) = R/(X) \oplus R/(2, X)$ then the second summand is pseudonull although $Ann(M) = (X)$ is principal.)

At the other extreme we may characterize pseudonull modules homologically. Let $e^k M = Ext_R^k(M, R)$, for all $k \geq 0$. The dual $M^* = e^0 M \cong Hom_R(M, R)$ is finitely generated and torsion free, and has the same rank as M. There is a natural *evaluation homomorphism* $ev_M : M \to M^{**}$, with kernel TM. The module M is *reflexive* if ev_M is an isomorphism. If $M = N^*$ then $(ev_N)^* ev_{N^*} = id_{N^*}$ and so

ev_M is a split monomorphism. Since M and M^* have the same rank, it must be an isomorphism. Thus a finitely generated R-module is reflexive if and only if it is a dual.

If M is a torsion module then $e^1 M = Hom_R(M, R_0/R)$, as follows on applying $Ext_R^*(M, -)$ to the exact sequence

$$0 \to R \to R_0 \to R_0/R \to 0.$$

Hence there is a similar evaluation $W_M : M \to e^1 e^1 M$.

THEOREM 3.9. *Let M be a finitely generated R-module. Then*

(1) *M is a torsion module if and only if $M^* = 0$;*
(2) *$Hom_R(M, R_0/R)$ is a torsion R-module with no nontrivial pseudonull submodule;*
(3) *$\mathrm{Ker}(W_{TM})$ is the maximal pseudonull submodule of M.*

Hence M is pseudonull if and only if $e^0 M = e^1 M = 0$.

PROOF. The first assertion is obvious. Let $m_1, \ldots, m_q$ generate M and let $f : M \to R_0/R$. If $f(m_i) = r_i/s_i$ mod R (with $s_i \neq 0$) then $(\Pi s_i) f = 0$ and so $Hom_R(M, R_0/R)$ is a torsion module. If $f_\wp = 0$ for all height 1 prime ideals $\wp$ then $f(m) \in \cap R_\wp = R$ for each m in M. Therefore $Hom_R(M, R_0/R)$ has no nonzero pseudonull submodule. It is an immediate consequence of the structure theorem for finitely generated modules over a principal ideal domain that $W_{TM,\wp} : TM_\wp \to e^1 e^1 TM_\wp$ is an isomorphism for each height 1 prime ideal $\wp$. Hence $\mathrm{Ker}(W_{TM})$ is pseudonull and the remaining assertions follow readily. $\square$

3.2. The Elementary Divisor Theorem

Let R be a factorial noetherian domain and M a finitely generated R-module which is the direct sum of cyclic submodules. If M satisfies the Elementary Divisor Theorem then all of its elementary ideals are principal. To establish the converse we shall begin with the special case of the direct sum of two cyclic modules.

LEMMA 3.10. *Let $M = (R/(\alpha\beta_1)) \oplus (R/(\alpha\beta_2))$ where $(\beta_1, \beta_2) = R$. Then $M \cong (R/(\alpha\beta_1\beta_2)) \oplus (R/(\alpha))$.*

PROOF. Let m and $n \in (R/(\alpha\beta_1)) \oplus (R/(\alpha\beta_2))$ generate the first and second summands, respectively. Since $(\beta_1, \beta_2) = R$, there are elements a, b in R such that $a\beta_1 + b\beta_2 = 1$. Let $m' = am + bn$ and $n' = \beta_2 m - \beta_1 n$. Then $m = \beta_1 m' + bn'$ and $n = \beta_2 m' - an'$, so m' and n' also generate M. Clearly $\alpha\beta_1\beta_2 m' = 0 = \alpha n'$, so there is an epimorphism from $(R/(\alpha\beta_1\beta_2)) \oplus (R/(\alpha))$ to M. Since these are both pure modules with 0th elementary ideal $(\alpha^2\beta_1\beta_2)$ this must be an isomorphism. □

In general we may show that if a direct sum of cyclic modules has all its elementary ideals principal then so does the direct sum of any two cyclic summands. The lemma may then be applied repeatedly.

THEOREM 3.11. *Let $M \cong \oplus_{i \geq 1}(R/(\xi_i))$ and suppose that the ideals $E_i(M)$ are all principal. Then M satisfies the Elementary Divisor Theorem.*

PROOF. We shall work through the case of three summands and then comment on how the argument may be extended.

Let $\alpha = \mathrm{hcf}(\xi_1, \xi_2, \xi_3)$, $\beta_1 = \mathrm{hcf}(\xi_2/\alpha, \xi_3/\alpha)$, $\beta_2 = \mathrm{hcf}(\xi_1/\alpha, \xi_3/\alpha)$ and $\beta_3 = \mathrm{hcf}(\xi_1/\alpha, \xi_2/\alpha)$. Then we may factor the elements ξ_i as $\xi_1 = \alpha\beta_2\beta_3\gamma_1$, $\xi_2 = \alpha\beta_1\beta_3\gamma_2$ and $\xi_3 = \alpha\beta_1\beta_2\gamma_3$. Note that α divides all the ξs, each β divides all but one of the ξ/αs, and each γ divides all but two of the $\xi/\alpha\beta$s. We then have $E_0(M) = (\alpha^3(\beta_1\beta_2\beta_3)^2\gamma_1\gamma_2\gamma_3)$. Let $J_k = \Delta_k(M)^{-1}E_k(M)$ for $k \geq 0$. Then $J_1 = (\beta_3\gamma_1\gamma_2, \beta_2\gamma_1\gamma_3, \beta_1\gamma_2\gamma_3)$ and $J_2 = (\beta_2\beta_3\gamma_1, \beta_1\beta_3\gamma_2, \beta_1\beta_2\gamma_3)$, where each of these triples have no common factor. In fact $J_1 = J_2 = R$, since the elementary ideals of M are principal, but the notation shall be useful.

Suppose that $(\xi_1, \xi_2) = \alpha\beta_3(\beta_2\gamma_1, \beta_1\gamma_2)$ is not principal. Then $(\beta_2\gamma_1, \beta_1\gamma_2)$ is contained in some maximal ideal m. Since every element of J_2 is divisible by either β_1 or $\beta_2\gamma_1$, we cannot have β_1 in m, for otherwise $J_2 \leq m$, contradicting $J_2 = R$. Similarly β_2 is not in m, so $(\gamma_1, \gamma_2) \leq m$. But every element of J_1 is divisible by either γ_1 or γ_2 and so $J_1 \leq m$, contradicting $J_1 = R$. Therefore we must have $(\xi_1, \xi_2) = (\alpha\beta_3)$.

By Lemma 3.10, $(R/(\xi_1)) \oplus (R/(\xi_2)) \cong (R/(\delta) \oplus (R/(\alpha\beta_3))$, where $\delta = \alpha\beta_1\beta_2\beta_3\gamma_1\gamma_2$. Applying the argument above, and then the lemma, gives $(R/(\alpha\beta_3)) \oplus (R/(\xi_3)) \cong (R/(\alpha\beta_1\beta_2\beta_3\gamma_3)) \oplus (R/(\alpha))$. One more similar step shows that M satisfies the Elementary Divisor Theorem.

In general, if $M \cong \oplus_{k\geq1}(R/(\xi_i))$ we may factor each element ξ_i as $\xi_i = \alpha(\beta \ldots)(\gamma \ldots) \ldots (\mu \ldots)\nu_i$, where α is the highest common factor of the ξ_is, each β is common to $n - 1$ of the quotients ξ_i/α, each γ is common to $n - 2$ of the $\xi/\alpha\beta$s, and so on. (In principle there may be $2^n - 1$ different factors, with 2^{n-1} occurring in any one ξ.) We then have

$$\Delta_0(M) = \alpha^n(\Pi\beta)^{n-1} \ldots (\Pi\mu)^2(\Pi\nu),$$

$$\Delta_1(M) = \alpha^{n-1}(\Pi\beta)^{n-2} \ldots (\Pi\mu),$$

and so on down to $\Delta_n(M) = \alpha$. (Each product is taken over all possible αs, and so on.) If we suppose that (ξ_1, ξ_2) is not principal, we may write $\xi_i = \xi\psi_i$ where $\xi = $ h.c.f.(ξ_1, ξ_2) and (ψ_1, ψ_2) is contained in some maximal ideal m. Assuming that $J_k = R$ for $k \geq 2$ we may then use $J_{n-1} = R$ to show that no 'β' factors of ψ_1 or ψ_2 are contained in m, and working down conclude that $(\nu_1, \nu_2) \leq m$. But then $J_1 \leq m$ and so the elementary ideals of M could not be principal. □

The Elementary Divisor Theorem also holds in the above sense for finitely generated modules over local domains, by Exercise 22 of Chapter VII of [**Bou**]. However, it does not hold for the direct sum $(R/(\xi)) \oplus (R/(\psi))$, if ξ and ψ have no common factor but generate a proper ideal $(\xi, \psi) \neq R$.

3.3. Extensions

In this section we shall consider relations between the determinantal invariants of a module and its sub- and quotient modules.

THEOREM 3.12. *Let* $0 \to K \to M \to C \to 0$ *be an exact sequence of R-modules, and let* r *be the rank of* C. *Then*

(1) $E_i(M) \geq E_j(K)E_{i-j}(C)$ *for all* $i \geq j$;

(2) *if* $p.d._R C \leq 1$ *then* $E_r(M) = E_0(K)E_r(C)$;

(3) *if* K *is a torsion module then* $\widetilde{E}_r(M) = \widetilde{E}_0(K)\widetilde{E}_r(C)$.

PROOF. Let $P(K)$ and $P(C)$ be presentation matrices for K and C, respectively. It is easy to see that M has a presentation matrix of the form $\begin{pmatrix} P(K) & 0 \\ * & P(C) \end{pmatrix}$. The first assertion is immediate. Since two ideals agree if and only if their localizations at maximal ideals of R agree we may assume that R is a local ring. In particular, projective R-modules are free. Therefore if $p.d._R C \leq 1$ we may assume that $P(C)$ is a $q \times (q+r)$ matrix of rank r. Then the only nonzero elements of $E_r(M)$ are those obtained by deleting r columns from $P(C)$ and taking the product of the resulting element of $E_r(C)$ with a subdeterminant of $P(K)$ of column index 0. The final assertion follows from the structure theorem for modules over discrete valuation rings, after localizing at height 1 prime ideals. $\square$

COROLLARY 3.12.1. *If R has global dimension ≤ 2 then* $E_r(M) = E_0(TM)E_r(M/TM)$.

PROOF. Since M/TM is torsion free, $p.d._R(M/TM) \leq 1$. $\square$

If R is an integral domain the *resultant* $Res(\theta, \psi)$ of $\theta, \psi \in R[u]$ is an element of R which is 0 if and only if θ and ψ have a common factor. If θ and ψ factor as $\theta = a\Pi_{i=1}^m (u - \xi_i)$ and $\psi = b\Pi_{j=1}^n (u - \zeta_i)$ in some field containing R then $Res(\theta, \psi) = a^n b^m \Pi(\zeta_i - \xi_j) = b^m \Pi\theta(\zeta_j)$. Clearly, $Res(\theta, \psi) = (-1)^{mn} Res(\psi, \theta)$ and depends only on the class of ψ in $R[u]/(\theta)$, and formation of resultants is compatible with change of coefficient domains. (See page 135 of [**Lan**].)

THEOREM 3.13. *Let R be a factorial noetherian domain and N a finitely generated torsion $R[u]$-module. Let $\theta \in R[u]$ be a monic polynomial. Then $Q = N/\theta N$ is finitely generated as an R-module, and is a torsion module if and only if $\Delta_0(N)(\xi) \neq 0$ for all nonzero roots ξ of θ in an algebraic closure Ω of R_0. If for each principal prime ideal $\wp$ of R the $R_\wp[u]$-ideal $E_0(N_\wp)$ is generated by the image of $\Delta_0(N)$ then $\Delta_0(Q) \doteq Res(\Delta_0(N), \theta)$.*

PROOF. If θ is monic of degree m then $R[u]/(\theta) \cong R^m$, and so Q is finitely generated over R. To verify the torsion criterion it suffices

to extend coefficients from R to Ω, where it becomes obvious. In the
second assertion we may assume that Q is a torsion R-module, for
otherwise both sides of the equation are 0. We may further assume
that R is a discrete valuation ring, since $\Delta_0(Q)$ is determined by
the localizations of Q at principal prime ideals of R. Since $E_0(N)$ is
principal and nonzero N has no nontrivial pseudonull submodules,
by Theorems 3.7 and 3.8. Hence multiplication by θ is injective on N,
for otherwise N would contain $R[u]/(\phi)$ for some nontrivial factor ϕ
of θ, which would divide $\Delta_0(N)$. There is a submodule $C \leq N$ which
is a direct sum of cyclic modules and such that $\Delta_0(C) \doteq \Delta_0(N)$.
Hence $P = N/C$ is pseudonull and so of finite length as an R-module.
In particular, $\Delta_0(P/\theta P) \doteq \Delta_0(K)$, where $K = \mathrm{Ker}(\theta id_P)$. We now
apply the Snake Lemma to the endomorphism of the short exact
sequence $0 \to C \to N \to P \to 0$ induced by multiplication by θ to
obtain an exact sequence

$$0 \to K \to C/\theta C \to Q \to P/\theta P \to 0.$$

Using Theorem 3.12 (for R-modules) several times, we conclude that
$\Delta_0(Q) \doteq \Delta_0(C/\theta C)$. Thus we are reduced to the case of cyclic
modules, which follows from basic properties of the resultant. $\square$

COROLLARY 3.13.1. *Let* $\theta, \psi \in R[u]$. *If* θ *or* ψ *is monic then*
$Res(\theta, \psi) \doteq \Delta_0(M)$, *where* $M = R[u]/(\theta, \psi)$. $\square$

Similar arguments apply for Laurent extensions $R[u, u^{-1}]$, provided the highest and lowest coefficients of θ (or ψ) are units.

The next result is from [**Le87**].

THEOREM 3.14. *Let* R *be a factorial noetherian domain and* N
a finitely generated torsion $R[u, u^{-1}]$-*module. Let* K *and* C *be the
kernel and cokernel of multiplication by* $u - 1$ *on* N. *Then* $\Delta_0(C) \doteq$
$\Delta_0(K)\varepsilon(\Delta_0(N))$ *(where* K *and* C *are considered as* R-*modules and*
$\varepsilon : R[u, u^{-1}] \to R$ *sends* u *to* 1*).*

PROOF. If $u - 1$ is injective this is a special case of Theorem 13.
If $u = 1$ then $K = C$ and $Ann(N) = (u - 1, J)$ for some ideal J in
R. Either $J = 0$, in which case $\Delta_0(N) \doteq u - 1$ and $\delta_0(K) = 0$, or
$J \neq 0$ and $\Delta_0(N) \doteq 1$. In either case the result holds.

If N is an extension of N_2 by the submodule N_1 and the lemma holds for N_1 and N_2 then it holds for N, by a Snake Lemma argument. In particular, we may show by induction on k that it holds if $(u-1)^k N = 0$. (Take $N_1 = (u-1)^{k-1}$ and $N_2 = N/N_1$.)

Since R is noetherian and N is finitely generated the increasing sequence of submodules $\{\text{Ker}(u-1)^k\}$ must be finite. Let N_1 be its maximal member. Then $(u-1)$ is nilpotent on N_1 and injective on N/N_1, so the result follows from our earlier considerations and another application of the Snake Lemma. $\qquad \square$

3.4. Reidemeister-Franz torsion

Let R be an integral domain, with field of fractions R_0. If b and c are ordered bases for a finitely generated free R-module M, with $b_i = \Sigma \Gamma_{ij} c_j$, let $[b/c] = \det(P)$ be the determinant of the change of basis matrix. (Our notation follows [**Tu86**]. In more general treatments of K-theoretic torsion $[b/c]$ denotes the image of the matrix P in $K_1(R)$. See [**Mi66**] or [**Tur**].)

Let C_* be a finitely generated free chain complex over R_0, and suppose that c_i and h_i are bases for C_i and $H_i(C_*)$, respectively. (Note that the empty set is the unique basis for the trivial module 0.) Choose partial bases b_i in C_i such that ∂b_i is a basis for $\text{Im}(\partial_i)$, and choose lifts $\tilde{h}_i$ of the bases h_i to $\text{Ker}(\partial_i)$. Then $\partial_{i+1}(b_{i+1})\tilde{h}_i b_i$ is a basis for C_i. The *Reidemeister-Franz torsion* of the based complex is $\tau(C_*; c_*, h_*) = \Pi[\partial_{i+1}(b_{i+1})\tilde{h}_i b_i / c_i]^{e(i)}$, where $e(i) = (-1)^{i+1}$, for all $i \in Z$. This element of $R_0^\times$ depends only on C_*, c_* and h_*, and not on the choice of partial bases b_* and lifts $\tilde{h}_i$.

If C_* is a finitely generated free chain complex over R with given bases c_i and $h_i \subset H_i(C_*)$ represents a basis for $R_0 \otimes_R H_i(C_*)$ we set $\tau(C_*; c_*, h_*) = \tau(R_0 \otimes_R C_*; c_*, h_*)$. If $R_0 \otimes_R C_*$ is acyclic we write just $\tau(C_*; c_*)$, and $\tau(C_*)$ for its image in $R_0^\times / R^\times$ (which is independent of the bases c_*).

The torsion is multiplicative with respect to extensions.

THEOREM. [**Mi66**] *Let* $0 \to A_* \to B_* \to C_* \to 0$ *be an exact sequence of finitely generated free R-chain complexes, with bases* a_*, $b_* = a_* \sqcup b'_*$, *and* c_*, *respectively, such that* b'_* *maps bijectively* to c_*. *Suppose also that the homology modules are free, with bases* h_*^A, h_*^B *and* h_*^C. *Then the exact sequence of homology* H_* *is an acyclic complex with basis* $h_* = h_*^A \sqcup h_*^B \sqcup h_*^C$, *and* $\tau(B_*; b_*, h_*^B) = \pm\tau(A_*; a_*, h_*^A)\tau(C_*; c_*, h_*^C)\tau(H_*; h_*)$. $\qquad\square$

This was first proven in [**Wh50**] for the case when all the complexes are acyclic. See Theorem 3.2 of [**Mi66**] for the general case.

The order of a torsion module is also multiplicative in short exact sequences, by Theorem 3.12, and together these results enable us to relate Reidemeister-Franz torsion and the orders of torsion modules.

THEOREM 3.15. *Let* $C_* = 0 \to C_n \to \cdots \to C_0 \to 0$ *be a finitely generated free chain complex over a factorial noetherian domain* R, *with bases* c_*, *and suppose that* $R_0 \otimes_R C_*$ *is acyclic. Then* $\tau(C_*) \doteq \Pi\Delta_0(H_i(C_*))^{e(i)}$, *where* $e(i) = (-1)^{i+1}$, *for all* $i \geq 0$.

PROOF. We shall argue by induction on the length n of C_*. As the result is clearly true if $n = 1$ we may assume that $n > 1$ and that the theorem holds for all such complexes of length less than n.

Let $Z_{n-2} = \mathrm{Ker}(\partial_{n-2})$ have rank r, and choose $c_1', \ldots, c_r' \in C_{n-1}$ whose images under ∂_{n-1} generate a free submodule of rank r. Let $j : C' \to C_{n-1}$ be the inclusion of the submodule of C_{n-1} generated by $\{c_1', \ldots, c_r'\}$. Let C_*' be the subcomplex of C_* such that $C_q' = C_q$ if $q < n - 1$, $C_{n-1}' = C'$ and $C_q' = 0$ if $q \geq n$. Let D_* be the complex

$$\ldots C' \xrightarrow{\ id\ } C' \ldots$$

concentrated in degrees n and $n - 1$, and let $C_*'' = C_* \oplus D_*$. Define a chain homomorphism $\alpha_* : C_*' \to C_*''$ by $\alpha_q = id_{C_q}$ if $q < n - 1$ and $\alpha_{n-1} = j \oplus id_{C'}$. Then α_* is a monomorphism with cokernel E_* again concentrated in degrees n and $n - 1$, and given by

$$C_n \oplus C' \xrightarrow{(\partial_n, j)} C_{n-1}.$$

Each of these complexes is free, and $R_0 \otimes_R C_*'$, $R_0 \otimes_R C_*''$ and $R_0 \otimes_R E_*$ are acyclic.

Equip each of the auxiliary complexes with the obvious bases. The multiplicativity of torsion implies that $\tau(C_*; c_*) = \tau(C_*''; c_*'') = \tau(C_*'; c_*')\tau(E_*; e_*)$. The inclusion of C_* into C_*'' is a chain homotopy equivalence, and so $H_q(C_*) \cong H_q(C_*'')$ for all q. The long exact sequence of homology determined by the exact sequence

$$0 \to C_*' \xrightarrow{\alpha_*} C_*'' \to E_* \to 0$$

breaks up into isomorphisms $H_q(C_*') = H_q(C_*)$ if $q < n - 2$ and an exact sequence $0 \to H_{n-1}(C_*) \to \ldots H_{n-2}(C_*) \to 0$. These modules are all R-torsion modules, and so $\Delta_0(H_{n-1}(E_*))\Delta_0(H_{n-2}(C_*'))^{-1} = \Delta_0(H_{n-1}(C_*))\Delta_0(H_{n-2}(C_*))^{-1}$, by Theorem 3.12. As C_*' and E_* are each of length $\leq n - 1$ the result now follows from the inductive hypothesis. □

This was first proven in [**Mi68**] for R a PID and in [**Tu86**] for any factorial noetherian domain. It may also be extended to any Krull domain by localization at height 1 primes and using Milnor's argument. (The result should then be viewed as identifying two fractional ideals.)

In §4.5 of Chapter VII of [**Bou**] it is shown that if R is an integrally closed noetherian domain then $M \mapsto \widetilde{E}_0(M)$ gives rise to an isomorphism $K_0(\mathcal{T}/\mathcal{T}') \cong D(R)$, where $\mathcal{T}/\mathcal{T}'$ is the category of finitely generated R-torsion modules modulo the full subcategory of pseudonull modules and $D(R)$ is the group of divisors (divisorial fractional ideals) of R. These different manifestations of $\widetilde{E}_0(M)$ are presumably related by a K-theoretic localization sequence.

3.5. Steinitz-Fox-Smythe invariants

Two nonzero ideals I and J of an integral domain R are in the same *ideal class* if they are isomorphic as R-modules, i.e., if there are nonzero a and b in R such that $aI = bJ$. Every finitely generated torsion-free R-module of rank 1 is isomorphic to an ideal, and so represents an ideal class. The set of ideal classes is a semigroup with product represented by multiplication of ideals.

Let M be a finitely generated R-module of rank r and let Q be a $p \times q$ presentation matrix for M. The ideal class of the ideal of R generated by all the subdeterminants of maximal rank from a fixed set of $q - r$ linearly independent rows of Q depends only on the module M, and not on the matrix Q or the set of rows chosen

[St12, FS64]. This invariant is the *SFS row class* $\rho(M)$; there is a similar column class $\gamma(M)$.

LEMMA 3.16. [St12] *Let M be a module of rank r. Then the product $\gamma(M)\rho(M)$ is the ideal class of $E_r(M)$.*

PROOF. Let Q be a presentation matrix for M. Suppose that the top left $(q-r) \times (q-r)$ submatrix D has nonzero determinant d. Given a set H of r linearly independent rows there is a matrix $M_H \in GL(q-r, R_0)$ which carries the first $q-r$ rows of Q onto H. Similarly given a set K of $q-r$ linearly independent columns there is a matrix N_K which carries the first $q-r$ columns of Q onto K. Hence if d_{HK} is a nonzero $(q-r) \times (q-r)$ subdeterminant corresponding to the intersection of such sets of rows and columns then $dd_{HK} = \det(M_H D)\det(DN_K)$ is the product of a subdeterminant from the first $q-r$ columns with one from the first $q-r$ rows. Thus the class of $dE_r(M)$ is $\gamma(M)\rho(M)$.   □

THEOREM 3.17. *The SFS row class $\rho(M)$ is the isomorphism class of the rank 1 torsion free module $(\wedge_r M)/T(\wedge_r M)$.*

PROOF. Let U be a $(q-r) \times q$ submatrix of maximal rank $q-r$ of a $p \times q$ presentation matrix Q. Let $\psi : (R^q)^r \to R$ be the function such that $\psi(v_1, \ldots, v_r)$ is the determinant of the $q \times q$ matrix whose first r rows are the vectors v_i and whose final $q-r$ rows are given by U. Then ψ is multilinear and alternating, and $\psi(v_1, \ldots, v_r) = 0$ if any of the arguments is in the image of Q, so ψ factors through $\wedge_r M$. The image of ψ is the ideal generated by the $(q-r) \times (q-r)$ subdeterminants of R, and ψ induces an isomorphism from $(\wedge_r M)/T(\wedge_r M)$ onto this ideal.   □

It follows immediately that $\rho(M/TM) = \rho(M)$. If N is another finitely generated R-module of rank s and $S = M \oplus N$ then $\rho(S) = \rho(M \oplus N)$ is the ideal class of the tensor product of $(\wedge_r M)/T(\wedge_r M)$ with $(\wedge_s N)/T(\wedge_s N)$. Thus ρ defines a homomorphism from the semigroup of finitely generated R-modules (with respect to direct sum) to the semigroup of ideal classes (with respect to product of ideals), i.e., $\rho(M \oplus N) = \rho(M)\rho(N)$. As $\rho(R^n) = 1$ (the principal ideal class) this invariant depends only on the stable isomorphism class of M (with respect to stabilization by forming direct sums with free modules), and $\rho(P)$ is the class of an invertible ideal if P is projective.

If ρ is the row class of an ideal I let ρ^* be the ideal class of $e^0 I = Hom_R(I, R)$. The homomorphism $\tau : \wedge_r(e^0 M) \to e^0(\wedge_r M)$, determined by $\tau(f_1 \wedge \cdots \wedge f_r)(m_1 \wedge \cdots \wedge m_r) = \det[f_i(m_j)]$ is natural, and is an isomorphism if R is a field. The definition of τ is compatible with localization and passage to a quotient with respect to an ideal. Therefore $\tau_\wp$ is an epimorphism for all prime ideals $\wp$, by Nakayama's Lemma. Thus τ is an epimorphism and so $\rho(e^0 M) = \rho(M)^*$.

There does not appear to be a simple general characterization of $\gamma(M)$ although it is easily seen that $\gamma(M \oplus R^n) = \gamma(M)$. If $p.d._R M \leq 1$ then $\gamma(M)$ is invertible. In particular, if M is projective then $\gamma(M) = \rho(e^0 M) = \rho(M)^*$. In this case $\rho(M)$ is the class of an invertible ideal and $\rho(M)^*$ is the class of the inverse ideal.

It is not generally true that $\gamma(M) = \rho(M)^*$ for every finitely generated module. If $R = \mathbb{Q}[x, y, z]$ and M is the ideal (x, y, z) then $\gamma(M) = \rho(M) =$ the class of M, which is not principal, but $e^0 M \cong R$ and $\rho(M)^* = 1$.

3.6. 1- and 2-dimensional rings

A noetherian ring is *1-dimensional* if it has a prime ideal which is not maximal, and if every such prime ideal $\wp$ is nilpotent, i.e., such that $\wp^n = 0$ for n large. Such rings arise naturally in the study of knot modules. (See Chapter 7.) A 1-dimensional noetherian domain R is *Dedekind* if it satisfies one (and hence all) of the following equivalent conditions:

(1) *R is integrally closed;*
(2) *the localization $R_{\mathfrak{m}}$ is a discrete valuation ring for each maximal ideal $\mathfrak{m}$.*
(3) *every nonzero ideal is invertible;*
(4) *every nonzero ideal factors uniquely as a product of maximal ideals.*

The ideal class semigroup is then a group. (See Chapter 9 of [**AM**].)

LEMMA 3.18. *A Dedekind domain with only finitely many maximal ideals is a principal ideal domain.*

PROOF. Let R be a Dedekind domain with finitely many maximal ideals $\{\mathfrak{m}_i \mid 1 \leq i \leq n\}$. Since $\mathfrak{m}_i^2 \subsetneq \mathfrak{m}_i$ we may choose $x_i \in \mathfrak{m}_i - \mathfrak{m}_i^2$, for $1 \leq i \leq n$. There are elements y_i such that $y_i - x_i \in \mathfrak{m}_i^2$ and $y_i - 1 \in \mathfrak{m}_j$, for all $1 \leq i \neq j \leq n$, by the Chinese Remainder Theorem (Exercise 9.9 of [**AM**]). Hence the prime factorization of the ideal $y_i R$ is $\mathfrak{m}_i$. It follows immediately that R is a PID. $\square$

THEOREM 3.19. *Let R be a Dedekind domain and M a finitely generated R-module. Then $M \cong (M/TM) \oplus TM$, and $p.d._R M \leq 1$. If M has rank $r > 0$ then $M/TM \cong R^{r-1} \oplus J$, where J is an ideal of R with ideal class $\rho(M)$. The torsion submodule TM is determined by the elementary ideals of M.*

PROOF. Since $E_r(M) \neq 0$ it is invertible. Therefore M/TM is projective and $p.d._R M \leq 1$, by Theorem 3.7. Hence $M \cong (M/TM) \oplus TM$, and so $E_i(TM) = E_{r+i}(M)$ for all $i \geq 0$. If $r > 0$ there is a nonzero homomorphism $\theta : M \to R$, since M/TM embeds in $R \otimes (M/TM) \cong R^r$. As $\theta(M)$ is an ideal it is projective, so $M \cong \mathrm{Ker}(\theta) \oplus \theta(M)$. We may assume by induction on r that $\mathrm{Ker}(\theta)/TM \cong R^{r-2} \oplus J'$ for some ideal J' of R. Let $\{\mathfrak{m}_i \mid i \in I\}$ be the finite set of maximal ideals which contain $\theta(M) \cap E_r(M)$. The localization R_S with respect to the multiplicative system $S = R - \cup_{i \in I} \mathfrak{m}_i$ is a PID, by Lemma 3.18. Hence $J'_S \cong R_S$ and so there is an epimorphism $\psi : J'_S \to (R/\theta(M))_S = R/\theta(M)$. Let $\tilde{\psi} : J' \to R$ be a homomorphism lifting $\psi|_{J'}$ (which exists since J' is projective). Define $\xi : J' \oplus \theta(M) \to R$ by $\xi(j', n) = \tilde{\psi}(j') + n$ for all $j' \in J'$ and $n \in \theta(M)$. Then ξ is an epimorphism, so $M/TM \cong R^{r-1} \oplus J$, where $J = \mathrm{Ker}(\xi)$. Clearly $J \cong \wedge_r(M/TM)$.

Since $TM = TM_S$ it is determined as a R_S-module by its elementary ideals $E_i(TM_S) = E_i(TM)_S = E_{r+i}(M)_S$. $\square$

In particular, Dedekind domains have global dimension 1.

THEOREM 3.20. *Let M be a finitely generated module over a 1-dimensional noetherian ring R. Then M is free of rank r if and only if $E_{r-1}(M) = 0$, $E_r(M) = R$ and $\wedge_r M \cong R$.*

PROOF. The conditions are clearly necessary. If $E_{r-1}(M) = 0$ and $E_r(M) = R$ then $M_\wp$ is free of rank r for all prime ideals $\wp$ of R, by an easy extension of Theorem 3.7. If $r = 0$ then $M = 0$, by part (1) of Theorem 3.1, so we may assume $r > 0$. Therefore $M \cong R^{r-1} \oplus J$ for some ideal J of R, by the Stable Range Theorem [**Ba64**]. But then $\wedge_r M \cong J$, so M is free if and only if $\wedge_r M \cong R$. $\square$

COROLLARY 3.20.1. *If I is an invertible ideal in a 1-dimensional noetherian domain R then I can be generated by two elements.*

PROOF. Let P be a projective module such that $I \oplus P$ is free, and let J be an ideal in the class $\rho(P)$. Then $I \oplus J \cong R^2$, by the Theorem, and so I is a quotient of R^2. $\square$

THEOREM 3.21. *Let R be an integrally closed noetherian domain such that all maximal ideals of R have height 2, and let M be a finitely generated R-module. Then M is pseudonull if and only if it has finite length.*

PROOF. Since a finitely generated module is pseudonull or of finite length if and only if its cyclic submodules are pseudonull or of finite length (respectively), it suffices to assume that $M \cong R/I$, for some ideal I. The ideal I is the intersection of powers of prime ideals, by the Primary Decomposition Theorem for noetherian rings. Clearly "R/I is pseudonull" and "R/I has finite length" are each equivalent to this decomposition being an intersection of powers of maximal ideals, since all maximal ideals of R have height 2. $\square$

THEOREM 3.22. *Let R be an integrally closed noetherian domain of global dimension 2, and let M be a finitely generated R-module. Then*

(1) *p.d.$_R M \leq 1$ if and only if M has no nontrivial pseudonull submodule;*

(2) *the maximal pseudonull submodule of M is isomorphic to $e^2 e^2 M$.*

PROOF. Let $K = \text{Ker}(W_{TM})$ and $L = \text{Im}(W_{TM})$. Then K is the maximal pseudonull submodule of M and $L \leq e^1 e^1 M$, by Theorem 3.9.

Applying the functor $e^* = Ext_R^*(-, R)$ to a short exact sequence of R-modules $0 \to A \to B \to C \to 0$ gives a nine-term exact sequence terminating with $\cdots \to e^2 C \to e^2 B \to e^2 A \to 0$, since R has global dimension 2. In particular, if $p.d._R B \leq 1$ then $p.d._R A \leq 1$ also. Therefore $p.d._R L \leq 1$.

If $p.d._R M \leq 1$ then $K = 0$, by Theorem 3.7. Conversely, if $K = 0$ then $TM = L$ and so $p.d._R TM \leq 1$. Since M/TM is finitely generated and torsion free it is a submodule of a free module, and so $p.d._R (M/TM) \leq 1$ also. Therefore $p.d._R M \leq 1$.

Since TM is an extension of L by K the e^*-sequence gives an isomorphism $e^2 TM \cong e^2 K$, while since M is an extension of M/TM by TM there is an isomorphism $e^2 M \cong e^2 TM$. Since $e^0 K = e^1 K = 0$ it is easily seen that $K \cong e^2 e^2 K$. Therefore $K \cong e^2 e^2 M$. $\square$

With more care one may show that there is a *natural* isomorphism $K \cong e^2 e^2 M$. More generally, if R has global dimension d and M is a finitely generated R-module then M has no nontrivial submodule of finite length if and only if $e^d M = 0$ and the maximal submodule of finite length is naturally isomorphic to $e^d e^d M$ [**LV68**].

3.7. Bilinear pairings

An *inner product module* over R is a finitely generated R-module X with a bilinear pairing $b : X \times X \to R$ which is symmetric and non-singular (the adjoint map $Ad(b) : X \to X^*$ given by $Ad(b)(x)(y) = b(x, y)$ is an isomorphism). In particular, X is reflexive. In the study of the stable classification of inner product modules it is usual to require also that X be projective. Here we shall see that projectivity follows from the existence of such a non-singular symmetric bilinear form exactly when R is regular of global dimension ≤ 2.

LEMMA 3.23. *Let R be a noetherian domain. The following are equivalent:*

(1) *every inner product module over R is projective;*

(2) *every reflexive module is projective;*

(3) *R has global dimension ≤ 2.*

PROOF. If M is reflexive then $M \oplus e^0 M$ supports an inner product, given by $b((m, \mu), (m', \mu')) = \mu(m') + \mu'(m)$. Hence (1) implies (2). Clearly (2) implies (1). If $Q : R^p \to R^q$ is a homomorphism with cokernel M then $M^* \cong \mathrm{Ker}(Q^*)$. Hence every dual is projective if and only if R has global dimension ≤ 2, by Schanuel's Lemma. $\square$

Suppose now that R is factorial, and has an involution $\bar{}$. Let $W_\varepsilon(R_0, R, -)$ denote the Witt group of ε-linking pairings over R (as defined in §8 of Chapter 2). If a torsion R-module M supports a primitive hermitean pairing then M has no nontrivial pseudonull submodule, by Theorem 3.9.

LEMMA 3.24. *If R is a principal ideal domain then primitive pairings on torsion modules are non-singular.*

PROOF. Let M be a torsion R-module. As M has finite length, and as $e^1 M$ is isomorphic to M, by the Structure Theorem for finitely generated modules over PIDs, any monomorphism from M to $e^1 M$ must be an isomorphism. $\square$

The Witt groups of greatest interest to the algebraists are based on non-singular pairings with an additional quadratic structure. If 2 is invertible in R any non-singular pairing can be endowed with such a quadratic structure. In particular, if R is a PID containing $\frac{1}{2}$ every primitive pairing on a R-torsion module is non-singular and has a unique quadratic structure.

LEMMA 3.25. *If a sesquilinear pairing $c : M_1 \times M_2 \to R_0/R$ is primitive on both sides then $\Delta_i(M_1) \doteq \overline{\Delta_i(M_2)}$, for all $i \geq 0$.*

PROOF. This follows on localizing with respect to the multiplicative systems $R - (\wp \cup \bar{\wp})$, for each height 1 prime $\wp$ of R. $\square$

LEMMA 3.26. *If (N, c) is a neutral ε-linking pairing over R then $\Delta_0(N) \doteq f\bar{f}$ for some nonzero $f \in R$.*

PROOF. Let $P < N$ be a submodule such that $P = P^\perp$. The pairing c induces a sesquilinear pairing of P and N/P into R_0/R which is primitive on both sides. Since $\Delta_0(P) \doteq \overline{\Delta_0(N/P)}$, by Lemma 3.25, $\Delta_0(N) \doteq \Delta_0(P)\Delta_0(N/P) \doteq \Delta_0(P)\overline{\Delta_0(P)}$. $\square$

THEOREM 3.27. *If (M, b) and (M', b') are Witt equivalent ε-linking pairings over R then $f\bar{f}\Delta_0(M) \doteq f'\overline{f'}\Delta_0(M')$ for some f, f' in $R - \{0\}$.*

PROOF. This follows from Theorem 3.12 and Lemma 3.26. $\square$

The *determinant class* $\delta(M, b)$ of a ε-linking pairing (M, b) is the class of $\Delta_0(M)$ modulo products with $uf\bar{f}$, where u is a unit of R and $f \neq 0$. By the theorem, δ determines a homomorphism from the Witt group $W_\varepsilon(R_0, R, -)$ to the group

$$\{f \in R_0^\times \mid \bar{f} = uf \ for \ some \ u \in R^\times\}/\{ug\bar{g} \mid u \in R^\times, g \in R_0^\times\}.$$

If R is a Dedekind domain with an involution the argument of §10 of Chapter 2 extends to give natural isomorphisms $W_\varepsilon(R_0, R, -) \cong \oplus_{\wp=\bar{\wp}}W_\varepsilon(R_0, R_\wp, -) \cong \oplus_{\wp=\bar{\wp}}W_\varepsilon(k_\wp, -)$, where the sum is taken over the maximal ideals which are invariant under the involution.

CHAPTER 4

The Maximal Abelian Cover

In this chapter we shall define the Crowell sequence for a group and extend to links various results of Crowell on the Alexander modules of classical knots. We shall focus particularly on the interactions between properties known to hold for boundary links, such as the Alexander module having rank μ, the first nonzero Alexander ideal being principal and the longitudes being in $(\pi_\omega)'$.

4.1. Metabelian groups and the Crowell sequence

Let G be a finitely generated group and $\varepsilon : \mathbb{Z}[G] \to \mathbb{Z}$ be the augmentation homomorphism of the group ring. If G is generated by $\{g_1, \ldots g_n\}$ the augmentation ideal $I = I(G) = \mathrm{Ker}(\varepsilon)$ is generated as a $\mathbb{Z}[G]$-module by $\{g_i - 1 \mid 1 \leq i \leq n\}$. The *augmentation sequence* for G is the sequence

$$0 \to I(G) \to \mathbb{Z}[G] \xrightarrow{\varepsilon} \mathbb{Z} \to 0.$$

Tensoring this sequence with a right $\mathbb{Z}[G]$-module M gives an exact sequence

$$0 \to H_1(G; M) \to M \otimes_{\mathbb{Z}[G]} I(G) \to M \to \varepsilon M = M/I(G)M \to 0.$$

Suppose instead that M is a left $\mathbb{Z}[G]$-module and let $Der(G; M) = Hom_{\mathbb{Z}[G]}(I(G), M)$. Then the functor $Hom_{\mathbb{Z}[G]}(-, M)$ gives an isomorphism $H^1(G; M) \cong Der(G; M)/P(G; M)$, where $P(G, M)$ is the image of $M = Hom_{\mathbb{Z}[G]}(\mathbb{Z}[G], M)$ under restriction. The group $Der(G; M)$ may be identified with the group of *derivations* or *crossed homomorphisms* $\tilde{f} : G \to M$ such that $\tilde{f}(gh) = \tilde{f}(g) + g\tilde{f}(h)$ for all $g, h \in G$, via $f \mapsto \tilde{f}$, where $\tilde{f}(g) = f(g - 1)$. These in turn correspond to splitting homomorphisms for the semidirect product $M \rtimes G$, sending $g \in G$ to $(\tilde{f}(g), g) \in M \rtimes G$. The subgroup $P(G, M)$

corresponds to the "principal" crossed homomorphisms $\tilde{f}_m$, where $\tilde{f}_m(g) = (g-1)m$ for all $g \in G$. If the commutator subgroup G' acts trivially on M these groups are left $\mathbb{Z}[G/G']$-modules,

Let $A(G) = \mathbb{Z}[G/G'] \otimes_{\mathbb{Z}[G]} I(G)$. Then $A(G) = I/II'$, where $I' = I'(G)$ is the ideal generated by $\{w - 1 \mid w \in G'\}$, and is finitely generated as a $\mathbb{Z}[G/G']$-module. The conjugation action of G/G' on G'/G'', given by $gG'.aG'' = gag^{-1}G''$ for $g \in G$ and $a \in G'$, makes G'/G'' into a $\mathbb{Z}[G/G']$-module, and the homomorphism $\delta : G'/G'' \to A(G)$ which sends aG'' to $[a - 1] = (a - 1) + II'$ is injective, with image I'/II'. Let $\phi_G([g - 1]) = gG' - 1$, for $g \in G$. The four-term exact sequence

$$Cr(G): \quad 0 \to G'/G'' \xrightarrow{\delta} A(G) \xrightarrow{\phi_G} \mathbb{Z}[G/G'] \xrightarrow{\varepsilon} \mathbb{Z} \to 0$$

is called the *Crowell sequence* for G. (This corresponds to the choice $M = \mathbb{Z}[G/G']$ in the above paragraph.) There is an equivalent short exact sequence

$$0 \to G'/G'' \to A(G) \to I(G/G') \to 0.$$

(Tensoring with $\mathbb{Z}$ leads to an isomorphism $\varepsilon A(G) \cong \varepsilon I(G/G')$.) If M is a $\mathbb{Z}[G/G']$-module then $Der(G; M) \cong Hom_{\mathbb{Z}[G/G']}(A(G), M)$.

A homomorphism $f : G \to H$ induces a homomorphism $Cr(f)$ of Crowell sequences, and the induced homomorphism $G/G'' \cong H/H''$ is an isomorphism if and only if $f^{ab} : G/G' \to H/H'$ and $Cr(f)$ are isomorphisms [**Cr61, Cr71**]. The Crowell sequence represents an element of $Ext^2_{\mathbb{Z}[G]}(\mathbb{Z}, G'/G'')$, which corresponds to the extension

$$1 \to G'/G'' \to G/G'' \to G/G' \to 1$$

via the isomorphism $Ext^2_{\mathbb{Z}[G]}(\mathbb{Z}, G'/G'') \cong H^2(G/G'; G'/G'')$. (The group G/G'' may be recovered from $Cr(G)$ as a quotient of a semidirect product $(G'/G'') \rtimes F(r)$. See §3 below for the cases with G/G' free abelian.)

If X is a finite complex with fundamental group G, basepoint $*$ and maximal abelian cover $p : X' \to X$ then the Crowell sequence may be identified with the nontrivial part of the long exact sequence of homology for the pair $(X', p^{-1}(*))$. In particular, a finite presentation for the group G determines a presentation for $A(G)$ as a module (with the same numbers of generators and relations) by the free differential calculus. (See §6 of Chapter 10 below.)

We shall henceforth assume that $G/G' \cong \mathbb{Z}^{\mu}$ is free abelian of finite rank $\mu > 0$, with a given basis. We may then identify $\mathbb{Z}[G/G']$ with $\Lambda_{\mu} = \mathbb{Z}[t_1, t_1^{-1}, \ldots, t_{\mu}, t_{\mu}^{-1}]$, and we shall let $I_{\mu} = I(\mathbb{Z}^{\mu}) = (t_1 - 1, \ldots, t_{\mu} - 1)$ be the augmentation ideal of Λ_{μ}. We shall also let $R\Lambda_{\mu} = R \otimes_{\mathbb{Z}} \Lambda_{\mu}$, for any PID R. Then $R\Lambda_{\mu}$ is a factorial noetherian domain of global dimension μ, if R is a field, and $\mu + 1$ otherwise, by Hilbert's Syzygy Theorem. In particular, its height 1 prime ideals are principal. Moreover projective $R\Lambda_{\mu}$-modules are free, by a theorem of Suslin. (See page 67 of [**Lam**].)

4.2. Free metabelian groups

The Crowell sequence for the free group $F(\mu)$ is a partial resolution for $\mathbb{Z}$ as a Λ_{μ}-module. This may be continued to a full resolution $K(\Lambda_{\mu})_*$ with j^{th} term $K(\Lambda_{\mu})_j = \wedge_j(\Lambda_{\mu})^{\mu}$ and differentials given by

$$\partial_k(e_{i_1} \wedge \cdots \wedge e_{i_k}) = \Sigma_{j=1}^{j=k}(-1)^{j-1}(t_{i_j} - 1)(e_{i_1} \wedge \ldots \widehat{e_{i_j}} \cdots \wedge e_{i_k}).$$

(This is the *Koszul complex* for Λ_{μ} with respect to the regular sequence $(t_1 - 1, \ldots, t_{\mu} - 1)$. It is also the equivariant cellular chain complex for R^{μ}, as the universal cover of $(S^1)^{\mu}$ with its standard minimal cell structure.) We may also construct this complex by taking the tensor product over $\mathbb{Z}$ of μ copies of $K(\Lambda_1)_*$.

Since all the differentials of $\mathbb{Z} \otimes_{\Lambda} K(\Lambda)_*$ are 0 we find $e^q\mathbb{Z} = 0$ if $q \neq \mu$, $e^{\mu}\mathbb{Z} = \mathbb{Z}$ and $Tor_q^{\Lambda}(\mathbb{Z}, \mathbb{Z}) \cong \mathbb{Z}^{\binom{\mu}{q}}$.

LEMMA 4.1. $E_1(I_{\mu}) = (I_{\mu})^{\mu-1}$.

PROOF. We shall induct on μ. The result is clear if $\mu = 1$, since $I_1 = (t_1 - 1) \cong \Lambda_1$. It is also clear from the Koszul complex that $E_1(I_{\mu}) \leq (I_{\mu})^{\mu-1}$. Let $\phi_j : \Lambda_{\mu-1} \to \Lambda_{\mu}$ be the ring homomorphism defined by $\phi_j(t_i) = t_i$ if $1 \leq i < j$ and $\phi_j(t_i) = t_{i+1}$ if $j \leq i < \mu$, for $1 \leq j \leq \mu$. Then $\lambda \mapsto (t_j - 1)\lambda$ determines an isomorphism of Λ_{μ}-modules $\Lambda_{\mu}/(\phi_j(I_{\mu-1})) \cong I_{\mu}/(\phi_j(I_{\mu-1}))$. There is an exact sequence

$$0 \to (\phi_j(I_{\mu-1})) \to I_{\mu} \to I_{\mu}/(\phi_j(I_{\mu-1})) \to 0.$$

Hence $E_1(I_\mu) \geq E_0(I_\mu/(\phi_j(I_{\mu-1})))E_1(\phi_j(I_{\mu-1})) = (\phi_j(I_{\mu-1}))^{\mu-1}$, by Theorem 3.12. Since each of the generating monomials of $(I_\mu)^{\mu-1}$ involves only $\mu - 1$ of the variables it follows that $E_1(I_\mu) \geq (I_\mu)^{\mu-1}$, proving the lemma. $\qquad\qquad\Box$

The Koszul complex gives a presentation with $\binom{\mu}{2}$ generators and $\binom{\mu}{3}$ relations for the Λ_μ-module $B(\mu) = F(\mu)'/F(\mu)''$. Hence $E_{\mu-1}(B(\mu)) \leq (I_\mu)^m$, where $m = \binom{\mu-1}{2}$. It can be shown that these ideals are equal [**CSt69**].

THEOREM 4.2. *Let G be a finitely generated group such that $G/G' \cong \mathbb{Z}^\mu$. If G maps onto $F(\mu)/F(\mu)''$ then $E_{\mu-1}(G) = 0$. Moreover $G/G'' \cong F(\mu)/F(\mu)''$ if and only if $E_{\mu-1}(G) = 0$ and $E_\mu(G) = \Lambda_\mu$.*

PROOF. An epimorphism $f : G \to H$ induces an epimorphism $A(f) : A(G) \to A(H)$. Therefore the image of $E_i(G)$ in $\mathbb{Z}[H/H']$ is contained in $E_i(H)$. In particular, if $G/G' \cong \mathbb{Z}^\mu$ and G maps onto $F(\mu)/F(\mu)''$ then $E_{\mu-1}(G) = 0$.

If $E_{\mu-1}(G) = 0$ and $E_\mu(G) = \Lambda_\mu$ then $A(G/G'') = A(G)$ is projective of rank μ, by Corollary 3.7.1. Hence $A(G/G'') \cong (\Lambda_\mu)^\mu = A(F(\mu)/F(\mu)'')$, since projective Λ_μ-modules are free. Therefore $G/G'' \cong F(\mu)/F(\mu)''$, by Corollary 2 of [**GM86**]. The converse is clear. $\qquad\qquad\Box$

This criterion for recognizing free metabelian groups derives from the one given by Groves and Miller, but has simpler hypotheses.

COROLLARY 4.2.1. *Let G be a finitely generated group such that $G/G' \cong \mathbb{Z}^2$. Then G maps onto $F(2)/F(2)''$ if and only if G'/G'' maps onto Λ_2.*

PROOF. The condition is clearly necessary. Suppose it holds, and let $H = (G/G'')/T(G'/G'')$. Then $H'/H'' \cong \Lambda_2$, so $E_1(H) = 0$ and $E_2(H) \geq E_1(I_2)E_1(H'/H'') = I_2$. Moreover $\varepsilon(E_2(H)) = \mathbb{Z}$, since $H/H' \cong \mathbb{Z}^2$. Therefore $E_2(H) = \Lambda_2$ and so $H \cong F(2)/F(2)''$. $\qquad\Box$

This corollary also follows from consideration of the possible extensions of $\mathbb{Z}^2$ by $F(2)'/F(2)'' \cong \Lambda_2$. For $H^2(\mathbb{Z}; \Lambda_2) \cong \mathbb{Z}$, and if

$G(n)$ is the group corresponding to $n \in \mathbb{Z}$ then $G(-n) \cong G(n)$ and $G(n)/G(n)' \cong \mathbb{Z}^2$ if and only if $n = \pm 1$.

THEOREM 4.3. *Let G be a finitely presentable group such that $G/G' \cong \mathbb{Z}^\mu$ and $\mathrm{def}(G) = \mu$. Then $E_{\mu-1}(G) = 0$, and G maps onto $F(\mu)/F(\mu)''$ if and only if $E_\mu(G)$ is principal.*

PROOF. Since $G/G' \cong \mathbb{Z}^\mu$ and $A(G)$ has a presentation of deficiency μ we have $E_{\mu-1}(G) = 0$ and $\varepsilon(E_\mu(G)) = \mathbb{Z}$, so $A(G)$ has rank μ. If G maps onto $F(\mu)/F(\mu)''$ then $A(G)$ maps onto $(\Lambda_\mu)^\mu$, so $A(G) \cong (\Lambda_\mu)^\mu \oplus TA(G)$. Adding μ relations to kill a basis for the free summand gives a square presentation matrix for $TA(G)$. Therefore $E_\mu(G) = E_\mu(A(G)) = E_0(TA(G))$ is principal. Conversely if $E_{\mu-1}(G) = 0$ and $E_\mu(G)$ is principal then $A(G) \cong (\Lambda_\mu)^\mu \oplus TA(G)$, by Theorem 3.7 and the fact that projective Λ_μ-modules are free. Let $K < G$ be the subgroup containing G'' and such that $K/G'' = T(G'/G'') = TA(G)$. Then $A(G/K) \cong A(G)/TA(G) \cong (\Lambda_\mu)^\mu$, and so G/K maps onto $F(\mu)/F(\mu)''$, by Theorem 4.2. $\square$

4.3. Link module sequences

Let $0 \to B \to A \to I_\mu \to 0$ be an exact sequence of finitely generated Λ_μ-modules, where $\varepsilon A \cong \varepsilon I_\mu = \mathbb{Z}^\mu$. We may construct the group corresponding to this sequence as follows. Let $f : (\Lambda_\mu)^\mu \to A$ be a homomorphism lifting $\phi_{F(\mu)}$, and let $S = B \rtimes (F(\mu)/F(\mu)'')$, where $F(\mu)/F(\mu)''$ acts on the Λ_μ-module B through $F(\mu)/F(\mu)'$. Let $\Gamma(f|_{B(\mu)})$ be the graph of $f|_{B(\mu)}$. Then $G = S/\Gamma(f|_{B(\mu)})$ is a metabelian group with $G/G' \cong \mathbb{Z}^\mu$ and $Cr(G)$ is isomorphic to the sequence we started with.

LEMMA 4.4. *Let G and H be finitely generated metabelian groups such that $G/G' \cong H/H' \cong \mathbb{Z}^\mu$. Fix an isomorphism $\alpha : G/G' \cong H/H'$. Then Cr determines a surjection from $\{f \in Hom(G, H) \mid f^{ab} = \alpha\}$ to the set of homomorphisms from $Cr(G)$ to $Cr(H)$ inducing $\mathbb{Z}[\alpha] : \mathbb{Z}[G/G'] \cong \mathbb{Z}[H/H']$.*

PROOF. Let $\gamma : Cr(G) \to Cr(H)$ be a homomorphism which induces $\mathbb{Z}[\alpha]$, and let $f : F(\mu) \to G$ be a homomorphism such that

f^{ab} is an isomorphism. We may choose a homomorphism $h : F(\mu) \to H$ such that $Cr(h) = \gamma Cr(f)$. Then $h|_{F(\mu)'} = \gamma|_{G'} f|_{F(\mu)'}$. Therefore h induces a homomorphism h_G from $G \cong G' \rtimes (F(\mu)/F(\mu)'')/\Gamma(f|_{F(\mu)'})$ to $H \cong H' \rtimes (F(\mu)/F(\mu)'')/\Gamma(h|_{F(\mu)'})$. It is easily verified that $Cr(h_G) = \gamma$. $\qquad\square$

If $G \cong F(\mu)/F(\mu)''$ then these homomorphism sets may each be identified with $(H')^\mu$ and the correspondance is bijective.

A *link module sequence* is an exact sequence of Λ_μ-modules

$$0 \to B \to A \to I_\mu \to 0$$

such that A has a finite presentation of deficiency 1. Clearly A has rank ≥ 1, and so $E_0(A) = 0$.

LEMMA 4.5. *Let* $0 \to B \to A \to I_\mu \to 0$ *be a link module sequence. Then either* $p.d._{\Lambda_\mu}B \leq p.d._{\Lambda_\mu}A = \mu - 1$ *or* $p.d._{\Lambda_\mu}A \leq p.d._{\Lambda_\mu}B = \mu - 2$ *or* $p.d._{\Lambda_\mu}A = p.d._{\Lambda_\mu}B \geq \mu$.

PROOF. We apply the functor $e^*(-) = Ext^*_{\Lambda_\mu}(-, \Lambda_\mu)$ to the link module sequence, and use the fact that $e^{q-1}I_\mu = e^q\mathbb{Z}$, for $q > 0$. $\qquad\square$

An immediate consequence of this lemma is that if A has rank 1 then B has a square presentation matrix if and only if $\mu \leq 3$. For if A has deficiency 1 and rank 1 then $p.d._{\Lambda_\mu}A \leq 1$ and B is a Λ_μ-torsion module. Since projective Λ_μ-modules are free, B has a square presentation matrix if and only if $p.d._{\Lambda_\mu}B \leq 1$.

THEOREM 4.6. *Let* $0 \to B \to A \to I_\mu \to 0$ *be a link module sequence, and let A have rank r. Then*

   (1) *if $\mu = 1$ then $A \cong B \oplus \Lambda_1$ and $E_1(A) = (\Delta_1(A))$, while if $\mu > 1$ then $E_1(A) = \Delta_1(A)I_\mu$;*
   (2) *if $r = 1$ then $B = TA$ has no nonzero pseudonull submodule and $Ann(B) = (\lambda_1(B)) = (\lambda_2(A))$;*
   (3) *if $E_{r-1}(B)$ and $E_r(A)$ are both principal then $\mu \leq 2$;*
   (4) *if $\mu = 2$ then $E_0(B) = (\Delta_1(A))$; if moreover $\mathbb{Z} \otimes_\Lambda A \cong \mathbb{Z}^2$ then $\mathbb{Z} \otimes_\Lambda B \cong \mathbb{Z}/l\mathbb{Z}$, where $l = |\varepsilon(\Delta_1(A))|$, and the sequence splits if and only if $l = 1$;*
   (5) *if $\mu > 2$ and $\mathbb{Z} \otimes_\Lambda A \cong \mathbb{Z}^\mu$ then the sequence does not split;*

(6) if $E_r(A) = (\Delta_r(A))$ and $\varepsilon(\Delta_r(A)) = \pm 1$ then A is $\mathbb{Z}$-torsion free.

PROOF. If $\mu = 1$ then $I_1 = (t_1 - 1) \cong \Lambda_1$, so the sequence splits. Suppose that $r = 1$. Then $B = TA$ and $A/TA \cong I_\mu$, so $\rho(A)$ is the class of I_μ. Moreover $\gamma(A) = 1$ since A has deficiency 1. Hence $\rho(A)$ is the class of $E_1(A)$, by Lemma 3.17. If $r > 1$ then $E_1(A) = (\Delta_1(A)) = 0$.

Since $p.d._\Lambda A \leq 1$ if $r = 1$, part (2) follows from Theorem 3.8.

If $E_{r-1}(B)$ and $E_r(A)$ are both principal then B/TB and A/TB are projective, and so

$$0 \to B/TB \to A/TB \to I_\mu \to 0$$

is a projective resolution for the augmentation module. Hence $\mu = 1 + p.d._{\Lambda_\mu} I_\mu \leq 2$.

Since $\mathbb{Z} \otimes_\Lambda A \cong \mathbb{Z}^\mu = \mathbb{Z} \otimes I_\mu$, the sequence can only split if $\mathbb{Z} \otimes_\Lambda B = 0$. Therefore $E_0(\mathbb{Z} \otimes_\Lambda B) \neq 0$ and so $r = 1$. From this and the other assumptions on A it follows that $Tor_1^\Lambda(\mathbb{Z}, A) \cong \mathbb{Z}^{\mu-1}$. On applying $Tor_*^\Lambda(\mathbb{Z}, -)$ to the link module sequence we find that $\mathbb{Z} \otimes_\Lambda B$ is a quotient of $Tor_1^\Lambda(\mathbb{Z}, I_\mu) \cong \mathbb{Z}^{\binom{\mu}{2}}$ and has rank at least $\binom{\mu}{2} - \mu + 1$. Hence $\mu \leq 2$.

If $\mu = 2$ then $p.d._\Lambda I_2 = 1$, so $E_1(A) = E_0(B)E_1(I_2)$, by Theorem 3.12. Hence $E_0(B) = (\Delta_1(A))$. Moreover $Ext_\Lambda^1(I_2, B) \cong B/I_2 B = \mathbb{Z} \otimes_\Lambda B$, which is cyclic of order $l = |\varepsilon(\Delta_1(A))|$. Therefore $\mathbb{Z} \otimes_\Lambda B = 0$ if and only if $l = 1$, and then $Ext_\Lambda^1(I_2, B) = 0$, so $A \cong B \oplus I_2$.

Suppose that (6) holds. Let p be an integral prime and suppose that $pm = 0$ for some $m \in A$. Then $Ann(\Lambda_\mu m)$ contains p and $Ann(TA)$, and hence $E_0(TA)$, by part (1) of Theorem 3.1. Hence if $Ann(\Lambda_\mu m) \leq (\delta)$, δ divides p and $\Delta_0(TA) = \Delta_r(A)$. Since $\varepsilon(\Delta_r(A))$ is $= \pm 1$, δ must be a unit and so $\Lambda_\mu m$ is pseudonull. It now follows from Theorems 3.7 and 3.8 that $m = 0$. $\qquad\square$

Parts (1) and (4-6) were originally proven in [**Cr65**].

Since $E_1(I_\mu) = (I_\mu)^{\mu-1}$ we have $E_{k-1}(B)(I_\mu)^{\mu-1} \leq E_k(A)$, by Theorem 3.12. The link module sequence together with the Crowell sequence for $F(\mu)$ give an exact sequence

$$0 \to B(\mu) \to B \oplus (\Lambda_\mu)^\mu \to A \to 0.$$

Using Theorem 3.12 again gives $E_{k-1}(B) = E_{k+\mu-1}(B \oplus (\Lambda_\mu)^\mu) \geq E_k(A).E_{\mu-1}(B(\mu))$, for any $k \geq 1$. Hence $E_{k-1}(B) \geq E_k(A).(I_\mu)^r$, where $r = \binom{\mu-1}{2}$. (In particular, $E_2(A) \leq E_1(B)$ if $\mu = 2$.) If $\mu > 1$ then $E_0(B) = \Delta_1(A)(I_\mu)^m$, where $m = \binom{\mu-2}{2}$, while if $k > 1$ then $E_{k-1}(B) \geq E_k(A).(I_\mu)^{r+k-\mu}$, where $r = \binom{\mu-1}{2}$. (See [**CSt69, Tr82'**].)

4.4. Localization of link module sequences

Let $S = 1 + I_\mu = \{f \in \Lambda_\mu \mid \varepsilon(f) = 1\}$ be the multiplicative system consisting of Laurent polynomials augmenting to 1. The next result is due to Massey, who used completion rather than localization.

THEOREM 4.7. [**Ma80**] *Let* $0 \to B \to A \to I_\mu \to 0$ *be a link module sequence such that* $\mathbb{Z} \otimes_\Lambda A \cong \mathbb{Z}^\mu$. *Then*

(1) A_S *has a presentation with* μ *generators and* $s < \mu$ *relations;*

(2) B_S *has a presentation with* $\binom{\mu}{2}$ *generators and* $\binom{\mu}{3} + s$ *relations. Moreover,* $\binom{\mu}{3}$ *of these generators are common to all such modules.*

PROOF. Let $g : A \to I_\mu$ be the epimorphism provided by the link module sequence, and choose a map $f : (\Lambda_\mu)^\mu \to A$ such that gf sends the ith standard generator to $t_i - 1 \in I_\mu$. Then $\mathbb{Z} \otimes_\Lambda f$ is an isomorphism and so Nakayama's Lemma implies that $f_S : (\Lambda_\mu S)^\mu \to A_S$ is an epimorphism. Let $K = \text{Ker}(f_S)$. Then $\mathbb{Z} \otimes_\Lambda K \cong Tor_1^\Lambda(\mathbb{Z}, A)$, which is generated by at most $\mu - 1$ elements, since A has deficiency ≥ 1 and $\mathbb{Z} \otimes_\Lambda A \cong \mathbb{Z}^\mu$. Therefore K is generated by at most $\mu - 1$ elements, by Nakayama's Lemma. (This argument applies for *any* epimorphism $F : (\Lambda_\mu S)^\mu \to A_S$.)

The homomorphism f also induces a commuting diagram of homomorphisms between the localized sequences

$$0 \to B(\mu)_S \to (\Lambda_\mu S)^\mu \to I_\mu S \to 0$$

and

$$0 \to B_S \to A_S \to I_{\mu S} \to 0,$$

from which (2) follows easily. □

In particular, $B_S = 0$ if $\mu = 1$ and B_S is cyclic if $\mu = 2$.

COROLLARY 4.7.1. *If $B = I_\mu B$ then $\mu \leq 2$.*

PROOF. Since localization is exact and $B_S = 0$ we have $A_S \cong I_{\mu S}$. Hence $p.d._\Lambda A \geq p.d._{\Lambda_S} I_{\mu S} = \mu - 1$. On the other hand, A has rank 1 and deficiency 1, so $p.d._\Lambda A \leq 1$. Therefore $\mu \leq 2$. □

COROLLARY 4.7.2. *If A has rank $r = \mu$ then $A_S \cong (\Lambda_{\mu S})^\mu$ and so $\mathbb{Z} \otimes_\Lambda TA = 0$.*

PROOF. Since $TA_S = 0$ we have $\mathbb{Z} \otimes_\Lambda TA = (\mathbb{Z} \otimes_\Lambda TA)_S = \mathbb{Z} \otimes_\Lambda TA_S = 0$. □

The higher Tors of such a module are also 0, by the following lemma.

LEMMA 4.8. [**Sa81"**] *Let M be a finitely generated Λ_μ-module such that $\mathbb{Z} \otimes_\varepsilon M = 0$. Then $Tor_i^\Lambda(\mathbb{Z}, M) = 0$, for all $i \geq 0$.*

PROOF. Since $M_S/I_{\mu S} M_S = \mathbb{Z} \otimes_\varepsilon M_S = 0$ we have $M_S = 0$, by Nakayama's Lemma. Therefore $Tor_i^\Lambda(\mathbb{Z}, M) = Tor_i^\Lambda(\mathbb{Z}, M)_S = Tor_i^\Lambda(\mathbb{Z}, M_S) = 0$, for all $i \geq 0$, since $\mathrm{Ker}(\varepsilon)$ acts trivially on $\mathbb{Z}$. □

Let $\widehat{\Lambda_\mu}$ be the I_μ-adic completion of Λ_μ. Then Λ_μ embeds in $\widehat{\Lambda_\mu} \cong \mathbb{Z}[[X_1, \ldots, X_\mu]]$ via $t_i \mapsto 1 + X_i$. This embedding extends to an embedding of $\Lambda_{\mu S}$ into $\widehat{\Lambda_\mu}$. In each of these rings the image of I_μ generates the Jacobson radical.

If $\widehat{M}$ is the I_μ-adic completion of a finitely generated Λ_μ-module M then $\widehat{M} \cong \widehat{\Lambda_\mu} \otimes_\Lambda M$. Tensor product with $\widehat{\Lambda_\mu}$ is a faithfully flat functor on finitely generated $\Lambda_{\mu S}$-modules, by Krull's Theorem (Theorem 10.17 of [**AM**]). Hence the completed link module sequence is also exact. Arguments involving I_μ-adic completion are usually equivalent to arguments involving localization with respect to S.

4.5. Chen groups

Let G be a finitely generated group. The *Chen groups* for G are the quotients $Ch(G; q) = G_q G''/G_{q+1} G''$, for $q \geq 2$. (See [**Ch51**]. The notation $Q(G; q)$ is used in [**Mu70, Hi78**] and [**Ma80**].) For later use, we shall set $G(\infty) = \cap_{q \geq 1}(G_q G'')$. Clearly $G'' G_\omega \leq G(\infty)$; in general, this inclusion is proper. Note that $G'' \leq G_4$.

LEMMA 4.9. *Let G be a finitely generated group with $G/G' \cong \mathbb{Z}^\mu$, and let $B = G'/G''$. Then $B/(I_\mu)^n B \cong G'/G_{n+2} G''$ for all $n \geq 0$. If moreover $f : G \to H$ is a homomorphism which induces isomorphisms on abelianization then the induced homomorphisms $Ch(f; q) : Ch(G; q) \to Ch(H; q)$ are isomorphisms if and only if $A(f)_S$ is an isomorphism.*

PROOF. An easy induction on n shows that $(I_\mu)^n B$ is the image of G_{n+2} in B. The second assertion is straightforward. $\square$

Thus the $(q + 2)^{th}$ Chen group is the qth summand of the associated graded module $Gr(B) = \oplus_{q \geq 0}((I_\mu)^q B_S/(I_\mu)^{q+1} B_S)$. (This is also $\oplus_{q \geq 0}((I_\mu)^q \hat{B}/(I_\mu)^{q+1} \hat{B})$.)

THEOREM 4.10. *Let G be a finitely generated group with $G/G' \cong \mathbb{Z}^\mu$. Then $E_{\mu-1}(A(G)) = 0$ if and only if $Ch(G; q) \cong Ch(F(\mu); q)$ for all $q \geq 1$.*

PROOF. Let $B = G'/G''$. If $f : F(\mu) \to G$ is a homomorphism which induces isomorphisms on abelianization then it induces epimorphisms on all Chen groups, and also induces an epimorphism $A(f)_S : (\Lambda_\mu S)^\mu \to A(G)_S$.

Then A has rank μ if and only if $A(f)_S$ is an isomorphism. Since $B/I^\mu B \cong G'/G_{n+2} G''$ this is so if and only if the induced maps on Chen groups are all isomorphisms. Since they are epimorphisms, and since these groups are finitely generated, this must be so if $Ch(G; q) \cong Ch(F(\mu); q)$ for all $q \geq 1$. $\square$

4.6. Applications to links

Let L be a μ-component 1-link with exterior $X = X(L)$ and group $\pi = \pi L$. Then $A(L) = A(\pi)$, $\alpha(L) = rank_{\Lambda_\mu} A(L)$, $E_k(L) =$

$E_k(A(L))$ and $\Delta_k(L) = \Delta_k(A(L))$ are the *Alexander module, Alexander nullity, k^{th} Alexander ideal* and *k^{th} Alexander polynomial* of L, respectively. Let $\lambda_k(L) = \lambda_k(A(L))$, for $k \geq 1$, and $B(L) = \pi'/\pi''$. Since X is a compact bounded 3-manifold and $\chi(X) = 0$, it is homotopy equivalent to a finite 2-complex with one 0-cell $*$, $n + 1$ 1-cells and n 2-cells, for some $n \geq 0$. Let $p' : X' \to X$ be the maximal abelian cover of X. The meridians of L determine an isomorphism of the covering group $Aut(X'/X)$ with Z^μ. The equivariant chain complex of X' is homotopy equivalent to a finite free complex D_* with $D_0 = \Lambda_\mu$, $D_1 = (\Lambda_\mu)^{n+1}$ and $D_2 = (\Lambda_\mu)^n$, while the relative complex of the pair $(X', p'^{-1}(*))$ is chain homotopy equivalent to the complex $\ldots 0 \to D_2 \to D_1 \to 0$. Hence there is an exact sequence

$$0 \to H_2(X; \Lambda_\mu) \to (\Lambda_\mu)^n \xrightarrow{\partial_2} (\Lambda_\mu)^{n+1} \to A(L) \to 0.$$

LEMMA 4.11. *Let L be a μ-component 1-link. Then*

(1) $\mathbb{Z} \otimes_\Lambda A(L) \cong \mathbb{Z}^\mu$, $\varepsilon(E_\mu(L)) = \mathbb{Z}$ *and* $1 \leq \alpha(L) \leq \mu$;

(2) $H_2(X; \Lambda_\mu)$ *is torsion free of rank* $\alpha(L) - 1$, *and*
$$p.d._\Lambda H_2(X; \Lambda_\mu) = \max\{0, p.d._\Lambda A(L) - 2\} \leq \mu - 1;$$

(3) *if* $\alpha(L) = 2$ *then* $H_2(X; \Lambda_\mu) \cong \Lambda_\mu$.

PROOF. It is immediate that $\alpha(L) = \min\{k \mid E_k(L) \neq 0\} \geq 1$. Since we may obtain the cellular chain complex for $(X, *)$ by tensoring that for $(X', p'^{-1}(*))$ over Λ_μ with $\mathbb{Z}$ we have $\mathbb{Z} \otimes_\Lambda A(L) \cong H_1(X, *; \mathbb{Z}) \cong \mathbb{Z}^\mu$. Hence $\varepsilon(E_\mu(L)) = \mathbb{Z}$ and so $\alpha(L) \leq \mu$.

Part (2) follows on applying Schanuel's Lemma to the above sequence.

Suppose now that $\alpha(L) = 2$. Then $H_2(X; \Lambda_\mu)$ is torsion free of rank 1. Thus if u and v belong to $H_2(X; \Lambda_\mu)$. there are α and β in Λ_μ such that $\alpha u = \beta v$. We may assume that α and β have no common factor. Since Λ_μ is factorial $v = \alpha w$ for some w in $(\Lambda_\mu)^n$ which must actually be in $H_2(X; \Lambda_\mu)$ by the exactness of the above sequence and the fact that $(\Lambda_\mu)^{n+1}$ is torsion free. Hence every 2-generator submodule of $H_2(X; \Lambda_\mu)$ is cyclic. As this module is finitely generated and torsion free it is free of rank 1. □

In general, $H_2(X; \Lambda_\mu)$ is free if and only if $p.d._\Lambda A(L) \leq 2$. If $\mu \leq 2$ then $p.d._\Lambda A(L) \leq \mu$. Can we ever have $p.d._\Lambda A(L) = \mu + 1$?

THEOREM 4.12. *Let L be a μ-component 1-link with group $\pi = \pi L$. Then*

(1) *if $\mu = 1$ then $A(L) \cong (\pi'/\pi'') \oplus \Lambda_1$ and $E_1(L) = (\Delta_1(L))$, while if $\mu > 1$ then $E_1(L) = \Delta_1(L)I_\mu$;*

(2) *if $\alpha(L) = 1$ then $\pi'/\pi'' = TA(L)$ has no nonzero pseudonull submodule and $Ann(\pi'/\pi'') = (\lambda_2(L))$;*

(3) *if $E_{\alpha(L)-1}(\pi'/\pi'')$ and $E_{\alpha(L)}(L)$ are both principal then $\mu \leq 2$;*

(4) *if $\mu = 2$ then $|\varepsilon(\Delta_1(L))| = |\ell_{12}|$, the absolute value of the linking number, and $A(L) \cong (\pi'/\pi'') \oplus I_2$ if and only if $\ell_{12} = \pm 1$;*

(5) *if $\mu > 2$ then the link module sequence $0 \to \pi'/\pi'' \to A(L) \to I_\mu \to 0$ does not split;*

(6) *if $\alpha(L) = \mu$ and $E_\mu(L) = (\Delta_\mu(L))$ then A is $\mathbb{Z}$-torsion free;*

(7) *$\widetilde{\alpha_k}T(\pi'/\pi'') = (\lambda_{\alpha(L)+k}(L))$, for each $k \geq 1$.*

PROOF. Most of this theorem follows immediately from Theorem 4.6. If $\mu = 2$ then π/π_3 has a presentation

$$\langle x, y \mid x, y \rightleftharpoons [x, y], \ [x, y]^{\ell_{12}} = 1 \rangle,$$

by Theorem 1.4, and so $\mathbb{Z} \otimes_\pi B(L) = \pi'/\pi_3 \cong \mathbb{Z}/\ell_{12}\mathbb{Z}$. Hence $|\varepsilon(\Delta_1(L))| = |\ell_{12}|$. Part (7) follows from Theorem 3.4. □

Wirtinger presentations were used to prove (1) [**To53**], (2) and (3) in the knot-theoretic case ($\mu = 1$) [**Cr64**], and (4) and (5) [**Cr65**]. The higher Alexander ideals need not be principal, even for knots. (For instance, $B(9_{46}) = (\Lambda_1/(t - 2)) \oplus (\Lambda_1/(2t - 1))$, so $E_2(9_{46}) = (3, t + 1)$.)

COROLLARY 4.12.1. *The group π is a semidirect product $\pi' \rtimes \mathbb{Z}^\mu$ if and only if $\mu = 1$ or $\mu = 2$ and $\ell_{12} = \pm 1$.*

PROOF. The conditions are necessary by (4) and (5) of the theorem. If they hold the inclusion of a meridian or a boundary component (respectively) induces a splitting $\pi \cong \pi' \rtimes \mathbb{Z}^\mu$. □

We may use Theorem 4.7 to show that 1, $\mathbb{Z}$ and $\mathbb{Z}^2$ are the only nilpotent 1-link groups. If K is a knot the lower central series of

$\pi = \pi K$ terminates at π_2. Therefore π is nilpotent if and only if K is trivial. In general, if πL is nilpotent so are all of its quotient groups. Thus we may assume that $\mu \leq 3$, so $B(L)_S$ has a presentation of deficiency ≥ 0 as a $\Lambda_{\mu S}$-module. As $(I_\mu)^n B(L) = 0$ for n large, $B(L)_S = B(L)$ and is finitely generated as an abelian group. If $\mu > 1$ this implies $B(L) = 0$, and so $\pi L = \mathbb{Z}^\mu$. Since $H_2(\pi L; \mathbb{Z})$ is a quotient of $H_2(X(L); \mathbb{Z}) \cong \mathbb{Z}^{\mu-1}$ we must have $\mu \leq 2$.

Since loops at $*$ in X lift to paths in X' with endpoints in $p'^{-1}(*)$ the meridians and longitudes of L give rise to *meridianal* and *longitudinal elements* m_i and ℓ_i in $A(L)$. The next result is the analogue of Theorem 1.4 for $A(L)$, and may be deduced from it by considering the I_μ-adic completion. We shall instead give a homological argument close to that of Levine.

THEOREM 4.13. [Le83] *Let L be a μ-component 1-link with group $\pi = \pi L$. Then $A(L)_S$ is generated by a set of meridianal elements m_i, subject only to the relations $(T_i - 1)m_i = (t_i - 1)\Sigma_{j=1}^{j=\mu} \lambda_{ij} m_j$, where $T_i = \Pi_{j=1}^{j=\mu} t_j^{\ell_{ij}}$ and $\ell_i = \Sigma_{j=1}^{j=\mu} \lambda_{ij} m_j$, for $1 \leq i \leq \mu$. Moreover $\varepsilon(\lambda_{ij}) = \ell_{ij}$, and any one of the relations is redundant.*

PROOF. We shall follow the notation of Theorem 1.4. Since the meridians freely generate the nilpotent quotients G/G_n, the inclusion of meridians induces isomorphisms $(\Lambda_{\mu S})^\mu \cong H_1(W, *; \Lambda_{\mu S}) = A(G)_S$, by Theorems 4.7 and 4.10. The module $H_2(X, W; \Lambda_{\mu S})$ is freely generated by the classes d_i corresponding to lifts of the 2-cells D_i, for $1 \leq i \leq \mu$, by excision. The image of d_i under the connecting homomorphism δ is $(T_i - 1)m_i - (t_i - 1)\ell_i$, where m_i and ℓ_i are classes corresponding to meridianal and longitudinal loops on $W \cap \partial X(L_i)$.

Since $A(L)_S$ is freely generated by the meridians, we have $\ell_i = \Sigma_{j=1}^{j=\mu} \lambda_{ij} m_j$ in $A(L)_S$, for some coefficients $\lambda_{ij} \in \Lambda_{\mu S}$. On considering the images under the homomorphism $\phi_{\pi S} : A(L)_S \to \Lambda_{\mu S}$ of the localised Crowell sequence, we obtain the equations $(T_i - 1)(t_i - 1) = (t_i - 1)\Sigma_{j=1}^{j=\mu} \lambda_{ij}(t_j - 1)$, for $1 \leq i \leq \mu$. Dividing by $t_i - 1$ and considering Taylor expansions about $t_1 = 1, \ldots, t_\mu = 1$ now gives $\varepsilon(\lambda_{ij}) = \ell_{ij}$. The final assertion follows since $\cup \partial D_i = \partial N$ in W, and so $\Sigma \delta(d_i) = 0$. $\square$

If $0 \to B_S \to A_S \to I_{\mu S} \to 0$ is a sequence of $\Lambda_{\mu S}$-modules such that $\mathbb{Z} \otimes A_S \cong \mathbb{Z}^{\mu}$, A has generators m_i with images $\phi(m_i) = t_i - 1$ in $I_{\mu S}$ and $TA_S = 0$ then there is a μ-component 1-link L with all linking numbers 0 and such that the sequence

$$0 \to (B(L)/TA(L))_S \to (A(L)/TA(L))_S \to I_{\mu S} \to 0$$

is isomorphic to the given sequence, via an isomorphism carrying meridianal elements to the m_i [**Le83**].

COROLLARY 4.13.1. [**Tr83**] *Let* $d_i = $ h.c.f.$\{\ell_{ij} \mid 1 \le j \le \mu\}$, *for* $1 \le i \le \mu$. *Then* $E_{\mu-1}(L) \equiv (d_1(t_1 - 1), \ldots, d_\mu(t_\mu - 1))$ *mod* $(I_\mu)^2$. $\square$

See [**Tr84**] for refinements of this result, and a more precise description of the presentation matrix, in terms of Milnor invariants.

COROLLARY 4.13.2. [**Tr84**] *If* $\mu \ge 2$ *then* $\Delta_1(L) \in (I_\mu)^{\mu-2}$. *If moreover all the linking numbers are 0 then* $\Delta_1(L) \in (I_\mu)^{2\mu-3}$.

PROOF. The entries in the above presentation matrix are all in $I_{\mu S}$, and so $E_1(L) \le (I_{\mu S})^{\mu-1}$. If all the linking numbers are 0 the entries are in $(I_{\mu S})^2$ and so $E_1(L) \le (I_{\mu S})^{2\mu-2}$. Since $E_1(L) = \Delta_1(L)I_\mu$ the result follows. $\square$

In particular, if $\Delta_1(L) = 1$ then $\mu \le 2$ and $\pi' = \pi''$. If all the linking numbers are 0 and μ is even the exponent can be improved to $2\mu - 2$. These estimates are best possible, in general [**Tr84**]. (However, if L can be obtained by surgery on the trivial link then $\Delta_1(L) \in (I_\mu)^{3\mu-4}$ [**Pl88**].)

The localized invariants are trivial for homology boundary links.

The ith longitudinal element ℓ_i is *unlinked* if the ith longitude is in π', i.e., if $\ell_{ij} = 0$ for all $j \ne i$. It is then the image of this longitude in $B(L) = \pi'/\pi''$, and $(t_i - 1)\ell_i = 0$. Henceforth we shall abbreviate "longitudinal element" to "longitude". The *longitudinal polynomial* of ℓ_i is the greatest common divisor in $\Lambda_\mu/(t_i - 1) \cong \Lambda_{\mu-1}$ of the annihilator of the image of ℓ_i in $A(L)/U$, where U is the submodule generated by the other unlinked longitudes. (If ℓ_i is not unlinked we define the longitudinal polynomial to be 0.)

4.7. Chen groups, nullity and longitudes

If L is a μ-component boundary 1-link with group $\pi = \pi L$ then $E_{\mu-1}(L) = 0$, $E_\mu(L)$ is principal, $H_2(X; \Lambda_\mu) \cong (\Lambda_\mu)^{\mu-1}$ and the longitudes of L are in $(\pi_\omega)' < \pi''$. (In particular, if $\mu = 1$ then $H_2(X; \Lambda_1) = 0$ and $\pi_\omega = \pi'$.) These properties all follow from the fact that the maximal abelian and maximal free covering spaces X' and X_ω of $X = X(L)$ may be obtained by splitting X along a family of disjoint Seifert surfaces. We may then apply Alexander duality in S^3 and a Mayer-Vietoris argument. In this section we shall relate these properties for more general links.

THEOREM 4.14. *Let L be a μ-component 1-link with group $\pi = \pi L$, and let P be the submodule of $A(L)$ generated by the longitudes. Then the following are equivalent:*

(1) $E_{\mu-1}(L) = 0$;
(2) $Ch(\pi; q) \cong Ch(F(\mu); q)$, *for all $q \geq 1$;*
(3) *the longitudes of L are in $\pi(\infty)$;*
(4) *there is an $f \in Ann(P)$ such that $\varepsilon(f) = 1$;*
(5) *the linking numbers are all 0 and*
$$\pi(\infty)/\pi'' = T(\pi'/\pi'') = TA(L).$$

Moreover, P is then pseudonull.

PROOF. The equivalence (1) $\Leftrightarrow$ (2) was proven in Theorem 4.10.

Let $\theta : F(\mu) \to \pi$ be the homomorphism determined by a choice of meridians. The induced homomorphisms $\theta_q : F(\mu)/F(\mu)_q \to \pi/\pi_q$ are epimorphisms. Hence the homomorphisms $Ch(\theta; q)$ are also epimorphisms. The longitudes of L are all in π' if and only if $Ch(\theta; 3)$ is an isomorphism, by Theorem 1.4. We assume henceforth that this is so. Then $(t_i - 1)\ell_i = 0$, for all $1 \leq i \leq \mu$.

Let $[w_{i,q}]$ be the image of ℓ_i in $Ch(F(\mu); q)$, and let $B = \pi'/\pi''$ and $B(\mu) = F(\mu)'/F(\mu)''$. Then $Ch(\theta; q)([w_{i,q}])$ is the image of ℓ_i in $\pi'/\pi_q\pi'' = B/(I_\mu)^{q-2}B$. If $Ch(\pi; q) \cong Ch(F(\mu); q)$ then $Ch(\theta; q)$ is an isomorphism, since $Ch(F(\mu); q)$ is a finitely generated abelian group. Hence $(t_i-1)[w_{i,q}]$ is in $(I_\mu)^{q-2}B(\mu)$, for all i. There is a $k \geq 0$ such that if $(t_i - 1)w \in (I_\mu)^t B(\mu)$ for some i, t then $w \in (I_\mu)^{t-k}B(\mu)$, by the Artin-Rees Lemma (Proposition 10.9 of [**AM**]).

Thus if $Ch(\pi;q) \cong Ch(F(\mu);q)$ for all $q \geq 1$ then ℓ_i is in $\cap_{q \geq 0}(I_\mu)^q B = \cap_{q \geq 2}((\pi_q \pi'')/\pi'') = \pi(\infty)/\pi''$, and so the longitudes are in $\pi(\infty)$. Conversely, if the longitudes are in $\pi(\infty)$ then $Ch(\theta;q)$ is an isomorphism for all $q \geq 2$. Thus (2) $\Leftrightarrow$ (3).

The ideal $E_\mu(L)$ contains some δ such that $\varepsilon(\delta) = 1$, since $\varepsilon(E_\mu(L)) = \mathbb{Z}$. Let $D = \{\delta^n \mid n \geq 0\}$. If $E_{\mu-1}(L) = 0$ then $A(L)_D$ is a projective $\Lambda_{\mu D}$-module, by Corollary 3.7.1, so some power δ^N annihilates $TB = TA(L)$. Moreover $T(\pi'/\pi'')$ is annihilated by $1+j$, where $j = \delta^N - 1 \in I_\mu$. Since

$$\cap_{n \geq 0}((I_\mu)^n B) = \{g \in B \mid (1+j)g = 0 \ for \ some \ j \in I_\mu\},$$

by Krull's Theorem (Theorem 10.17 of [**AM**]), we get $\pi(\infty)/\pi'' = TB = TA(L)$. Thus (1) implies (4) and (5).

If (4) holds then $1 - f \in I_\mu$, so $P = I_\mu P$. Hence $P \leq \pi(\infty)/\pi'' = \cap_{q \geq 2}((I_\mu)^q B)$, and so (4) implies (3). Similarly, (5) implies (3) since the longitudes represent elements of $T(\pi'/\pi'')$.

Since $Ann(P)$ contains both $\Pi_{i=1}^{i=\mu}(t_i - 1)$ and δ^N (for some $N \gg 0$), which have no nontrivial common factor, P is pseudonull. $\square$

The equivalence of (1-3) when $\mu = 2$ is due to Murasugi [**Mu70**]. If the linking numbers are all 0 but $\Delta_1(L) \neq 0$ then P is nontrivial, but $A(L)$ has no nontrivial pseudonull submodule, by part (2) of Theorem 4.12. (In this case the longitudinal polynomials are all 0.)

Similarly, Theorem 1.4 implies that $\pi/\pi_q \cong F(\mu)/F(\mu)_q$ for all $q \geq 1$ if and only if all the longitudes are in G_ω. (For a homology boundary 1-link we can see this more directly. As $\pi/\pi_\omega \cong F(\mu)$, which has no non-cyclic abelian subgroups, the longitudes are in π_ω.)

THEOREM 4.15. *Let L be a μ-component 1-link with group $\pi = \pi L$ and such that $E_{\mu-1}(L) = 0$, and let P be the submodule of $A(L)$ generated by the images of the longitudes. Then the following are equivalent:*

(1) *there is an epimorphism from π onto $F(\mu)/F(\mu)''$;*
(2) *$E_\mu(A(L)/P)$ is principal;*
(3) *$\pi/\pi(\infty) \cong F(\mu)/F(\mu)''$;*
(4) *$A(L) \cong (\Lambda_\mu)^\mu \oplus TA(L)$.*

PROOF. Suppose that $f : \pi \to F(\mu)/F(\mu)''$ is an epimorphism. Then $A(L) \cong (\Lambda_\mu)^\mu \oplus TA(L)$ and $\mathrm{Ker}(f) = \pi(\infty)$, since $B(\mu)$ is torsion free as a Λ_μ-module. Moreover π/π'' is a semidirect product $(\pi(\infty)/\pi'') \rtimes F(\mu)/F(\mu)''$, and so $\pi'/\pi'' \cong B(\mu) \oplus (\pi(\infty)/\pi'')$.

Let M be the closed 3-manifold obtained via 0-framed surgery on L, and let $p_M : M' \to M$ be its maximal abelian cover. Then $H_1(M;\mathbb{Z}) \cong \pi/\pi'$ (since the longitudes are in π') and $H_1(M;\Lambda_\mu) = H_1(M';\mathbb{Z}) \cong (\pi'/\pi'')/P \cong B(\mu) \oplus T$, where $T = (\pi(\infty)/\pi'')/P$ is a torsion module. As $H_2(M;\Lambda_\mu) \cong \overline{H^1(M;\Lambda_\mu)} \cong e^0 B(\mu)$, it has projective dimension at most 1. Therefore the only nonzero entries in the E_2 page of the UCSS $E_2^{pq} = e^q H_p(M;\Lambda_\mu) \Rightarrow H^{p+q}(M;\Lambda_\mu)$ are $E_2^{0\mu} \cong \mathbb{Z}$, $E_2^{1q} = e^q B(\mu) \oplus e^q T$ (for $0 \leq q \leq \mu+1$), $E_2^{20} = e^0 H_2(M;\Lambda_\mu)$ and $E_2^{21} = e^1 H_2(M;\Lambda_\mu)$, which is 0 if $\mu = 2$ and $\cong \mathbb{Z}$ if $\mu > 2$. (Note that if $\mu = 1$ or 2 then $B(\mu) \cong (\Lambda_\mu)^{\mu-1}$, while if $\mu \geq 3$ then $e^q B(\mu) = 0$ for $q \neq 0$ or $\mu - 2$.)

Let δ and D be as in Theorem 4.14. Then $\mathbb{Z}_D = \mathbb{Z}$ and $T_D = 0$, so $e^q(T)_D = 0$ also, since localization is exact. Since $H^3(M;\Lambda_\mu) \cong \mathbb{Z}$ we may conclude from the localized spectral sequence that the maps between the copies of $\mathbb{Z}$ in the positions $(0,\mu)$, $(1,\mu-2)$ and $(2,1)$ are what they should be. Therefore it follows from the unlocalized spectral sequence that $e^q T = 0$ if $q > 2$ and there is an exact sequence

$$0 \to e^1 T \to H^2(M;\Lambda_\mu) \to e^0 H_2(M;\Lambda_\mu) \to e^2 T \to 0.$$

Poincaré duality gives isomorphisms $H^2(M;\Lambda_\mu) \cong \overline{B(\mu)} \oplus \overline{T}$ and $e^0 H_2(M;\Lambda_\mu) = e^0 \overline{e^0 B(\mu)}$. Hence there is an exact sequence

$$0 \to B(\mu) \xrightarrow{\ \alpha\ } e^0 e^0 B(\mu) \to e^2 \overline{T} \to 0$$

and $e^1 T \cong \overline{T}$. On dualizing this sequence it follows that $e^0 \alpha$ is an isomorphism, since $e^q e^2 \overline{T}$ is 0 for $q < 2$. Hence $e^0 e^0 \alpha$ is an isomorphism. But $ev_{B(\mu)} : B(\mu) \to e^0 e^0 B(\mu)$ is an isomorphism and hence $e^0 e^0 \alpha = \alpha$. Therefore $e^2 \overline{T} = 0$. (If $\mu \leq 2$ this follows more easily, for $e^2 T$ is then a pseudonull module with a short free resolution.) Hence $e^0 T = 0$ and so $p.d._\Lambda T \leq 1$. Therefore $E_\mu(A(L)/P) = E_0(T)$ is invertible, by Theorem 3.7, and so is principal, since Λ_μ is factorial. Thus (1) implies (2).

If $E_\mu(A(L)/P)$ is principal then $A(L)/TA(L) = A(\pi/\pi(\infty))$ is projective, and therefore free, by Theorem 3.7. Hence $\pi/\pi(\infty) \cong F(\mu)/F(\mu)''$, by Theorem 4.2. Thus (2) implies (3), which in turn clearly implies (4).

If $A(L) \cong (\Lambda_\mu)^\mu \oplus TA(L)$ then $TA(L) \leq B(L)$. Let K be the preimage of $TA(L)$ in π and $G = \pi/K$. Then $G/G'' \cong F(\mu)/F(\mu)''$, by Theorem 4.2, and so π maps onto $F(\mu)/F(\mu)''$. This proves the theorem. $\square$

COROLLARY 4.15.1. *If the above conditions hold then P is the maximal pseudonull submodule of $A(L)$, $Ann(TA(L)/P) = (\lambda_\mu(L))$, $A(L)/P$ is $\mathbb{Z}$-torsion free and $E_\mu(L) = \Delta_\mu(L)E_0(P)$.*

PROOF. Since $E_\mu(A(L)/P)$ is principal and P is pseudonull it follows from Theorems 3.7 and 3.12 that $E_\mu(L) = E_0(TA(L)) = \Delta_\mu(L)E_0(P)$. The remaining assertions are immediate consequences of Theorem 3.8 and part (6) of Theorem 4.12. $\square$

COROLLARY 4.15.2. *$E_{\mu-1}(L) = 0$ and $E_\mu(L)$ is principal if and only if π maps onto $F(\mu)/F(\mu)''$ and the longitudes of L lie in π''. If these conditions hold then $H_2(X; \Lambda_\mu)$ is free.* $\square$

The conditions on π and the longitudes in this corollary each imply that $E_{\mu-1}(L)$ is 0; otherwise all four conditions are independent. The connection between longitudes and principality for 2-component homology boundary links is due to Crowell and Brown [**Cr76**].

There are examples of 3-component homology boundary links for which the module $H_2(X; \Lambda_\mu)$ is not free. (See §7 of Chapter 8.)

Boundary 1-links may also be characterized as links whose components are separated by a system of connected closed surfaces C_i such that each component of the link is null-homologous in $S^3 - \cup C_i$. (We may take $C_i = \partial R_i$, where the R_i are disjoint regular neighbourhoods in S^3 of a set of orientable spanning surfaces for the link. See [**Sm66, Sm70**].) By Alexander duality, each component of C lifts to X'. Hence the projection $p' : X' \to X$ induces an epimorphism $H_2(p'; \mathbb{Z})$. We show next that in general this projection is an epimorphism if and only if $A(L)$ has rank μ.

THEOREM 4.16. *Let L be a μ-component 1-link with group $\pi = \pi L$. The homomorphism $H_2(p) : H_2(X(L); \Lambda_\mu) \to H_2(X(L); \mathbb{Z})$ has cokernel $Tor_1^\Lambda(\mathbb{Z}, A(L))$, and is an epimorphism if and only if $E_{\mu-1}(L) = 0$.*

PROOF. The Cartan-Leray spectral sequence of the projection $p : (X', p'^{-1}(*)) \to (X, *)$ gives an exact sequence

$$\cdots \to \mathbb{Z} \otimes_\Lambda H_2(X; \Lambda) \to H_2(X; \mathbb{Z}) \to Tor_1^\Lambda(\mathbb{Z}, A(L)) \to 0.$$

It is clear from the proof of Theorem 4.7 that $Tor_1^\Lambda(\mathbb{Z}, A(L)) = 0$ if and only if $A_S \cong (\Lambda_\mu S)^\mu$, where $S = 1 + I_\mu$. This is in turn equivalent to $E_{\mu-1}(L) = 0$, by Lemma 4.9 and Theorem 4.10. $\square$

If $\alpha(L) = \mu = 2$ then $H_2(X(L); \Lambda_2) \cong \Lambda_2$, by Lemma 4.11. See also Theorem 11.12 for a related characterization of 2-component boundary links among homology boundary links.

4.8. *I*-equivalence

Let $\mathcal{L}$ be an *I*-equivalence between μ-component n-links $L(0)$ and $L(1)$, and let $X(\mathcal{L}) = S^{n+2} \times [0, 1] \setminus \mathcal{L}$. Then the inclusions of $X(L(0))$ and $X(L(1))$ into $X(\mathcal{L})$ induce isomorphisms on homology, so $\pi L(0)/\pi L(0)_q \cong \pi L(1)/\pi L(1)_q$, for all $q \geq 1$, by Theorem 1.3.

Let $Cr(\pi)_S$ denote the localization of the Crowell sequence for π with respect to the multiplicative system $S \subseteq \Lambda_\mu \setminus \{0\}$.

THEOREM 4.17. *Let L be a μ-component 1-link with group $\pi = \pi L$ and let $S = 1 + I_\mu$. Then*

(1) *the sequence $Cr(\pi)_S$ is invariant under I-equivalence;*
(2) *if $\Delta_1(L) = \delta u$, where $\varepsilon(u) = \pm 1$ and no factor of δ augments to ± 1, then (δ) is invariant under I-equivalence;*
(3) *$\alpha(L)$ is invariant under I-equivalence;*
(4) *$Tor_1^\Lambda(\mathbb{Z}, A(L))$ is invariant under I-equivalence;*
(5) *If $L(0)$ and $L(1)$ are concordant 1-links then there are f_0, f_1 in S such that $f_0 \overline{f_0} \Delta_\alpha(L(0)) \doteq f_1 \overline{f_1} \Delta_\alpha(L(1))$ in Λ_μ.*

PROOF. Part (1) follows from Lemma 4.9. It implies that the principal ideal generated by the image of $\Delta_1(L)$ in $\Lambda_\mu S$ is invariant under I-equivalence. As an element of Λ_μ becomes a unit in $\Lambda_\mu S$ if and only if it augments to ± 1, this in turn implies (2). It also implies (3), as $\alpha(L)$ is the rank of $A(L)_S$.

Let $T = Tor_1^\Lambda(\mathbb{Z}, A(L))$. Then $T = T_S$, since $I_\mu T = 0$. Since localization is an exact functor we have $T = T_S = Tor_1^{\Lambda_\mu S}(\mathbb{Z}_S, A(L)_S)$, which is clearly invariant under I-equivalence.

Let $\mathcal{L}$ be a concordance from $L(0)$ to $L(1)$, with exterior Z, as in Theorem 2.4. The argument of that theorem shows that the Blanchfield pairing on $\hat{t}H_1(\partial Z; \Lambda_\mu)$ is neutral (without localization) and so $\Delta_0(\hat{t}H_1(\partial Z; \Lambda_\mu)) = f\overline{f}$ for some $f \neq 0$ in Λ_μ, by Lemma 3.26. Moreover the Mayer-Vietoris sequence of the triple $(\partial Z', X_0', X_1')$ shows that the kernel and cokernel of the natural homomorphism from $\hat{t}H_1(X_0; \Lambda_\mu) \oplus \hat{t}H_1(X_1; \Lambda_\mu)$ to $\hat{t}H_1(\partial Z; \Lambda_\mu)$ are annihilated by $\Pi = \Pi_{1 \leq i \leq \mu}(t_i - 1)$. Write $f = gh$ and $\Delta_\alpha(L(i)) = u_i\delta_i$, where g, u_0 and u_1 are in S and h, δ_0 and δ_1 has no factors augmenting to ± 1. Then we see immediately that $u_0 u_1 = g\overline{g}$. Moreover $\overline{u_1} \doteq u_1$, since the Alexander polynomials are essentially invariant under conjugation. (See Theorem 5.1.) As $\delta_0 \doteq \delta_1$, by part (1), we see that $\Delta_\alpha(L(0))u_1\overline{u_1} \doteq \Delta_\alpha(L(1))g\overline{g}$. $\square$

Parts (1) and (2) are essentially from [**Ma80**], while (2) and (3) were proven for PL I-equivalences in [**Kw77**], and (5) is from [**Kw78**].

COROLLARY 4.17.1. *The linking numbers are invariant under I-equivalence.*

PROOF. We may assume $\mu = 2$, and then the result follows from part (4) of Theorem 4.12 and part (2) of Theorem 4.17. $\square$

COROLLARY 4.17.2. [**Kw78**] *If L is a slice link then $\alpha(L) = \mu$ and $\Delta_\mu(L) \doteq f\overline{f}$ for some $f \in S$.* $\square$

Conversely, any such product $f\overline{f}$ (with $f \in \Lambda_\mu$ and $\varepsilon(f) = \pm 1$) is realized as $\Delta_\mu(L)$ for some μ-component slice link L [**Na78**].

These invariants are invariants of F-isotopy, by Theorem 1.9. A stronger result holds for strict F-isotopy.

THEOREM 4.18. *Let $L(0)$ and $L(1)$ be μ-component 1-links which are strictly F-isotopic. Then $A(L(0)) \cong A(L(1))$ as Λ_μ-modules.*

PROOF. It is sufficient to assume that $L(1)$ is obtained from $L(0)$ by a strict elementary F-isotopy on the ith component. The result is then an immediate consequence of a Mayer-Vietoris sequence argument with coefficients Λ_μ. $\qquad\square$

4.9. The sign-determined Alexander polynomial

The cellular chain modules of a regular covering of a finite cell complex pair (X, Y) have natural bases, determined by lifting cells. (We consider the attaching maps for the cells to be part of the data of a cell complex, and so each cell is naturally oriented, by our conventions on orientations for discs and spheres.) The bases are well defined up to order and the action of the covering group. In particular, if $p : \hat{X} \to X$ is the covering associated to an epimorphism $\theta : \pi_1(X) \to Z^\mu$, $\hat{Y} = p^{-1}(Y)$ and $H_*(X, Y; (\Lambda_\mu)_0) = 0$ then the torsion of $C_*(\hat{X}, \hat{Y})$ with respect to such bases is well defined up to multiplication by elements of $\Lambda_\mu^\times$, and so we obtain an invariant $\tau(X, Y; \theta) \in (\Lambda_\mu)_0^\times / \Lambda_\mu^\times$. If $H_*(X, Y; (\Lambda_\mu)_0) \neq 0$ then $\tau(X, Y; \theta)$ is defined to be 0. If θ is the Hurewicz epimorphism to $H_1(X; \mathbb{Z})/(torsion)$ we shall write just $\tau(X, Y)$. This is in fact an invariant of simple homotopy type. The multiplicativity of the torsion gives $\tau(X; \theta) = \tau(X, Y; \theta)\tau(Y; \theta)$. There is a similar "Mayer-Vietoris" equation $\tau(X \cup Y; \theta)\tau(X \cap Y; \theta) = \tau(X; \theta)\tau(Y; \theta)$.

THEOREM 4.19. *Let L be a μ-component 1-link. Then $\tau(X(L)) \doteq (t-1)^{-1}\Delta_1(L)$ if $\mu = 1$ and $\tau(X(L)) \doteq \Delta_1(L)$ if $\mu > 1$.*

PROOF. This is an immediate consequence of Theorem 3.15. $\qquad\square$

The group $\mathbb{Z}^\mu$ is the group of *positive* units of Λ_μ. Turaev showed how to refine the torsion to a "sign-determined" torsion, with values in $(\Lambda_\mu)_0^\times / \mathbb{Z}^\mu$. An *h-orientation* of a pair of finite complexes (X, Y) is an orientation for the $\mathbb{Q}$-vector space $\oplus H_i(X, Y; \mathbb{Q})$, in other words the orbit of a (graded) basis under the action of automorphisms $\oplus \alpha_i$ such that $\Pi \det(\alpha_i) > 0$. Clearly every nonempty finite complex has exactly two h-orientations. Our conventions on orienting spheres and links determine a canonical h-orientation of a 1-link exterior $X(L)$, represented by the image of any base-point in $H_0(X;\mathbb{Q})$, the meridians in $H_1(X;\mathbb{Q})$ and the oriented boundaries of $X(L_i)$ for $1 \le i < \mu$ in $H_2(X;\mathbb{Q})$.

Let h_* be an h-orientation for (X,Y). A basis c'_* for $C_*(\hat{X},\hat{Y})$ determines a basis e_* for $C_*(X,Y) = \mathbb{Z} \otimes_\Lambda C_*(\hat{X},\hat{Y})$. Let $a_i = \Sigma_{q \le i} c_q(X,Y)$ be the number of cells of dimension at most i and $b_i = \Sigma_{q \le i} b_q(X,Y)$ be the corresponding sum of Betti numbers, and let $N(C) = \Sigma a_i b_i$. Then $\xi = (-1)^{N(C)} \tau(C_*(X,Y); e_*, h_*)$ is in $\mathbb{Q}^\times$. The *sign-determined* torsion $\tau_+(X,Y;\theta)$ is the image of $\mathrm{sign}(\xi)\tau(C_*(\hat{X},\hat{Y}); e'_*)$ in $(\Lambda_\mu)_0^\times/\mathbb{Z}^\mu$ if $H_*(X,Y;(\Lambda_\mu)_0) = 0$ and is 0 otherwise. (The factor $(-1)^{N(C)}$ is needed in order that the sign-determined torsion be invariant under cellular subdivision.) The sign-determined torsion satisfies a product formula strengthening Theorem 3.15. (See Lemma 3.4.2 of [**Tu86**].)

The *sign-determined* Alexander polynomial of a μ-component 1-link L is defined to be $A_+(L) = \tau_+(X(L))$. This apparently minor improvement has proved very useful, both in providing new proofs of known properties on Alexander polynomials of links and extending such results [**Tu86, Tu88**], in extending the Casson-Walker invariant to all closed orientable 3-manifolds [**Les**], and most recently in identifying the Seiberg-Witten invariants of 3-manifolds in terms of Alexander polynomials [**MT96, Tu98**]. Moreover the interpretation of the Alexander polynomial as a torsion invariant places it in a wider context. Other normalizations of the Alexander polynomial have been given in [**Ha83**] (following an idea of [**Co70**]) and [**Les**] (specializing Turaev's idea to link exteriors in 3-manifolds).

In Chapter 5 we shall give a further variant of the Reidemeister-Franz torsion (also due to Turaev). If L is a classical link the corresponding torsion of $X(L)$ is closely related to $\Delta_\alpha(L)$, the first nonzero Alexander polynomial of L. However, in general the higher Alexander polynomials and the Alexander ideals do not appear to have similar interpretations.

4.10. Higher dimensional links

The preceding sections have concentrated on the classical case. While "knot modules" have been studied in detail (see [**Le77, Lev**] and Chapter 6 below), relatively little is known about the homology of the maximal abelian covers of higher dimensional links with $\mu > 1$ beyond the following analogue of Theorem 2.1.

THEOREM 4.20. [**Ko82, Sa84**] *Let L be a μ-component n-link with exterior $X = X(L)$. Then*

(1) $H_j(X; \Lambda_\mu)$ *and* $H^j(X; \Lambda_\mu)$ *are finitely generated, for all j, and are 0 if $j > n + 1$;*

(2) $\mathbb{Z} \otimes_\Lambda H_j(X; \Lambda_\mu) = 0$ *and so $H_j(X; \Lambda_\mu)$ is a torsion module, for $2 \leq j \leq n$.*
   If moreover either $n > 1$ or $n = 1$ and $\alpha(L) = \mu$ then

(3) $\mathbb{Z} \otimes TH_1(X(L); \Lambda_\mu) = 0$;

(4) $H_{n+1}(X; \Lambda_\mu)$ *is torsion free of rank $\mu - 1$;*

(5) $H_1(X; \Lambda_\mu)$ *has rank $\mu - 1$ (i.e., $E_{\mu-1}(\pi L) = 0$).*

PROOF. The first assertion and the fact that $H_{n+1}(X; \Lambda_\mu)$ is torsion free hold since X is a compact $(n+2)$-manifold with nonempty boundary and Λ_μ is a noetherian ring. A choice of meridians for L determines a map $f : V = \vee^\mu S^1 \to X = X(L)$, which we may assume is an embedding, and which is homologically n-connected, i.e. $H_q(X, V; \mathbb{Z}) = 0$ for all $q \leq n$.

Let $S = 1 + I_\mu = \{f \in \Lambda_\mu \mid \varepsilon(f) = 1\}$. If C_* is a finitely generated free Λ_μ-chain complex such that $H_0(\mathbb{Z} \otimes_\Lambda C_*) = 0$ then $H_0(C_*)_S = H_0(C_{*S}) = 0$, by Nakayama's Lemma. If $H_j(\mathbb{Z} \otimes_\Lambda C_*) = 0$ for $j \leq n$ an easy induction then shows that $H_j(C_*)_S = H_j(C_{*S}) = 0$ for $j \leq n$ also. Applying this argument to the cellular chain complex of the maximal abelian covering of the pair (X, V) we see that $H_j(X, V; \mathbb{Z})_S = 0$ for $j \leq n$. Since $H_1(f; \mathbb{Z})$ is an isomorphism and the Chen groups $Ch(\pi L; q)$ are all free $A(f)_S$ is an isomorphism, by Lemma 4.9. Hence $H_1(f; \Lambda)_S$ is also an isomorphism. Therefore $TH_1(X(L); \Lambda_\mu)_S = 0$ and $H_j(X; \Lambda)_S = 0$ for $2 \leq j \leq n$. Hence $H_1(X(L); \Lambda_\mu)$ has rank $\mu - 1$ and $H_j(X(L); \Lambda_\mu)$ is a torsion module for $2 \leq j \leq n$. Since $\chi(X(L)) = (1 - \mu)(1 - (-1)^n)$ it follows that

$H_{n+1}(X(L); \Lambda_\mu)$ also has rank $\mu - 1$. Since $\mathbb{Z} \otimes_\Lambda M \cong \mathbb{Z} \otimes_\Lambda M_S$ for any finitely generated Λ_μ-module this implies the theorem. $\qquad \square$

In particular, if L is a knot ($\mu = 1$) then $H_j(X; \Lambda_1) = 0$ for all $j > n$ and $\mathbb{Z} \otimes_\Lambda H_j(X; \Lambda_1) = 0$, for $1 \le j \le n$. (See Chapter 7 below for more on such "knot modules".)

COROLLARY 4.20.1. $\varepsilon(\Delta_0(H_j(X; \Lambda_\mu))) = \pm 1$, for $2 \le j \le n$. $\qquad \square$

The modules of homology boundary links satisfy a further constraint, due to Sato. A Λ_μ-module M is *of type BL* if there is a finitely generated abelian group A and an exact sequence

$$0 \to \Lambda_\mu \otimes A \xrightarrow{\ f\ } \Lambda_\mu \otimes A \to M \to 0,$$

where $\varepsilon \otimes f$ is an automorphism of A.

LEMMA 4.21. [Sa86] *Let A be a finitely generated group and f an endomorphism of $\Lambda_\mu \otimes A$. If $\varepsilon \otimes f$ is injective then so is f.*

PROOF. The corresponding result for field coefficients follows easily from determinantal characterizations of the rank. In the integral case one reduces firstly to the p-primary part of the $\mathbb{Z}$-torsion submodules and then inducts on n, where p^n is the exponent of this group. $\qquad \square$

Let L be a homology boundary n-link and let $d_0, \ldots, d_{n+2}$ be the maps in the Mayer-Vietoris sequence for $H_*(X; \Lambda_\mu)$ determined by a family of disjoint Seifert hypersurfaces U_i for L as in §2 of Chapter 2. The maps $\varepsilon \otimes d_j$ are monomorphisms if $1 \le i \le n - 1$, and are isomorphisms if $2 \le i \le n - 1$, by the Mayer-Vietoris sequence for $H_*(X; \mathbb{Z})$. Therefore d_j is a monomorphism in this range, by Lemma 4.21. Thus $H_j(X; \Lambda_\mu)$ is of type BL for $2 \le j \le n - 1$. If moreover L is a boundary link then $\varepsilon \otimes d_1$ is also an isomorphism, $TH_1(X; \Lambda_\mu)$ and $H_n(X; \Lambda_\mu)$ are of type BL, and $H_{n+1}(X; \Lambda_\mu) \cong (\Lambda_\mu)^{\mu-1}$. Conversely, given $n > 2$ and a family of modules H_j of type BL for $1 \le j < [n/2]$ and such that H_1 is $\mathbb{Z}$-torsion free there is a μ-component boundary n-link L such that $H_j(X; \Lambda_\mu) \cong H_j$ [Sa86].

Modules of type BL clearly have projective dimension ≤ 2. There is a purely algebraic proof for the case of field coefficients.

THEOREM 4.22. [**Ko89**] *Let L be a μ-component homology bound-ary n-link with exterior $X = X(L)$, and let $\mathbb{F}$ be a field. Then $H_i(X; \mathbb{F}\Lambda_\mu)$ has a square presentation matrix, for $2 \leq i \leq n - 1$.*

PROOF. Let $C_i = C_i(X^\omega; \mathbb{F})$ and $D_i = C_i(X'; \mathbb{F})$ be the chain complexes associated to the coverings X^ω and X', respectively, and let $\rho_i : C_i \to D_i$ be the natural homomorphism (reduction of coefficients). Then $\rho_i(\mathrm{Ker}(\partial_i^C)) \leq \mathrm{Ker}(\partial_i^D)$, for all i. Since $\mathbb{F}[F(\mu)]$ has global dimension 1 the submodules $\partial_i(C_i) \leq C_{i-1}$ are projective, and so $C_i \cong \mathrm{Ker}(\partial_i^C) \oplus \partial_i^C(C_i)$. Reduction of coefficients gives a corresponding splitting of D_i, and so $\rho_i(\mathrm{Ker}(\partial_i^C))$ is a direct summand. If $2 \leq i \leq n$ then $H_i(X; \mathbb{F}\Lambda_\mu) = H_i(D_*)$ is a finitely generated torsion $\mathbb{F}\Lambda_\mu$-module. Hence $\rho_i(\mathrm{Ker}(\partial_i^C)) = \mathrm{Ker}(\partial_i^D)$. Thus $\mathrm{Ker}(\partial_i^D)$ and $\mathrm{Im}(\partial_i^D)$ are projective in this range, and so $p.d_{\mathbb{F}\Lambda_\mu} H_i(X; \mathbb{F}\Lambda_\mu) \leq 1$ if $2 \leq i \leq n - 1$. Since projective $\mathbb{F}\Lambda_\mu$-modules are free, $H_i(X; \mathbb{F}\Lambda_\mu)$ then has a square presentation matrix. $\qquad\square$

The examples constructed from the parafree groups $G(i, j)$ of Baumslag show that in general higher dimensional links are not homology boundary links. Kobelskii asserted that $TH_1(X; \Lambda_\mu)$ and $H_j(X; \Lambda_\mu)$ are of type BL for any n-link L and $2 \leq j \leq n - 1$ [**Ko82**]. In [**Ko89**] he extended the conclusion of Theorem 4.22 to all 2-component n-links, by means of the following lemma, which he proved for $\mu = 2$ using a carefully chosen presentation for π.

LEMMA 4.23. *Let π be a finitely presentable group such that $\pi/\pi' \cong \mathbb{Z}^\mu$ and $H_2(\pi; \mathbb{Z}) = 0$. Then the abelianization homomorphism from π to $\mathbb{Z}^\mu$ factors through $F(\mu)/F(\mu)''$.*

PROOF. Let $A = A(\pi)$, $B = B(\pi) = \pi'/\pi''$ and ξ_π be the class of $Cr(\pi)$ in $\mathrm{Ext}_\Lambda^1(I_\mu, B)$. Let ξ_F be the class of $Cr(F(\mu))$ in $\mathrm{Ext}_\Lambda(I_\mu, B(\mu))$. Let $S = 1 + I_\mu$ and let $\xi_{\pi S}$ be the class of $Cr(\pi)_S$ in $\mathrm{Ext}_\Lambda^1(I_\mu S, B(\mu)_S) = \mathrm{Ext}_\Lambda^1(I_\mu, B(\mu))_S \cong \mathrm{Ext}_\Lambda^1(I_\mu, B(\mu))$. Then $A_S \cong (\Lambda_\mu S)^\mu$, by Theorems 1.3, 4.10 and Nakayama's Lemma. Since B_S is the kernel of an epimorphism from A_S to $I_2 S$ we have $B_S \oplus (\Lambda_\mu S)^\mu \cong B(\mu)_S \oplus (\Lambda_\mu S)^\mu$, by Schanuel's Lemma. Since A_S is free of rank μ it follows easily that $\xi_{\pi S} = \xi_{FS}$.

The image of B in B_S is contained in $\tilde{B} = s^{-1}B(\mu)$, for some $s \in S$. Let $\tilde{A}$ be the pushout of A and $\tilde{B}$ under B, and let $\tilde{\xi}$ be the class of the extension

$$0 \to \tilde{B} \to \tilde{A} \to I_\mu \to 0$$

in $Ext^1_\Lambda(I_\mu, \tilde{B})$. The images of ξ_π, ξ_F and $\tilde{\xi}$ in $Ext^1_\Lambda(I_{\mu S}, B(\mu)_S)$ agree, and so $\tilde{\xi} = \xi_F$.

There is a homomorphism from π/π'' to $F(\mu)/F(\mu)''$ which induces an isomorphism on abelianization and the pushout homomorphism from ξ to $\tilde{\xi}$, by Lemma 4.4. $\square$

In the rest of his argument he uses the ring $\mathbb{F}B = \mathbb{F}[G]/\mathbb{F}I'(G)^2$, where $G = F(2)/F(2)''$ and $\mathbb{F}I'(G)$ is the kernel of the abelianization homomorphism from $\mathbb{F}[G]$ to $\mathbb{F}\Lambda_2$. The key step is to show that if $\widehat{E}_1$ and $\widehat{E}_2$ are finitely generated free $\mathbb{F}B$-modules, $\psi_1 : \widehat{E}_1 \to E_1 = \mathbb{F}\Lambda_2 \otimes_{\mathbb{F}B} \widehat{E}_1$ and $\psi_2 : \widehat{E}_2 \to E_2 = \mathbb{F}\Lambda_2 \otimes_{\mathbb{F}B} \widehat{E}_2$ are the canonical epimorphisms and there is a commutative square

$$\begin{array}{ccc} \widehat{E}_1 & \xrightarrow{\widehat{\phi}} & \widehat{E}_2 \\ \psi_2 \downarrow & & \psi_2 \downarrow \\ E_1 & \xrightarrow{\phi} & E_2 \end{array}$$

such that $\omega \mathrm{Ker}(\phi) \le \psi_1(\mathrm{Ker}(\widehat{\phi}))$ for some $\omega \in \mathbb{F}\Lambda_2$ (cf. part (2) of Theorem 4.20) then $\mathrm{Im}(\phi)$ is projective. From this his main result follows easily. To establish the key step he exploits the fact that $\mathbb{F}B$ is a (non-commutative) extension of $\mathbb{F}\Lambda_2$ by the ideal $\mathbb{F}I'(G)/\mathbb{F}I'(G)^2$, which is free as a left (or right) $\mathbb{F}\Lambda_2$-module. He also applies the fact that if M is a $\mathbb{F}\Lambda_2$-module such that $p.d._{\mathbb{F}\Lambda}M > 1$ then it has a nontrivial pseudonull submodule to $M = \mathrm{Cok}(\phi)$. (See part (1) of Theorem 3.22.) It is not clear how to extend this part of his argument to the general case.

CHAPTER 5

Sublinks and Other Abelian Covers

The Torres conditions relate the Alexander invariants of a link to those of its sublinks. Similar reductions give invariants of intermediate abelian covering spaces. We shall consider in some detail the homology of abelian branched coverings of homology 3-spheres, branched over links.

5.1. The Torres conditions

Let L be a μ-component n-link and let $L(\hat{j})$ be the $(\mu - 1)$-component sublink obtained by deleting the jth component L_j. Let $\psi : \Lambda_\mu \to \Lambda_{\mu-1}$ be the epimorphism (of $\Lambda_{\mu-1}$-modules) determined by $\psi(t_i) = t_i$ if $1 \leq i < \mu$ and $\psi(t_\mu) = 1$. If M is a Λ_μ-module let $\psi M = M/(t_\mu - 1)M$.

The two conditions of Torres for classical links [**To53**] may be stated as follows;

(1) If $\mu = 1$ then $\overline{\Delta_1(L)} = t^{2a}\Delta_1(L)$ for some $a \in \mathbb{Z}$;
 if $\mu > 1$ then $\overline{\Delta_1(L)} = (-1)^\mu (\prod_{i=1}^{\mu} t_i^{b_i})\Delta_1(L)$,
 where $b_i \equiv 1 - \Sigma_{j=1}^{j=\mu} \ell_{ij}$ mod (2).
(2) If $\mu = 1$ then $\psi(E_1(L)) = \mathbb{Z}$;
 if $\mu > 1$ then $\psi(E_1(L)) = (\prod_{i=1}^{\mu-1} t_i^{\ell_{i\mu}} - 1)E_1(L(\hat{\mu}))$.

Condition (1) does not depend on the choice of representative for the first Alexander polynomial, while deleting other components of L leads to conditions similar to (2). Duality establishes (1) up to multiplication by units, as in the following theorem.

THEOREM 5.1. [**Bl57**] *Let L be a μ-component 1-link. Then* $\overline{\Delta_i(L)} \doteq \Delta_i(L)$, *for all $i \geq 0$, and* $\varepsilon(\Delta_j(L)) = \pm 1$, *for all $j \geq \mu$.*

PROOF. Let $\wp = (p)$ be a height 1 prime ideal of Λ_μ, generated by an irreducible element p. If $\overline{\wp} = \wp$ then p^a divides $\overline{\Delta_i(L)}$ if and only if it divides $\Delta_i(L)$. So we may assume that $\overline{\wp} \neq \wp$. In particular, the monomials $t_j - 1$ are units in the PID $\Lambda_{\mu\wp}$. On localizing the long exact sequence of homology with coefficients Λ_μ for the pair $(X, \partial X)$ and on observing that $\Pi(t_j - 1)$ annihilates $H_*(\partial X; \Lambda_\mu)$ we conclude that $H_q(X; \Lambda_{\mu\wp})$ is isomorphic to $H_q(X, \partial X; \Lambda_{\mu\wp})$, for all $q \geq 0$. The UCSS gives an exact sequence

$$0 \to e^1 H_1(X; \Lambda_{\mu\wp}) \to H^2(X; \Lambda_{\mu\wp}) \to e^0 H_2(X; \Lambda_{\mu\wp}) \to 0.$$

Let $T = TH_1(X; \Lambda_\mu)$. Then $e^1 T_\wp = e^1 H_1(X; \Lambda_{\mu\wp}) \cong T\overline{H_1(X; \Lambda_{\mu\overline{\wp}})} = \overline{T_{\overline{\wp}}}$, by Poincaré duality and the facts that $e^0 H_2(X; \Lambda_{\mu\wp})$ is torsion free and the rings $\Lambda_{\mu\wp}$ are PIDs. The result now follows from the Elementary Divisor Theorem, which implies that $N \cong e^1 N$, for N any finitely generated torsion module over a PID.

The second assertion follows from part (1) of Lemma 4.11.   □

Every $f \in \Lambda_\mu$ such that $\varepsilon(f) = \pm 1$ and $\bar{f} \doteq f$ is realizable as $\Delta_\mu(L)$ for some μ-component boundary 1-link [**Fr06**]. If L is a 1-link then $\overline{E_i(L)} = E_i(L)$, for $i = 0$ or 1, by Theorem 4.12.(1), but this is not true in general [**Tu89**]. The SFS row invariants may be used to show that there are knots for which $H_1(X; \Lambda)$ is not isomorphic to $\overline{H_1(X; \Lambda)}$ [**FS64**].

The more precise assertion of the first Torres condition follows easily from Theorem 5.1 and the second Torres condition. (See below.) We shall derive the later condition from excision and a Wang sequence.

LEMMA 5.2. [**Sa81**] *Let L be a μ-component n-link, and let $X = X(L)$, $Y = X(L(\hat{\mu}))$ and $\Pi = \prod_{i=1}^{\mu-1} t_i^{\ell_{i\mu}}$. Then*

(1) *if $n = 1$ there is an ideal J containing $\Pi - 1$ and an exact sequence $0 \to \Lambda_{\mu-1}/J \to \psi A(L) \to A(L(\hat{\mu})) \to 0$;*

(2) *if $n \geq 2$ and $2 \leq k \leq n$ there is an exact sequence*

$$0 \to \psi H_k(X; \Lambda_\mu) \to H_k(X; \Lambda_{\mu-1}) \to Tor_1^{\Lambda_\mu}(\Lambda_{\mu-1}, H_{k-1}(X; \Lambda_\mu)) \to 0,$$

and $H_k(X; \Lambda_{\mu-1}) \cong H_k(Y; \Lambda_{\mu-1})$ if $2 < k \leq n$.

PROOF. Suppose first that $n = 1$. Let $\hat{X}$ be the covering space of X induced by the maximal abelian cover $p_Y : Y' \to Y$ of Y, and let $* \in X$ be a basepoint. The cover $r : (X', p'^{-1}(*)) \to (\hat{X}, p_Y^{-1}(*))$ is infinite cyclic, and the Wang sequence for r gives an isomorphism $H_1(X, *; \Lambda_{\mu-1}) \cong \psi A(L)$. Now $Y \setminus X$ is a regular neighbourhood of L_μ, so $H_1(Y, X; \Lambda_{\mu-1}) = 0$ and $H_2(Y, X; \Lambda_{\mu-1}) \cong \Lambda_{\mu-1}/(\Pi - 1)$. The exact sequence of homology for the triple $(Y', \hat{X}, p_Y^{-1}(*))$ gives an exact sequence

$$H_2(Y, X; \Lambda_{\mu-1}) \to \psi A(L) \to A(L(\hat{\mu})) \to 0,$$

which reduces to the claimed form.

If $n \geq 2$ the argument is similar, but simpler as $H_q(Y, X; \Lambda_{\mu-1}) = 0$ for $2 < q \leq n + 1$. $\qquad\square$

It follows that if $n > 1$ and $\Delta_{0,q}(L) = \Delta_0(H_q(X(L); \Lambda_\mu))$, for all q, then $\psi(\Delta_{0,q}(L))$ divides $\Delta_{0,q}(L(\hat{\mu}))$ for $2 \leq q \leq n$. When $n = 2k - 1 \geq 3$ and $q = k$ this divisibility can be sharpened; see Theorem 5.6 below.

THEOREM 5.3. *Let L be a μ-component 1-link, and let $\Pi = \prod_{i=1}^{\mu-1} t_i^{\ell_{i\mu}}$ and $\alpha = \alpha(L)$. Then*

(1) *if $\mu = 1$ then $\psi(E_1(L)) = \mathbb{Z}$, while if $\mu > 1$ then $\psi(E_1(L)) = (\Pi - 1)E_1(L(\hat{\mu}))$;*

(2) *if $k > 1$ then $E_{k-1}(L(\hat{\mu})) + (\Pi - 1)E_k(L(\hat{\mu})) \leq \psi(E_k(L))$ and $\psi(E_k(L)) \leq E_{k-1}(L(\hat{\mu})) + I_{\mu-1}E_k(L(\hat{\mu}))$;*

(3) *$\alpha(L(\hat{\mu})) \geq \alpha - 1$, with equality only if $\ell_{i\mu} = 0$ for $i < \mu$;*

(4) *$\psi(\Delta_\alpha(L))$ divides $\Delta_{\alpha-1}(L(\hat{\mu}))$;*

(5) *if $\alpha(L(\hat{\mu})) = \alpha$ then $\psi(\Delta_\alpha(L))$ divides $(\Pi - 1)\Delta_\alpha(L(\hat{\mu}))$; if $\alpha(L(\hat{\mu})) > \alpha$ then $\psi(\Delta_\alpha(L)) = 0$.*

PROOF. We retain the notation from Lemma 5.2. If $\mu = 1$ then $\psi = \varepsilon$, so $\psi(E_1(L)) = \mathbb{Z}$, by Lemma 4.11, while the other assertions are trivial. We may assume henceforth that $\mu > 1$. If $\Delta_1(L(\hat{\mu})) = 0$ then both sides of the equation in part (1) are 0, so it is trivially true. Otherwise $H_2(Y; \Lambda_{\mu-1}) = 0$ and there is an exact sequence

$$0 \to \Lambda_{\mu-1}/(\Pi - 1) \to \psi A(L) \to A(L(\hat{\mu})) \to 0.$$

Moreover $A(L(\hat{\mu}))$ has a short free resolution and $\psi(E_1(L)) = E_1(\psi A(L))$, so (1) now follows from Theorem 3.12.

In general, there is an ideal J containing $\Pi - 1$ and an exact sequence

$$0 \to \Lambda_{\mu-1}/J \to \psi A(L) \to A(L(\hat{\mu})) \to 0.$$

Hence if P is a presentation matrix for $A(L(\hat{\mu}))$ then $\psi A(L)$ has a presentation matrix of the form $\left(\begin{smallmatrix} P & K^* \\ 0 & J^* \end{smallmatrix} \right)$ where K^* and J^* are column vectors and the entries of J^* generate the ideal J. Hence $\psi(E_k(L)) = E_k(\psi A(L))$ contains the products $E_0(\Lambda_{\mu-1}/J)E_k(L(\hat{\mu}))$ and $E_1(\Lambda_{\mu-1}/J)E_{k-1}(L(\hat{\mu}))$. Therefore

$$E_{k-1}(L(\hat{\mu})) + (\Pi - 1)E_k(L(\hat{\mu})) \le \psi(E_k(L)),$$

since $\Pi - 1 \in E_0(\Lambda_{\mu-1}/J)$ and $E_1(\Lambda_{\mu-1}/J) = \Lambda_{\mu-1}$. Similarly $\psi(E_k(L)) \le E_{k-1}(L(\hat{\mu})) + (K, J)E_k(L(\hat{\mu}))$. Since $\mathbb{Z} \otimes \psi A(L) \cong \mathbb{Z}^\mu$ and $\mathbb{Z} \otimes A(L(\hat{\mu})) \cong \mathbb{Z}^{\mu-1}$, the column $\left(\begin{smallmatrix} K^* \\ J^* \end{smallmatrix} \right)$ is in the span of the columns of $\left(\begin{smallmatrix} P \\ 0 \end{smallmatrix} \right)$ mod $I_{\mu-1}$. Hence

$$\psi(E_k(L)) \le E_{k-1}(L(\hat{\mu})) + I_{\mu-1}E_k(L(\hat{\mu})).$$

If $\Delta_{\alpha-1}(L(\hat{\mu})) \ne 0$ the rank of $\psi A(L)$ is strictly greater than that of $A(L(\hat{\mu}))$, and the sequence of Lemma 5.2 reduces to

$$0 \to \Lambda_{\mu-1} \to \psi A(L) \to A(L(\hat{\mu})) \to 0.$$

In particular, $\psi(E_\alpha(L)) \ne 0$ and so $\psi(\Delta_\alpha(L)) \ne 0$. Let $C = \psi T A(L)$. Since $\Delta_\alpha(L) = \Delta_0(T A(L))$, by Theorem 3.4, it follows that C is a torsion $\Lambda_{\mu-1}$-module, and $\psi(\Delta_\alpha(L))$ divides $\Delta_0(C)$, by Theorem 3.14. Since C is isomorphic to a submodule of $\psi A(L)$, by the Snake Lemma, it maps injectively to $A(L(\hat{\mu}))$. Therefore $\Delta_0(C)$ divides $\Delta_{\alpha-1}(L(\hat{\mu}))$, and so $\psi(\Delta_\alpha(L))$ divides $\Delta_{\alpha-1}(L(\hat{\mu}))$.

The remaining assertions of (3), (4) and (5) follow immediately from Lemma 5.2. $\square$

The result of part (2) is due to Traldi, who gave examples of 1-links such that both inclusions are strict, and examples such that one or both are equalities [**Tr82**].

Let $\nu_k(t) = (t^k - 1)/(t - 1)$, for $k \in \mathbb{Z}$.

COROLLARY 5.3.1. *If* $\mu = 2$ *then* $\psi(\Delta_1(L)) \doteq \nu_{\ell_{12}}(t)\Delta_1(L_1)$, *while if* $\mu > 2$ *then* $\psi(\Delta_1(L)) \doteq (\Pi - 1)\Delta_1(L(\hat{\mu}))$.

PROOF. These follow from (1) since $E_1(L) = \Delta_1(L)I_\mu$ if $\mu > 1$, by Theorem 4.12, and $\psi(I_2) = (t-1)$ while $\psi(I_\mu) = I_{\mu-1}$ if $\mu > 2$. $\square$

When $\mu = 2$ this implies part (4) of Theorem 4.12.

Since the ideal $(\Delta_1(L))$ is self conjugate, by Theorem 5.1, there is a unit $u \in \Lambda_\mu^\times$ such that $\overline{\Delta_1(L)} \doteq u\Delta_1(L)$. If $\mu = 1$ then $\overline{\Delta_1(L)} = u\Delta_1(L)$ and $\varepsilon(\Delta_1(L)) = \pm 1 \not\equiv 0 \mod (2)$ easily imply that $u = t^{2a}$ for some $a \in \mathbb{Z}$. In general $\overline{\Delta_1(L)} = u\Delta_1(L)$ implies that $\psi(\overline{\Delta_1(L)}) = \psi(u)\psi(\Delta_1(L))$. The Corollary then gives $\nu_{\ell_{12}}(t)\overline{\Delta_1(L_1)} = t_1^{\ell_{12}-1}\psi(u)\nu_{\ell_{12}}(t)\Delta_1(L_1)$ if $\mu = 2$, while if $\mu > 2$ we have $(\Pi - 1)\overline{\Delta_1(L(\hat{\mu}))} = \Pi\psi(u)(\Pi - 1)\Delta_1(L(\hat{\mu}))$. If sufficiently many linking numbers are nonzero a simple induction now gives the first Torres condition.

Otherwise, adjoin a new component K_0 such that $\ell_{0i} = \mathrm{lk}(K_0, L_i)$ is nonzero, for $1 \leq i \leq \mu$, and let $L^+ = K_0 \cup L$ be the resulting $(\mu+1)$-component link. We may now use the above argument to conclude that $\Delta_1(L^+)(t_0, \ldots t_\mu)$ satisfies the first Torres condition, so that $\overline{\Delta_1(L^+)} = (-1)^{\mu+1}(\prod_{i=0}^\mu t_i^{c_i})\Delta_1(L^+)$, with $c_i \equiv 1 - \Sigma_{j=0}^{j=\mu}\ell_{ij} \mod (2)$. On applying the second Torres condition to the link L obtained by deleting the component K_0 of L^+ we have $(\prod_{i=1}^\mu t_i^{\ell_{0i}} - 1)\Delta_1(L) = (-1)^{\mu+1}(\prod_{i=1}^\mu t_i^{c_i})(\prod_{i=1}^\mu t_i^{\ell_{0i}} - 1)\Delta_1(L)$. As the common factor is nonzero we get $-\overline{\Delta_1(L)} = (-1)^{\mu+1}(\prod_{i=1}^\mu t_i^{b_i})\Delta_1(L)$, where $b_i = c_i + \ell_{0i} \equiv 1 - \Sigma_{j=1}^{j=\mu}\ell_{ij} \mod (2)$. This proves the first Torres condition. (This argument is due to Fox and Torres [FT54].)

Using knots constructed by imbedding once-punctured surfaces in S^3 with suitable self-linking, Seifert showed that any integral Laurent polynomial $\delta \in \Lambda_1$ such that $\bar{\delta} = \pm t^n \delta$ and $\varepsilon(\delta) = \pm 1$ could be realized as the first Alexander polynomial of a knot [Se34]. Levine used surgery to construct a knot with $\pi'/\pi'' \cong \Lambda_1/(\delta)$, and hence (by taking knot sums) obtained a similar characterization of the family of Alexander polynomials $\Delta_i(K)$ ([Le65] - see also Theorem 7.10 below). The natural question is whether the Torres conditions suffice also if $\mu > 1$. We shall see (in Chapter 8) that in general further conditions are needed if $\mu = 2$ and $|\ell_{12}| > 1$.

THEOREM 5.4. *Let L be a μ-component 1-link. If $\alpha(L) = 2$ and $\alpha(L(\hat{\mu})) = 1$ then $Ann(\ell_\mu) = (t_\mu - 1, \theta)$ for some $\theta \in \Lambda_{\mu-1}$ and ℓ_μ generates $\mathrm{Ker}(t_\mu - 1)$.*

PROOF. We retain the notation from Lemma 5.2. The linking numbers $\ell_{i\mu}$ are all 0, by part (3) of Theorem 5.3. Hence the solid torus $Y - X$ lifts to $\hat{X}$, since it is a regular neighbourhood of L_μ, and so $H_3(Y, X; \Lambda_{\mu-1}) \cong \Lambda_{\mu-1}$, The connecting homomorphism to $H_2(X; \Lambda_{\mu-1})$ is injective, with image generated by the fundamental class of a lift D_μ of $\partial X(L_\mu)$ to Y'. The class $[D_\mu]$ has image ℓ_μ in $H_1(X; \Lambda_\mu)$ under the Wang sequence homomorphism.

Now $H_2(X; \Lambda_\mu) \cong \Lambda_\mu$, by Lemma 4.11, and $H_2(X; \Lambda_{\mu-1})$ is freely generated by $[D_\mu]$ as a $\Lambda_{\mu-1}$-module. Hence $Ann(\ell_\mu) = (t_\mu - 1, \theta)$, where $\theta[D_\mu]$ generates the image of $H_2(r)$ in $H_2(X; \Lambda_{\mu-1})$. The final assertion is clear from exactness of the Wang sequence. $\square$

5.2. Torsion again

Turaev introduced a further modification of the torsion, which takes Poincaré duality into account. Let $(X, \partial X)$ be an oriented $(2k + 1)$-dimensional Poincaré duality pair, and let X' be the maximal free abelian cover of X. Let $\theta : \pi_1(X) \to \mathbb{Z}^\mu$ be an epimorphism, and suppose that $H_*(\partial X; (\Lambda_\mu)_0) = 0$. Let C_* be the cellular chain complex of X', and let c_* be the basis of C_* determined by a choice of lifts of the cells of X. Choose bases h_i of $H_i(X; (\Lambda_\mu)_0)$, for $1 \leq i \leq 2k$, and let v_i be the matrix of the intersection pairing of $H_i(X; (\Lambda_\mu)_0)$ with $H_{2k+1-i}(X; (\Lambda_\mu)_0)$ with respect to these bases, for $1 \leq i \leq k$. Let $e(i) = (-1)^{i+1}$. Then varying the choice of bases changes $\tau(C_*; c_*, h_*)\Pi_{1 \leq i \leq k}(\det(v_i))^{-e(i)}$ by multiplication by elements $u g \bar{g}$, where $u \in \Lambda_\mu^\times$ and $g \in (\Lambda_\mu)_0^\times$, so the image $\omega(X; \theta)$ of this product in the quotient of $(\Lambda_\mu)_0^\times$ by these subgroups is well defined. If θ is the Hurewicz epimorphism to $H_1(X; \mathbb{Z})/(torsion)$ we shall just write $\omega(X)$. If X is homologically oriented $\omega(X)$ may be refined to a sign-determined invariant $\omega_+(X)$, as in §9 of Chapter

4. Changing the homological orientation of X changes the sign of $\omega_+(X)$, while changing the sign of the fundamental class affects all the intersection pairings, and thus multiplies $\omega_+(X)$ by $(-1)^s$, where s is the sum of the ranks of $H_i(X; \Lambda_\mu)$, for $0 \leq i \leq k$.

An argument related to that of Theorem 3.15 shows that $\omega(X)$ is the image of $\Delta_0(T_k)\Delta_0(J)^{e(k)}\Pi_{0 \leq i < k}\Delta_0(H_i(\partial X; \Lambda_\mu))^{e(i)}$, where T_k is the Λ_μ-torsion submodule of $H_k(X; \Lambda_\mu)$, J is the image of $H_k(\partial X; \Lambda_\mu)$ in T_k. (See Theorem 5.1.1 of [**Tu86**].)

Let L be a μ-component $(2k-1)$-link with $\mu \geq 1$ and let $X = X(L)$ and $\omega(L) = \omega(X)$. Then $\partial X \cong \mu S^1 \times S^{2k-1}$. Let $\Pi_S = \prod_{i \in S}(t_i - 1)$, for each $S \subseteq \{1, \ldots, \mu\}$. If $k = 1$ then $\Delta_0(H_0(\partial X; \Lambda_\mu)) = \Delta_0(H_1(\partial X; \Lambda_\mu)) = \Pi_I$, where $I = \{i \mid \ell_{ij} = 0, \forall j \neq i\}$. Since J is a quotient of $H_1(\partial X; \Lambda_\mu)$ it follows from the above formula that $\omega(L)$ is the image of $\Delta_\alpha(L)\Pi_\beta$, for some $\beta \subseteq I$. Similar formulae apply for the torsions $\omega(X; \theta)$ associated to intermediate abelian covers of X. (In Theorem 5.3.1 of [**Tu86**] it is shown that for the maximal abelian cover $\beta = \{i \mid \nu(i) \equiv \sum_{j \neq i} \ell_{ij} \ mod \ (2)\}$, where the $\nu(i)$ are the exponents in the equation $\overline{\Delta_\alpha(L)} = (-1)^m \prod_{i=1}^\mu t_i^{\nu(i)}\Delta_\alpha(L)$, following from Theorem 5.1.) In particular, if $k = \mu = 1$ then $\Delta_1(L)/(t-1)$ represents $\omega(L)$, as in Theorem 4.19. If $k \geq 2$ then $H_i(\partial X; \Lambda_\mu) = 0$ for $1 \leq i \leq k$ (so $J = 0$ also), and $H_k(X; \Lambda_\mu)$ is a Λ_μ-torsion module, by Theorem 4.20. Hence $\Delta_0(H_k(X; \Lambda_\mu)) \neq 0$ and $\Delta_0(H_k(X; \Lambda_\mu))\Pi_{1 \leq i \leq \mu}(t_i - 1)$ represents $\omega(L)$.

The ω-torsion may be used to strengthen the divisibility results of part (4) of Theorem 3, and to extend them to higher dimensional links. In each case the major part of the argument is in relating $\omega(L)$ with $\omega(L(\hat{\mu}))$. Once the connection is made and representatives for each are chosen as above, the results follow by elementary considerations of factorization in $\Lambda_{\mu-1}$.

THEOREM 5.5. [**Tu86**] *Let L be a μ-component 1-link such that $\alpha = \alpha(L) \geq 2$. Then there are a $\lambda \in \Lambda_{\mu-1}$ and a $\beta \subseteq \{1, \ldots, \mu-1\}$ such that $(\prod_{i=1}^{\mu-1} t_i^{\ell_{i\mu}} - 1)\Delta_\alpha(L(\hat{\mu})) \doteq (\prod_{i \in \beta}(t_i - 1))\lambda\overline{\lambda}\psi(\Delta_\alpha(L))$.*

PROOF. Let $\Pi = \prod_{i=1}^{\mu-1} t_i^{\ell_{i\mu}}$, and let $\phi : \pi L \rightarrow \mathbb{Z}^{\mu-1}$ be the composition of abelianization and projection onto the first $\mu - 1$ factors of $\mathbb{Z}^\mu$. We may assume that $(\Pi - 1)\Delta_\alpha(L(\hat{\mu})) \neq 0$, for otherwise the result is trivially true, as we may take $\lambda = 0$ and $\beta = \emptyset$. Hence $\alpha(L(\hat{\mu})) = \alpha(L)$ and $\psi(\Delta_\alpha(L))$ divides $(\Pi - 1)\Delta_\alpha(L(\hat{\mu}))$, by parts (3) and (4) of Theorem 5.3. In particular, $\psi(\Delta_\alpha(L)) \neq 0$. Then $\omega(X(L); \phi)$ is the image of $\Pi_\gamma \psi(\Delta_\alpha(L))$ and $\omega(L(\hat{\mu}))$ is the image of $\Pi_\xi \Delta_\alpha(L(\hat{\mu}))$, where γ and $\xi \subseteq \{1, \ldots, \mu - 1\}$, as observed above.

On applying the multiplicativity of torsion to the short exact sequence of chain complexes arising from the maximal abelian covering of the pair $(X(L(\hat{\mu})), X(L))$, we obtain $\omega(L(\hat{\mu}))(\Pi - 1) = \omega(X(L); \phi)$. Taking into account the indeterminacy in the definition of the torsions, we get an equation $k\bar{k}\Pi_\xi \Delta_\alpha(L(\hat{\mu}))(\Pi - 1) = uh\bar{h}\Pi_\gamma \psi(\Delta_\alpha(L))$, where $u \in \Lambda_\mu^\times$, $h, k \in \Lambda_\mu - \{0\}$, and $h\bar{h}$ and $k\bar{k}$ are relatively prime. Since $\psi(\Delta_\alpha(L))$ divides $\Delta_\alpha(L(\hat{\mu}))(\Pi - 1)$, by Theorem 5.3, $k\bar{k}$ divides Π_ξ, and so $k = 1$. Moreover Π_ξ divides $h\bar{h}\Pi_\beta$. Therefore $t_i - 1$ divides h for all $i \in \xi - \gamma$. Let $\beta = (\gamma - \xi) \cup (\xi - \gamma)$ and $\lambda = h/\Pi_{\xi-\gamma}$. Then $(\Pi - 1)\Delta_\alpha(L(\hat{\mu})) \doteq \lambda\bar{\lambda}\Pi_\beta \psi(\Delta_\alpha(L))$. $\qquad\square$

If $\alpha(L(\hat{\mu})) = \alpha - 1$ then $\psi(\Delta_\alpha(L))$ divides $\Delta_{\alpha-1}(L(\hat{\mu}))$, by part (4) of Theorem 5.3, and so is nonzero. As $\Pi - 1$ is now 0, by part (3) of Theorem 5.3, we must have $\lambda = 0$ in Theorem 5.5, which thus holds for trivial reasons only. There is instead the following more substantial result.

THEOREM 5.6. [**Tu88**] *Let L be a μ-component $(2k-1)$-link with $\mu \geq 2$ and let $\alpha = \alpha(L)$.*

(1) *If $k = 1$ then there are $\sigma \in \Lambda_{\mu-1}$ and $\delta \subseteq \{1, \ldots, \mu - 1\}$ such that $\Delta_{\alpha-1}(L(\hat{\mu})) \doteq (\prod_{i \in \delta}(t_i - 1))\sigma\bar{\sigma}\psi(\Delta_\alpha(L))$.*

(2) *If $k \geq 2$ then there is a $\sigma \in \Lambda_{\mu-1}$ such that $\varepsilon(\sigma) = 1$ and*
$$\Delta_0(H_k(X(L(\hat{\mu})); \Lambda_{\mu-1})) \doteq \sigma\bar{\sigma}\psi(\Delta_0(H_k(X(L); \Lambda_\mu))).$$

PROOF. (Sketch.) Turaev shows that in each case $\omega(L)$ has a representative $\omega \in (\Lambda_\mu)_0$ such that $(t_\mu - 1)\omega = f/g$, where $f, g \in \Lambda_\mu$ and $\psi(g) \neq 0$, and such that $\psi((t_\mu - 1)\omega)$ represents $\omega(L(\hat{\mu}))$. The rest of the proof is similar to that of Theorem 5.4. (If $k > 1$ we use $\Delta(L) = \Delta_0(H_k(X; \Lambda_\mu))$ instead of $\Delta_\alpha(L)$ and $\omega(L) = \Delta(L)\prod_{i=1}^\mu (t_i - 1)$.) If $k \geq 2$ then $\varepsilon(\sigma) = 1$, by Theorem 4.20. $\qquad\square$

Part (1) was first proven by Levine, for the cases with $\alpha(L) = 2$. If moreover $\mu = 2$ then ϕ is the longitudinal polynomial of ℓ_2. (See also Theorem 8.5 below.) If $\alpha(L) = \mu$ then $\delta = \emptyset$ and $\varepsilon(\phi) = \pm 1$, for then $\alpha(L(\hat{\mu})) = \mu - 1$ and we can iterate the argument until we reach a knot polynomial. However, if $\mu > \alpha(L)$ then δ can be any subset of $\{1, \ldots, \mu - 1\}$ [**Le87**].

5.3. Partial derivatives

Let ∂_i denote partial differentiation with respect to the ith variable t_i. Although ∂_i is not Λ_μ-linear, the Leibniz formula implies immediately that $\partial_i(\lambda\Delta) \equiv \lambda\partial_i\Delta$ mod (Δ), for any $\lambda, \Delta \in \Lambda_\mu$. Thus the principal ideal generated by the image of $\partial_i\Delta$ in any quotient ring of $\Lambda_\mu/(\Delta)$ depends only on the principal ideal (Δ). In particular, the ideal generated by the image of $\psi(\partial_\mu\Delta_1(L))$ in $\Lambda_{\mu-1}/(\Delta_1(L(\hat{\mu})))$ is well defined, by the Torres conditions. Murasugi showed that it is invariant under homotopy of the μth component.

THEOREM 5.7. [**Mu80'**] *Let L_- and L_+ be two μ-component 1-links such that $L_-(\hat{\mu}) = L_+(\hat{\mu}) = K$, say, and such that $L_{-\mu}$ is homotopic to $L_{+\mu}$ in $X(K)$. Then the images of $\psi(\partial_\mu\Delta_1(L_-))$ and $\psi(\partial_\mu\Delta_1(L_+))$ generate the same ideal in $\Lambda_{\mu-1}/(\Delta_1(K))$.*

PROOF. We may assume that $L_\pm$ differ only by an elementary homotopy, involving the crossing of two arcs in a small ball B. Such a homotopy may be realized by performing a ± 1-framed surgery on the boundary Z of a 2-disc $D \subset B$ which meets $L_{-\mu}$ transversely in two points, with opposite orientations, and is otherwise disjoint from L_-. In particular, we may view $X(L_+)$ as the union of $X(L_- \cup Z)$ with a solid torus.

Let $A_\pm = A(L_\pm)$ and $M = H_1(X(L_- \cup Z), *; \Lambda_\mu)$. Note that $H_q(X(L_\pm), X(L_\pm \cup Z); \Lambda_\mu) \cong \Lambda_\mu$ if $q = 2$, and is 0 if $q = 1$, by excision. Hence there are exact sequences

$$\Lambda_\mu \xrightarrow{i_-} M \to A_- \to 0$$

and

$$\Lambda_\mu \xrightarrow{i_+} M \to A_+ \to 0.$$

A Mayer-Vietoris argument (with coefficients $\Lambda_{\mu-1} = \psi(\Lambda_\mu)$) applied to the triple $(X(L_- \cup Z); X(L_-) \setminus B, B \setminus L_- \cup Z)$ implies that the homomorphism $\psi(i_-)$ splits.

Let $w, z \in M$ represent meridians for $L_{-\mu}$ and Z, respectively. Then $i_-(1) = z$ and $i_+(1) = z + (u - 1)w$, for some unit $u \in \Lambda_\mu^\times$. The module M has a presentation matrix (P, ω, ζ), where the last two columns correspond to generators mapping to w and z, respectively. Hence A_- has a presentation matrix $\left(\begin{smallmatrix} P & \omega & \zeta \\ 0 & 0 & 1 \end{smallmatrix}\right)$, while A_+ has a presentation matrix $\left(\begin{smallmatrix} P & \omega & \zeta \\ 0 & u-1 & 1 \end{smallmatrix}\right)$. Hence A_- and A_+ have the simpler presentation matrices (P, ω) and (P, ω_+), respectively, where $\omega_+ = \omega - (u - 1)\zeta$. Moreover $\psi(P)$ is a presentation matrix for $A(K)$, and $\psi(\zeta) = 0$ since $\psi(i_-)$ splits.

The first elementary ideals are generated by the minors obtained by deleting 1 column and some rows. Suppose that P has n columns, and let Q be an $n \times (n - 1)$ submatrix of P. Then

$$\det(Q, \omega) - \det(Q, \omega_+) = (1 - u)\det(Q, \zeta),$$

and so

$$\psi(\partial_\mu \det(Q, \omega)) - \psi(\partial_\mu \det(Q, \omega_+)) = (1 - \psi(u))\det(\psi(Q), \psi(\partial_\mu \zeta)),$$

since $\psi(\zeta) = 0$. Expanding the latter determinant by its last column gives a linear combination of $(n - 1) \times (n - 1)$ minors of $\psi(P)$, and so a multiple of $\Delta_1(K)$. The result is now straightforward. $\square$

Murasugi compared the Jacobian matrices of Wirtinger presentations for the links $L_\pm$; otherwise this is essentially his argument. (It also follows easily from the Conway skein relation for the normalized Alexander polynomials.) We are grateful to Murasugi for an outline in English of his result, which was originally published in Japanese.

In Chapter 7 we shall use such ideals to show that the Torres conditions do not completely characterize link polynomials, if $\mu = 2$ and $|\ell| > 1$. The first derivative of the Alexander polynomial also plays a role in Jin's exposition of the Kojima-Yamasaki invariant [**KY79, Ji87**].

Let $I = (i_1, \ldots, i_r)$ be a multi-index with $1 \le i_j \le \mu$ for all $1 \le j \le |I| = r$, and let $I! = i_1! \ldots i_r!$. Let $\partial_I = \partial_{i_1} \ldots \partial_{i_r}$ be the corresponding r-fold partial derivative. If $|I| < \mu - 2$ then $\varepsilon(\partial_I \Delta_1(L)) = 0$, by Corollary 4.13.2, while if $|I| = \mu - 2$ and $g \in \Lambda_\mu$ then $\varepsilon \partial_I (g \Delta_1(L)) = \varepsilon(g) \varepsilon \partial_I (\Delta_1(L))$, so the integers $|\varepsilon \partial_I (g \Delta_1(L))|$ depend only on I and L. These integers are absolute values of sums of products of linking numbers, by Theorem 5.3. (Explicit formulae for the derivatives of total order $|I| \le \mu$ of a normalization of $\Delta_1(L)$ are given in Chapter 2 of [**Les**].) In particular, $|\varepsilon \partial_i^{\mu-2}(\Delta_1(L))| = |\prod_{j \ne i} \ell_{ij}|$. If $\mu = 2$ then $|\ell_{12}| = |\varepsilon(\Delta_1(L))|$, while if $\mu = 3$ then $\ell_{ij}^2 |\varepsilon \partial_k (\Delta_1(L))| = |\varepsilon \partial_i (\Delta_1(L)) \varepsilon \partial_j (\Delta_1(L))|$ for i, j, k distinct, and so the $|\ell_{ij}|$ are determined if all are nonzero. We can detect when at most one is zero. If L is a 4-component link such that $\ell_{12} = \ell_{13} = \ell_{34} = 1$ and $\ell_{23} = \ell_{14} = \ell_{24} = 2$ and γL is the link obtained by interchanging components 2 and 4 then the values of these derivatives are equal for L and γL, and so do not determine the linking numbers. Can they be recovered from a normalization of the Alexander polynomial?

5.4. The total linking number cover

Let L be a μ-component 1-link and let $p_\tau : X^\tau \to X$ be the total linking number cover of $X = X(L)$. The total linking number homomorphism $\tau : \pi = \pi L \to \mathbb{Z}$ induces an epimorphism from Λ_μ to Λ. The *reduced Alexander module* of L is $A_{red}(L) = \tau A(L) = \Lambda \otimes_{\Lambda_\mu} A(L)$. The *reduced Alexander polynomial* of L is $\Delta_{red}(L)(t) = \tau(\Delta_1(L))$. (Thus $\Delta_{red}(L)(t) = \Delta_1(L)(t, \ldots, t)$.) This is well defined up to a unit factor $\pm t^n$. In the knot theoretic case $\mu = 1$ it is the usual first Alexander polynomial, and we may assume that $\Delta_{red}(L)(t) = \Delta_1(L)(t)$ is symmetric and has augmentation 1.

The equivariant chain complex of $(X^\tau, p_\tau^{-1}(*))$ is chain homotopy equivalent to a complex

$$0 \to \Lambda^s \xrightarrow{\tau(\partial)} \Lambda^{s+1} \to 0 \to 0$$

where ∂ is the boundary map for the corresponding complex for $(X', p^{-1}(*))$. Thus $\tau A(L) = H_1(X, *; \Lambda) = H_1(X^\tau, p_\tau^{-1}(*); \mathbb{Z})$. The exact sequence of homology for the pair $(X^\tau, p_\tau^{-1}(*))$ gives a short exact sequence

$$0 \to H_1(X; \Lambda) \to \tau A(L) \to \Lambda \to 0,$$

so $\tau A(L) \cong \Lambda \oplus H_1(X; \Lambda)$. Hence $E_i(H_1(X; \Lambda)) = E_{i+1}(\tau A(L)) = \tau(E_{i+1}(L))$, for all $i \geq 0$.

The *reduced nullity* of L is $\kappa(L) = \min\{k \mid \tau(E_k(L)) \neq 0\}$. It is obvious that $1 \leq \alpha(L) \leq \kappa(L) \leq \mu$, and it is easily seen that $\kappa(L) = 1 + rank_\Lambda H_1(X; \Lambda)$.

THEOREM 5.8. *The following are equivalent:*

(1) $\kappa(L) = \mu$;

(2) *the longitudes of L are in* $\mathrm{Ker}(\tau)'$, *the commutator subgroup of* $\mathrm{Ker}(\tau)$;

(3) $H_1(X; \mathbb{Z}) \cong H_1(M(L); \mathbb{Z})$ *and* $H_1(X; \Lambda) \cong H_1(M(L); \Lambda)$, *where $M(L)$ is obtained by 0-framed surgery on L in S^3.*

PROOF. If $\tilde{L}$ is a ν-component sublink of L then $\kappa(L) - \kappa(\tilde{L}) \leq \mu - \nu$; hence $\kappa(\tilde{L}) = \nu$ if $\kappa(L) = \mu$. In particular, if (1) holds then the 2-component sublinks have reduced nullity 0, and so the linking numbers are 0, by the Torres conditions. Hence all the longitudes of L are in $\pi' \leq \mathrm{Ker}(\tau)$. The image of each longitude in $H_2(X; \Lambda)$ is annihilated by $(t-1)$, since each longitude commutes with a meridian. By an easy argument (as in Theorem 4.14) there is a $\delta \in \tau(E_\mu(L))$ such that $\varepsilon(\delta) = 1$ and which annihilates the Λ-torsion submodule of $H_1(X; \Lambda)$. Hence all the longitudes are in $\mathrm{Ker}(\tau)'$.

Conversely, all the longitudes are in π' if and only if the inclusion of X into $M(L)$ induces an isomorphism $H_1(X; \mathbb{Z}) \cong H_1(M(L); \mathbb{Z})$, in which case $H_2(M(L), X; \Lambda) \cong \mathbb{Z}^\mu$, by excision, and the connecting homomorphism to $H_1(X; \Lambda)$ is trivial if and only if either (2) or (3) is true. Thus (2) and (3) are equivalent. The remaining implication now follows on taking coefficients $\mathbb{Q}$ and appealing to duality, which implies that $H_2(M(L); \mathbb{Q}\Lambda) \cong \mathbb{Q} \oplus (\mathbb{Q}\Lambda)^r$, where r is the rank over $\mathbb{Q}\Lambda$ of $H_1(M(L); \mathbb{Q}\Lambda) \cong H_1(X; \mathbb{Q}\Lambda)$ and so equals $\kappa(L) - 1$. $\square$

The argument of Theorem 4.15 may be adapted to show that if $\kappa(L) = \mu$ then $H_1(X; \Lambda)/TH_1(X; \Lambda) \cong \Lambda^\mu$ if and only if $\tau(E_\mu(L))$ is principal. In particular, this is so if $A(L)$ maps onto $(\Lambda_\mu)^\mu$. We then have $E_\mu(L) = \Delta_\mu(L)E_0(P)$, by Corollary 4.15.1, and $\varepsilon(E_\mu(L)) \neq 0$. Hence the ideal $\tau(E_0(P))$ is principal, but is not contained in $(t-1)$. However, it contains $(t-1)^N$ for N large, since $\prod_{i=1}^\mu (t_i - 1)$ annihilates P. Therefore $\tau(E_0(P)) = \Lambda$, and so $\tau(E_\mu(L)) = (\tau(\Delta_\mu(L)))$. This provides another test for homology boundary links.

LEMMA 5.9. *Let L be a μ-component 1-link. Then $\Delta_{red}(L) \neq 0 \Leftrightarrow$ $\tau A(L)$ has rank 1 $\Leftrightarrow H_1(X; \Lambda)$ is a torsion module $\Leftrightarrow H_2(X; \Lambda) = 0$. If $\Delta_{red}(L) \neq 0$ then $H_1(X; \Lambda)$ has a square presentation matrix, and order ideal $\tau(E_1(L))$. If $\mu = 1$ then $\tau(E_1(L)) = (\Delta_{red}(L))$, while if $\mu > 1$ then $\tau(E_1(L)) = (t-1)(\Delta_{red}(L)) \leq (t-1)^{\mu-1}$.*

PROOF. The first assertion follows on considering the chain complex of $(X^\tau, p_\tau^{-1}(*))$ and its homology. If they hold $p.d._\Lambda \tau A(L) \leq 1$, and so $p.d.H_1(X; \Lambda) \leq 1$, by Theorems 3.8 and 3.22. Hence $H_1(X; \Lambda)$ has a square presentation matrix. The conditions on $\tau(E_1(L))$ follow from part (1) of Theorem 4.12 and Corollary 4.13.2. $\qquad\square$

If $h(t)$ generates the principal ideal $(t-1)^{1-\mu}\tau(E_1(L))$ then $h(t^{-1}) = t^{2m}h(t)$ for some m, by the first Torres condition. Thus the *Hosokawa polynomial* $\nabla(L)(t) = t^m h(t)$ is symmetric (i.e., $\overline{\nabla(L)} = \nabla(L)$), and well-defined up to sign. Clearly $\nabla(L) \doteq \Delta_1(L)$ if $\mu = 1$ and $\nabla(L) \doteq \Delta_{red}(L)/(t-1)^{\mu-2}$ if $\mu > 1$. Hosokawa showed that any such symmetric polynomial is realized by some μ-component link, for each $\mu \geq 1$, and computed $|\varepsilon(\nabla(L))|$ as a determinant in the linking numbers of L [**Hs58**].

THEOREM 5.10. *Let L be a μ-component 1-link. Then $|\varepsilon(\nabla(L))| = |E_1(N)|$, where N is the $\mu \times \mu$ matrix with $N_{ii} = -\Sigma_{j=1}^{j=\mu}\ell_{ij}$ and $N_{ij} = \ell_{ij}$ if $i \neq j$.*

PROOF. Let $R = (t-1)H_1(X; \Lambda)$. The Wang sequence for p_τ gives an isomorphism $H_1(X; \Lambda)/R \cong \mathbb{Z}^{\mu-1}$ and an exact sequence

$$H_2(X; \mathbb{Z}) \to H_1(X; \Lambda) \to R \to 0.$$

Hence there is an exact sequence

$$H_2(X; \mathbb{Z}) \xrightarrow{\ U\ } H_1(X; \Lambda)/R \to R/(t-1)R \to 0.$$

Let x_i be the image of the orientation class of $\partial X(L_i)$ in $H_2(X;\mathbb{Z})$, for $1 \leq i \leq \mu$. Then $H_2(X;\mathbb{Z})$ is generated by the x_i, subject only to the relation $\Sigma x_i = 0$. Since $\tau A(L) \cong H_1(X;\Lambda) \oplus \Lambda$ and $\varepsilon A(L) = \mathbb{Z} \otimes_\Lambda A(L) \cong \mathbb{Z}^\mu$, with basis the images of the meridianal elements, we see that $H_1(X;\Lambda)/R = \mathbb{Z} \otimes_\Lambda H_1(X;\Lambda)$ is isomorphic to $\mathbb{Z}^{\mu-1}$, and has basis the images of the elements $m_i - m_1$, for $1 < i \leq \mu$. The matrix of U with respect to these bases is a minor of the matrix N. Since the sums of the rows and of the columns of N are each 0, the $(\mu-1) \times (\mu-1)$ minors of N all have the same absolute value. Therefore $|\varepsilon(\nabla(L))| = |\Delta_0(R/(t-1)R)| = |E_1(N)|$.   □

If the choice of Alexander polynomials is normalized (as in [**Les**] or [**Tu86**]) so that the second Torres condition is an equality of Laurent polynomials this formula follows from the equation

$$\varepsilon(\nabla(L)) = lim_{t\to 1}(\Delta_{red}(L)(t)/(t-1)^{\mu-2})$$

$$= \varepsilon((\frac{d}{dt})^{\mu-2}\Delta_{red}(L))/(\mu-2)! = \Sigma_{|I|=\mu-2}\varepsilon\partial_I(\Delta_1(L))/I!.$$

The summands may be computed inductively using the Torres conditions, and compared with the terms in the expansion of a principal $(\mu - 1) \times (\mu - 1)$ minor of N.

If all the linking numbers are 0 then $\Delta_{red}(L) = (t-1)^{2\mu-3}f(t)$, for some $f \in \Lambda$, by Corollary 4.13.2. The first Torres condition implies that $f(t^{-1}) = (-1)^{\mu-1}t^{2m+\mu-1}f(t)$, for some m, and so $f(1) = 0$ if μ is even. Kidwell used a Seifert surface argument to prove that $|f(1)|$ is the determinant of a $(\mu-1) \times (\mu-1)$ skew symmetric matrix derived from Hosokawa's matrix, and so is 0 if μ is even and a perfect square if μ is odd [**Ki78**]. Are there any other constraints on f?

5.5. Murasugi nullity

The total $\mathbb{Z}/2\mathbb{Z}$-linking number defines a homomorphism from πL to $\{\pm 1\}$ which sends each meridian to -1. Let $\tilde{\varepsilon} : \Lambda_\mu \to \tilde{\mathbb{Z}} = \Lambda_\mu/(t_1+1,\ldots,t_\mu+1)$ be the ring epimorphism determined by $\tilde{\varepsilon}(t_i) = -1$, for all $1 \leq i \leq \mu$. (As rings $\tilde{\mathbb{Z}}$ and $\mathbb{Z}$ are isomorphic, but they are distinct as Λ_μ-algebras.) The *Murasugi nullity* $\eta(L)$ of a μ-component 1-link L is the rank of $\tilde{\mathbb{Z}} \otimes_\Lambda A(L)$ [**Mu65'**].

Since $\tilde{\varepsilon} \equiv \varepsilon \bmod (2)$, it is easily seen that $\alpha(L) \le \kappa(L) \le \eta(L) \le \mu$. Moreover $\eta(L) = 1 + s$, where s is the number of factors $(t + 1)$ dividing $\Delta_{red}(L)$, since $\tilde{\varepsilon}$ factors through τ. (In particular, $\eta(L) = 1$ if and only if L has nonzero *determinant* $|\Delta_{red}(L)(-1)|$.) Theorem 4.17 implies that $\eta(L)$ is invariant under I-equivalence and that the power of 2 dividing $\tilde{\varepsilon}(E_i(L))$ is an invariant of I-equivalence, for all i [**Tr85**]. It is also invariant under changes of orientation, since $\tilde{\varepsilon}(t_i^{-1}) = \tilde{\varepsilon}(t_i)$. However, $\eta(Wh) = 1$ and so $\eta(L)$ is not a homotopy invariant.

THEOREM 5.11. *Let L be a μ-component $Z/2Z$-homology bound-ary 1-link with group $\pi = \pi L$. Then $\eta(L) = \mu$.*

PROOF. Let $f : \pi \to *^\mu(Z/2Z)$ be an epimorphism such that composition with abelianization sends meridians to standard gener-ators. Then the composite of f with the homomorphism which sends each standard generator to -1 is the total $Z/2Z$-linking number ho-momorphism. There is also an epimorphism from $\mathbb{Z}[(Z/2Z)^\mu] = \Lambda_\mu/(t_1^2 - 1, \ldots, t_\mu^2 - 1)$ to $\tilde{\mathbb{Z}}$ which sends the standard generators to -1. Hence $A(f) : A(\pi) \to A(*^\mu(Z/2Z))$ gives rise to an epimorphism $\tilde{\mathbb{Z}} \otimes_\Lambda A(\pi) \to \tilde{\mathbb{Z}} \otimes_\Lambda A(*^\mu(Z/2Z)) \cong \tilde{\mathbb{Z}}^\mu$, and so $\eta(L) = \mu$. $\square$

COROLLARY 5.11.1. *Given integers $1 \le \alpha \le \eta \le \mu$ there is a μ-component 1-link L with Alexander nullity $\alpha(L) = \alpha$ and Murasugi nullity $\eta(L) = \eta$.*

PROOF. Let $\mu' = \mu - \alpha + 1$ and $\eta' = \eta - \alpha + 1$. Let L' be a μ'-component link, all of whose linking numbers are odd, and let L'' be the link obtained by replacing each of the first η' components by the $(2, 1)$-cable about that component (i.e., by the boundary of a Möbius band whose centreline is that component). The link L obtained by adjoining a trivial $(\alpha - 1)$-component link to L'' is a μ-component link with $\alpha(L) = \alpha$ and $\eta(L) = \eta$. $\square$

The Murasugi nullity $\eta(L)$ and signature $\sigma(L) = \sigma_L(-1)$ may be computed from any diagram for L, in terms of the associated Goeritz matrix (as modified in [**GL78**] and [**Tr85**]). These invariants have natural interpretations in terms of 2-fold branched covers; see §8 below.

5.6. Fibred links

A μ-component n-link L is *fibred* if there is a fibre bundle projection $\tilde{\tau} : X \to S^1$ such that the induced map of fundamental groups is the total linking number homomorphism τ. (The latter condition is redundant if $\mu = 1$.) The fibre $F = \tilde{\tau}^{-1}(1)$ is a compact connected oriented $(n + 1)$-manifold with boundary $\partial F \cong \mu S^n$.

In the knot theoretic case $\mu = 1$ each fibre is a Seifert surface for the knot. The exterior of the Hopf link Ho is homeomorphic to $S^1 \times S^1 \times [0, 1]$, and so fibres over the circle in many ways. However, there is only one bundle projection inducing τ, up to composition with an isotopy of $X(Ho)$. We shall see that if $n > 1$ there are no fibred n-links with more than one component.

Let $e : \mathbb{R} \to S^1$ be the universal cover of S^1. Since $\mathbb{R}$ is contractible the pullback $e^*\tilde{\tau}$ is a trivial bundle, and so there is a commutative diagram

$$
\begin{array}{ccc}
F \times \mathbb{R} & \xrightarrow{\;\;E\;\;} & X \\
{\scriptstyle pr_2}\big\downarrow & & \big\downarrow{\scriptstyle \tilde{\tau}} \\
\mathbb{R} & \xrightarrow{\;\;e\;\;} & S^1
\end{array}
$$

where pr_2 is projection onto the second factor and E is a covering map. The covering group of E is generated by a map $H : (v, r) \mapsto (h(v), r + 1)$, where h is a homeomorphism of F, and E induces a homeomorphism from the mapping torus of h to X. The map h is called the *characteristic map* of the bundle, since it determines the bundle up to isomorphism. Although h is only defined up to pseudoisotopy, the induced map on homology is well defined, and induces multiplication by t on the homology of $F \times \mathbb{R}$, considered as an infinite cyclic cover of X.

THEOREM 5.12. *Let L be a μ-component fibred n-link.*

(1) *If $n = 1$ then $\Delta_{red}(L) \neq 0$ and the leading coefficient is ± 1.*
(2) *If $n > 1$ then $\mu = 1$.*

PROOF. Since $H_1(X; \Lambda) = H_1(F; \mathbb{Z})$ it is finitely generated as an abelian group. Therefore it must be a Λ-torsion module and so, if $n = 1$, $\Delta_{red}(L) \neq 0$ and its leading coefficient must be ± 1. If

$n > 1$ then $E_{\mu-2}(H_1(X;\Lambda)) = \tau(E_{\mu-1}(L)) = 0$, by Theorem 1.3 and Theorem 4.10, so $H_1(X;\Lambda)$ can only be a torsion module if $\mu = 1$. $\square$

If $n \geq 4$ an n-knot K is fibred if and only if $X(K)'$ has the homotopy type of a finite complex and a torsion invariant in $Wh(\pi K)$ is 0, by the Farrell fibration theorem. This criterion holds also for $n = 3$, if πK is "good", in the sense of [**FQ**]. As our primary interests are links we shall concentrate on the case $n = 1$. (See §5 of [**KW78**] for more on higher dimensional fibred knots.) A classical link L is fibred if and only if $\text{Ker}(\tau)$ is finitely generated, by Stallings' fibration theorem for 3-manifolds [**St62**]. The Whitehead link ($\ell = 0$) and the $(2, 2\ell)$-torus link ($\ell \neq 0$) are fibred, and so there are no restrictions on the linking numbers of fibred 2-component links. A fibred 2-component link has linking number 0 if and only if the boundary of the fibre is a union of longitudes, for the linking number is the image of either longitude under τ when $\mu = 2$.

Since the action of $h_1 = H_1(h;\mathbb{Z})$ determines the module structure on $H_1(X;\Lambda) = H_1(F;\mathbb{Z})$ its characteristic and minimal polynomials generate $E_0(H_1(X;\Lambda))$ and $Ann(H_1(X;\Lambda))$, respectively. Therefore if $\mu = 1$ the characteristic polynomial is $\Delta_1(L)$ and the minimal polynomial is $\lambda_2(L)$, while if $\mu > 1$ they are $(t-1)\Delta_{red}(L)$ and $(t-1)\Delta_{red}(L)/\Delta_1(H_1(X;\Lambda))$, respectively. (The minimal polynomial divides $(t-1)\tau(\lambda_2(L))$ since $E_1(H_1(X;\Lambda)) = \tau(E_2(L))$.)

The most interesting class of fibred links is the class of links of isolated singularities of plane algebraic curves. If f is a square-free holomorphic function such that $f(0) = 0$ the pair $(S_\epsilon^3, S_\epsilon^3 \cap f^{-1}(0))$ cut out by the sphere of radius ϵ centred at 0 determines a link $L(f)$ whose ambient isotopy type is independent of ϵ, for ϵ small. If ϵ is sufficiently small then $p(w, z) = f(w, z)/|f(w, z)|$ definies a fibration of $S_\epsilon^3 - f^{-1}(0)$ over S^1, and so L is fibred, with fibre F, say [**Mil**]. The characteristic automorphism h of the fibration is the *monodromy* of f at O. There is in fact no loss of generality in assuming that f is a square-free polynomial in $\mathbb{C}[w, z]$. We shall call such links *algebraic links*. We shall give a self-contained account of the principal algebraic invariants of such singularities in Chapter 10.

Here we shall summarize the main topological properties of algebraic links. The topology is well understood, as $L(f)$ is an "iterated torus link", with one component for each branch of f at O. The branches are the irreducible factors of f in $\mathbb{C}[[w,z]]$, and L_i is determined by the Puiseux expansion of the ith branch. The link $L(f)$ is determined up to isotopy (among all such links) by the extended Puiseux data for the factors of f, which provide a template for constructing the link iteratively by forming satellites. (This corresponds to the canonical decomposition of $X(L(f))$ along essential tori.)

The linking numbers of an algebraic link are all nonzero. For if f and g are irreducible and relatively prime then $\text{lk}(L(f), L(g)) = \dim_{\mathbb{C}} \mathbb{C}[[X, Y]]/(f, g)$, and is strictly positive. The link $L(f)$ is in fact determined among all such links by its component knots and their linking numbers, and hence by its multivariable Alexander polynomial $\Delta_1(L(f))$ [**Ym84**].

Mayer-Vietoris arguments show that the reduced Alexander polynomial of an iterated torus link is a product of cyclotomic polynomials [**SW77**]. If L is an algebraic link the characteristic polynomial of h_1 can be computed in terms of the Puiseux expansion of f [**A'C73**], and $\varepsilon(\nabla(L)) \neq 0$ [**Du75**]. In particular, the Borromean ring link Bo is not algebraic (although it is fibred [**Go75**]).

If L is a knot the algebraic monodromy h_1 has finite order, and determines the isotopy type of the knot [**Le72**]. However the isotopy class of h has finite order only if L is a torus knot. In general, $(h_1^{m!} - 1)^2 = 0$ for m sufficiently large, by Theorem 13.6 of [**EN**]. If L has $r > 1$ components and there is a prime p and a root of unity ζ_{p^m} of order p^m such that $(t - \zeta_{p^m})^{r-1}$ divides $\Delta_{red}(L)$ then h_1 has infinite order [**SW77**]. The trace of the monodromy is 1 [**A'C73**], and $H_1(F; \mathbb{Z})$ has a basis for which the Seifert matrix is upper triangular with entries -1 on the diagonal [**Du74**]. The Blanchfield pairing on $H_1(X; \mathbb{C}\Lambda[(t-1)^{-1}])$ may be described in terms of the local Gauss-Manin connection [**Ba85**]. If L has more than one component the monodromy may not determine the link. (The pair of polynomials $(w^{14} - z^{11})(w^{21} - z^{44})$ and $(w^{28} - z^{33})(w^7 - z^{22})$ gives a simple counterexample [**Gr74**].)

Yamamoto showed that the linking numbers of an algebraic link are determined by the multivariable polynomial $\Delta_1(L) \in \Lambda_\mu$ [**Ym84**]. As the linking numbers are nonzero, $\Delta_1(L)$ also determines each $\Delta_1(L_i)$, and hence each component knot L_i, by Theorem 1. Therefore $\Delta_1(L)$ is a complete invariant, since the component knots and their linking numbers determine the link, by a theorem of Zariski. (See also [**GDC99, CS00**].)

If an algebraic knot K is concordant to a sum of $n \geq 0$ nontrivial algebraic knots $\natural_{i=1}^n K_i$ then $n = 1$ and K is isotopic to K_1 [**Li79**]. In particular, concordant algebraic knots are isotopic, while the sum of two nontrivial algebraic knots is never algebraic. (However, see page 122 of [**EN**] for an example of a nontrivial linear relation between concordance classes of algebraic knots.)

In principle, all the topological invariants of $L(f)$ can be computed from the Puiseux data (or, equivalently, from the resolution graph Γ_f). This has largely been done in [**EN**], where these links are considered in the wider context of "graph links" - those whose exteriors have JSJ decompositions with no hyperbolic pieces. However, graph links need not be fibred, and not all iterated torus links are algebraic. (See Theorem 9.4 of [**EN**].) It is an open question whether every fibred classical link arises from an isolated singularity of a *real* polynomial map $F : \mathbb{R}^4 \to \mathbb{R}^2$. See [**BS98**].

5.7. Finite abelian covers

Let G be a finite group acting effectively on a closed 3-manifold M, and let $p : M \to M/G$ be the orbit map. Let $tr : H_1(M/G; \mathbb{Z}) \to H_1(M; \mathbb{Z})$ be the transfer homomorphism. Then p_*tr is multiplication by $|G|$ and $tr \circ p_* = \Sigma_{g \in G} g_*$. (See Chapter IV.2 of [**Bre**].)

Let $p_k : X_k \to X$ be the *finite cyclic cover* corresponding to the subgroup $\tau^{-1}(k\mathbb{Z})$ of π. The *k-fold cyclic branched covering space of S^3, branched over L* is the closed 3-manifold $M_k = X_k \cup \mu D^2 \times S^1$. (Thus we may obtain M_k from X_k by adjoining μ 2-cells to kill the kth powers of meridians and then closing off with μ 3-cells.)

There are Wang sequences relating the homology of X_k to that of X^τ, since p_τ factors through X_k. In low dimensions, the Wang sequence reduces to

$$\ldots H_1(X;\Lambda) \xrightarrow{t^k-1} H_1(X;\Lambda) \to H_1(X_k;\mathbb{Z}) \to \mathbb{Z} \to 0.$$

Let $H = H_1(X;\Lambda)$. The image of the kth power of a meridian in $H_1(X_k;\mathbb{Z})$ determines a splitting for the final epimorphism, so $H_1(X_k;\mathbb{Z}) \cong \mathbb{Z} \oplus H/(t^k-1)H$. In particular, $H/(t-1)H \cong \mathbb{Z}^{\mu-1}$.

Recall that $\nu_k(t) = (t^k - 1)/(t - 1)$. The exact sequence of coefficient modules

$$0 \to \mathbb{Z} \xrightarrow{\nu_k} \Lambda/(t^k-1) \to \Lambda/(\nu_k) \to 0$$

gives rise to an exact sequence

$$H_1(X;\mathbb{Z}) \xrightarrow{tr} H_1(X_k;\mathbb{Z}) \to H_1(X;\Lambda/(\nu_k)) \to 0,$$

where tr is the transfer. The image of the transfer is generated by the lifts of the kth powers of the meridians and so $H_1(M_k;\mathbb{Z}) \cong H_1(X;\Lambda/(\nu_k))$. On the other hand the latter group may be described as a quotient of the homology of X^τ, via the short exact sequence of chain complexes

$$0 \to C_* \xrightarrow{\nu_k} C_* \to (\Lambda/(\nu_k)) \otimes C_* \to 0,$$

where C_* is the singular chain complex for X^τ. Hence $H_1(M_k;\mathbb{Z}) \cong H/\nu_k H$.

On extending coefficients to $\mathbb{C}$ (and noting that $\mathbb{C}\Lambda$ is a PID) we see that the rank of $H_1(M_k;\mathbb{Z}) = \beta_1(M_k;\mathbb{C})$ is the number of roots of $\nabla(L)$ (multiplicities counted) which are nontrivial k^{th} roots of unity [Su74]. In particular, $\eta(L) = \beta_1(M_2;\mathbb{C}) + 1$ and $\beta_1(M_k;\mathbb{C})$ is periodic as a function of k.

If A is an abelian group let $|A|$ be its order, if this is finite, and let $|A| = 0$ otherwise. Since $\Delta_0(H) = (t-1)^{\mu-1}\nabla(L)$, the quotient $H_1(M_k;\mathbb{Z}) = H/\nu_k H$ is finite if and only if $Res(\nabla(L),\nu_k) \neq 0$, in which case

$$|H_1(M_k;\mathbb{Z})| = k^{\mu-1}|Res(\nabla(L),\nu_k)|,$$

by Theorem 3.13. If $\mu = 1$ this reduces to $|Res(\Delta_{red}(L),\nu_k)|$, while if $\mu > 1$ it is $k|Res(\Delta_{red}(L),\nu_k)|$, since $|Res(t-1,\nu_k)| = k$.

Tensoring the exact sequence

$$0 \to \Lambda/(\nu_k) \to \Lambda/(t^k-1) \to \Lambda/(t-1) \to 0$$

with H and setting $R = (t-1)H$ gives an exact sequence

$$0 \to R/\nu_k R \to H/(t^k - 1)H \to \mathbb{Z} \otimes_\Lambda H = \mathbb{Z}^{\mu-1} \to 0.$$

Therefore $R/\nu_k R$ is the torsion subgroup of $H/(t^k - 1)H$ if it is finite. In other words, if $\beta_1(M_k) = 0$ then the torsion subgroup of $H_1(X_k; \mathbb{Z})$ has order $|Res(\nabla(L), \nu_k)|$.

The above discussion is based on the approach of Sakuma [**Sk79**]. It may be extended in several ways, at the cost of inverting the order of the covering group. (The rest of this section is taken from [**HS97**].) Let L be a μ-component link in a homology 3-sphere Σ. Given an epimorphism $\phi : \pi L \to A$, we shall let $X_\phi(L)$ be the corresponding covering space of $X(L) = \Sigma \setminus L$; if A is finite let $M_\phi(L)$ be the corresponding branched covering space of Σ, branched over L. Let L_ϕ be the sublink of L whose components have meridians mapped nontrivially by ϕ.

We shall assume henceforth that A is a finite abelian group, of order $|A|$, and that $R = \mathbb{Z}[|A|^{-1}]$. Let ζ_n be a primitive n^{th} root of unity, and $\theta_n(t) \in \Lambda$ be the cyclotomic polynomial which is the minimal polynomial of ζ_n over $\mathbb{Q}$. Let $\varphi(n) = [\mathbb{Q}(\zeta_n) : \mathbb{Q}]$ be Euler's totient function. Then $\theta_n(t)$ has degree $\varphi(n)$. We shall use the same symbol to denote a character $\chi : A \to S^1$ of order $n(\chi) = n$ and the corresponding epimorphism to Z/nZ. The ring of algebraic numbers generated by R and the values of χ is $R_\chi = R[\zeta_n] \cong R\Lambda/(\theta_n)$. All tensor products shall be taken over $R[A]$, unless otherwise indicated.

LEMMA 5.13. *Let P be a set of characters $\chi : A \to S^1$ such that every subgroup $B \leq A$ with A/B cyclic is the kernel of exactly one character in P. Then $R[A] \cong \oplus_{\chi \in P} R_\chi$.*

PROOF. For each $\chi \in P$ let $e_\chi = |A|^{-1} \Sigma \chi'(a)a$, where the sum runs over all $a \in A$ and characters χ' with $Ker(\chi') = Ker(\chi)$. Then the e_χ are mutually orthogonal idempotents and $\Sigma_{\chi \in P} e_\chi = 1$. $\square$

LEMMA 5.14. *Let $\chi : A \to S^1$ be a character. Then*

(1) $R_\chi \otimes H_1(X_\phi(L); \mathbb{Z}) \cong R_\chi \otimes H_1(X_{\chi\phi}(L); \mathbb{Z})$;
(2) $R_\chi \otimes H_1(M_\phi(L); \mathbb{Z}) \cong R_\chi \otimes H_1(M_{\chi\phi}(L_{\chi\phi}); \mathbb{Z})$.

PROOF. Let C_* be the singular chain complex for $X_\phi(L)$, considered as a $\mathbb{Z}[A]$-complex. Then $\mathbb{Z}[A/\mathrm{Ker}(\chi)] \otimes C_*$ is the singular chain complex for $X_{\chi\phi}(L)$ and $R_\chi \otimes C_* = R_\chi \otimes \mathbb{Z}[A/\mathrm{Ker}(\chi)] \otimes C_*$. Since R_χ is a direct summand of a localization of $\mathbb{Z}[A]$ it is a flat $\mathbb{Z}[A]$-module and so (1) follows.

Let $K = \mathrm{Ker}(\chi)$ and let $p : M_\phi(L) \to M_{\chi\phi}(L_{\chi\phi}) = M_\phi(L)/K$ be the orbit map. Let $q = |K|^{-1}\mathrm{tr}$. Then $p_*q = id$ and $qp_* = |K|^{-1}\Sigma_{g \in K}g$ has image 1 in R_χ. Hence $R_\chi \otimes p$ is an isomorphism. $\square$

LEMMA 5.15. *Let* $\chi : A \to S^1$ *be a character. Then*

$$R_\chi \otimes H_1(M_{\chi\phi}(L_{\chi\phi}); \mathbb{Z}) \cong R_\chi \otimes H_1(X_{\chi\phi}(L_{\chi\phi}); \mathbb{Z}).$$

PROOF. The inclusion of $X_{\chi\phi}(L_{\chi\phi})$ into $M_{\chi\phi}(L_{\chi\phi})$ induces an epimorphism from $R_\chi \otimes H_1(X_{\chi\phi}(L_{\chi\phi}); \mathbb{Z})$ to $R_\chi \otimes H_1(M_{\chi\phi}(L_{\chi\phi}); \mathbb{Z})$, with kernel generated by lifts of multiples of meridians. Since each meridian of $L_{\chi\phi}$ has nontrivial image under $\chi\phi$ each such generator of this kernel is annihilated by $t^d - 1$, for some proper divisor d of the order of χ. But the images of such terms in R_χ are invertible, and so the kernel is 0. $\square$

LEMMA 5.16. *Let* $\chi : A \to S^1$ *be a character and let* $\tilde{\chi} : \pi L \to \mathbb{Z}$ *be an epimorphism lifting* $\chi\phi$. *Then*

  (1) $H_1(X(L); R) \cong R \otimes A(L)$;
  (2) *if* $\chi \neq 1$ *then* $R_\chi \otimes A(L)$ *is isomorphic to*

$$(R_\chi \otimes H_1(X_{\chi\phi}(L); \mathbb{Z})) \oplus R_\chi \cong (R\Lambda \otimes_{\tilde{\chi}} A(L))/\phi_{n(\chi)}(R\Lambda \otimes_{\tilde{\chi}} A(L));$$

  (3) $R\Lambda \otimes_{\tilde{\chi}} A(L) \cong H_1(X_{\tilde{\chi}}(L); R) \oplus R\Lambda$.

Moreover $H_1(X_{\tilde{\chi}}(L); R)$ *has a square presentation matrix as an* $R\Lambda$-*module and* $\tilde{\chi}(E_1(L))$ *is a principal ideal, generated by* $\tilde{\chi}(\Delta_1(L))$ *if* $\mu = 1$ *and by* $(t - 1)\tilde{\chi}(\Delta_1(L))$ *if* $\mu > 1$.

PROOF. The link exterior $X(L)$ is homotopy equivalent to a finite 2-complex with one 0-cell and Euler characteristic 0. The first two assertions and the first part of (3) follow from the exact sequence of homology for covering spaces of $X(L)$, together with the facts that R_χ is a flat $\mathbb{Z}[A]$-module and that $R_\chi \otimes R = 0$ unless $\chi = 1$.

The cellular chain complex for $X_{\tilde{\chi}}(L)$ (with coefficients R) is a free $R\Lambda$-chain complex C_* with $C_0 \cong R\Lambda$, $C_1 \cong (R\Lambda)^{g+1}$ and $C_2 \cong (R\Lambda)^g$, for some $g \geq 0$. Since $\mathrm{Im}(\partial_1) = \mathrm{Ker}(\varepsilon) = (t-1)$ is free of rank 1 and projective $R\Lambda$-modules are free $\mathrm{Ker}(\partial_1) \cong (R\Lambda)^g$, and so $H_1(X_{\tilde{\chi}}(L); R)$ has a square presentation matrix. In particular, $\tilde{\chi}(E_1(L)) = E_1(R\Lambda \otimes_{\tilde{\chi}} A(L)) = E_0(H_1(X_{\tilde{\chi}}(L); R))$ is principal.

We may also see this directly. If $\mu = 1$ then $E_1(L)$ is principal, and so $\tilde{\chi}(E_1(L)) = (\tilde{\chi}(\Delta_1(L)))$. If $\mu > 1$ then $E_1(L) = \Delta_1(L)I_\mu$. We may assume that $\tilde{\chi}$ maps the ith meridians to t^{d_i} in $\mathbb{Z}$. Since $\tilde{\chi}$ is an epimorphism h.c.f.$(d_1, \ldots, d_\mu) = 1$ and so $(t^{d_1} - 1, \ldots, t^{d_\mu} - 1) = (t - 1)$. Hence $\tilde{\chi}(E_1(L)) = (\tilde{\chi}(\Delta_1(L))(t-1))$. $\qquad\square$

It follows from these lemmas that the first homology of $X_\phi(L)$ and $M_\phi(L)$ with coefficients R are determined as modules over $R[A]$ by the homology of intermediate cyclic covers with coefficients R_χ, and that these are essentially direct summands of quotients of the localized Alexander module.

LEMMA 5.17. *Let H be a finitely generated $R\Lambda$-module. Then $H/\theta_n H$ is determined as a module by the elementary ideals of H and an ideal class invariant. In particular, the $\mathbb{Z}$-torsion subgroup T has order $|T| = |R\Lambda/(E_r(H), \theta_n)|$, where $r = \min\{j \mid E_j(H) \not\subseteq (\theta_n)\}$. Hence $|Res(\Delta_r(H), \theta_n)|$ divides $|T|$ in R.*

PROOF. The module $H/\theta_n H$ is finitely generated and of rank r as a module over $R[\zeta_n]$, and its $\mathbb{Z}$-torsion subgroup is also its $R[\zeta_n]$-torsion submodule. The first assertion then follows from Theorem 3.19, since $R[\zeta_n]$ is a Dedekind domain.

Let I be the set of maximal ideals of $R[\zeta_n]$ which contain $Ann(T)$, and let S be the multiplicative system $S = R[\zeta_n] \setminus \cup_{m \in I} m$. Since T_S and $R[\zeta_n]_S/E_0(T)_S$ are $R[\zeta_n]_S$-torsion modules they have finite composition series. The simple $R[\zeta_n]_S$-modules are the quotients $R[\zeta_n]_S/m_S \cong R[\zeta_n]/m$ for $m \in I$, and the number of simple factors isomorphic to a given simple module in any such composition series for $R[\zeta_n]_S/E_0(T_S)$ is the same as for T_S. Hence $|T| = |T_S| = |R[\zeta_n]_S/E_0(T_S)| = |R[\zeta_n]/E_0(T)| = |R\Lambda/(E_r(H), \theta_n)|$.

The final observation is clear, since $(E_r(H), \theta_n) \leq (\Delta_r(H), \theta_n)$.
$\qquad\qquad\qquad\qquad\qquad\qquad\qquad\qquad\qquad\qquad\qquad\qquad\square$

If H is a Λ-module with a square presentation matrix and θ_n does not divide $\Delta_0(H)$ then $r = 0$ and $H/\theta_n H = T$. In this case $|H/\theta_n H| = |Res(\Delta_0(H), \theta_n)| = |\prod_{(k,n)=1} \Delta_0(H)(\zeta_n^k)|$. (The product is taken over all primitive n^{th} roots of unity.)

THEOREM 5.18. *Let L be a link in a homology 3-sphere Σ and let $\phi : \pi L \to A$ be an epimorphism to a finite abelian group. Then*

(1) $H_1(X_\phi(L); \mathbb{Z}[|A|^{-1}]) \cong \oplus_{\chi \in P}(R_\chi \otimes H_1(X_{\chi\phi}(L); \mathbb{Z}))$;
(2) $H_1(M_\phi(L); \mathbb{Z}[|A|^{-1}]) \cong \oplus_{\chi \in P}(R_\chi \otimes H_1(X_{\chi\phi}(L_{\chi\phi}); \mathbb{Z}))$.

In particular, $H_1(X_\phi(L); \mathbb{Z}[|A|^{-1}])$ and $H_1(M_\phi(L); \mathbb{Z}[|A|^{-1}])$ are determined by the Alexander ideals of L together with the Steinitz-Fox-Smythe row class invariants corresponding to characters $\chi \in P$ such that R_χ is not a principal ideal domain.

PROOF. The direct sum decompositions follow from Lemmas 5.13 to 5.15, and the further assertions then follow from Lemmas 5.16 and 5.17.
$\qquad\qquad\qquad\qquad\qquad\qquad\qquad\qquad\qquad\qquad\qquad\qquad\square$

Let $null(L; \chi) = \max\{d \mid \chi\phi(E_d(L)) = 0\}$ and $N = null(L; \chi) + 1$. Then $E(L; \chi) = \chi\phi(E_N(L))$ is an ideal in the ring $\mathbb{Z}_\chi = \mathbb{Z}[\zeta_{n(\chi)}]$.

COROLLARY 5.18.1. *Let T_X and T_M be the $\mathbb{Z}$-torsion subgroups of $H_1(X_\phi(L); \mathbb{Z})$ and $H_1(M_\phi(L); \mathbb{Z})$, respectively. Then*

(1) $\beta_1(X_\phi(L)) = \Sigma_{\chi \in P}\varphi(n(\chi))null(L; \chi)$;
(2) $\beta_1(M_\phi(L)) = \Sigma_{\chi \in P}\varphi(n(\chi))null(L_{\chi\phi}; \chi)$;
(3) $|T_X| = \prod_{\chi \in P}|\mathbb{Z}_\chi/E(L; \chi)|$ and
$|T_M| = \prod_{\chi \in P}|\mathbb{Z}_\chi/E(L_{\chi\phi}; \chi)|$,
up to powers of divisors of $|A|$.

Moreover if χ has order n and $\tilde{\chi} : \pi L \to \mathbb{Z}$ is an epimorphism lifting $\chi\phi$ then $|\mathbb{Z}_\chi/E(L; \chi)| = |\Lambda/(\tilde{\chi}(E(L; \chi)), \theta_n)|$ and is divisible by $|Res(\tilde{\chi}(\Delta_N(L)), \theta_n)|$, where $N = null(L; \chi) + 1$. $\qquad\square$

Note in particular that $H_1(M_\phi(L); \mathbb{Z})$ is finite if and only if $\chi\phi(\Delta_1(L_{\chi\phi})) \neq 0$ in $\mathbb{Q}_\chi$, for all $\chi \in P$.

The above argument does not work for integral homology, since $\mathbb{Z}[A]$ does not decompose as a direct sum of Dedekind domains. However, the natural homomorphism from $\mathbb{Z}[A]$ to $\oplus_{\chi \in P} \mathbb{Z}_\chi$ is injective, and its cokernel F is a finite $\mathbb{Z}[A]$-module with exponent dividing $|A|$. The long exact sequence of homology derived from the coefficient module sequence gives rise to a homomorphism from $H_1(X_\phi(L); \mathbb{Z})$ to $\oplus_{\chi \in P}(\mathbb{Z}_\chi \otimes H_1(X_{\chi\phi}(L); \mathbb{Z}))$ with cokernel $H_1(X(L); F)$ and kernel a quotient of $H_2(X(L); F)$. Mayberry and Murasugi give a formula for the order of $H_1(M_\phi(L); \mathbb{Z})$, when it is finite, without localization. In our terms their formula reads approximately as follows: $|H_1(M_\phi(L); \mathbb{Z})| = D(\phi) \prod_{\chi \in P} |Res(\tilde{\chi}(\Delta_1(L)), \theta_{n(\chi)})|$, where $D(\phi)$ is an integer defined in [**MM82**] which divides n and depends only on the homomorphism from $\mathbb{Z}^\mu = \pi/\pi'$ to A induced by ϕ. See [**Po04**] for an argument based on Reidemeister torsion.

5.8. Cyclic branched covers

Suppose now that $A = Z/kZ$. Then $H_1(M_\chi(L); \mathbb{Z})$ is annihilated by ν_k. The cover is meridian-cyclic or ξ-cyclic in the terminology of [**MM82**] if χ maps each meridian of L to a generator of A; it is strictly cyclic if all meridians have the same image. The argument for the latter case at the beginning of §7 extends to show that $H_1(M_\chi(L); \mathbb{Z}) \cong H_1(X_{\tilde{\chi}}(L); \mathbb{Z})/\nu_k H_1(X_{\tilde{\chi}}(L); \mathbb{Z})$, where $\tilde{\chi} : \pi L \to \mathbb{Z}$ is an epimorphism lifting χ. (In the strictly cyclic case this is the total linking number cover up to sign, i.e., $\tilde{\chi} = \pm\tau$.) It is finite if and only if $Res(\tilde{\chi}(\Delta_1(L)), \nu_k) \neq 0$.

Suppose that $\tilde{\chi}(\Delta_1(L)) \neq 0$. Let J be the image of $H_1(X; \Lambda_\mu)$ in $H_1(X_{\tilde{\chi}}(L); \mathbb{Z})$. Then $H_1(X_{\tilde{\chi}}(L); \mathbb{Z})/J \cong \mathbb{Z}^{\mu-1} = (\Lambda/(t - 1))^{\mu-1}$, by an iterated Wang sequence argument. Hence $H_1(M_\chi(L); \mathbb{Z})$ is an extension of $(Z/nZ)^{\mu-1} = (\Lambda/(t - 1, \nu_n))^{\mu-1}$ by $J/\nu_n J$. Since $H_1(X_{\tilde{\chi}}(L); \mathbb{Z})$ has a square presentation matrix with nonzero determinant it is a torsion Λ-module of projective dimension ≤ 1. Hence J is also a torsion Λ-module of projective dimension ≤ 1 and so also has a square presentation matrix with nonzero determinant. If $\nabla_{\tilde{\chi}}(L)$ is the latter determinant and $\mu > 1$ then $\nabla_{\tilde{\chi}}(L) = \tilde{\chi}(\Delta_1(L))/(t-1)^{\mu-1}$. (Otherwise $\nabla_{\tilde{\chi}}(L) = \Delta_1(L)$.) If $n = p^r$ is a prime power then

$Res(\nabla_{\tilde\chi}(L), \nu_n) \equiv \varepsilon(\nabla_{\tilde\chi}(L))^{p^r - p} \bmod (p)$, and so $J/\nu_n J$ is finite of order prime to p if $(\varepsilon(\nabla_{\tilde\chi}(L)), p) = 1$. Moreover if $s < s'$ then $(\theta_{p^s}, \theta_{p^{s'}}) = (\theta_{p^s}, p)$. Hence $J/\nu_n J \cong \oplus_{s=1}^{s=r}(J/\theta_{p^s}J)$ and Theorem 3.19 applies to give the structure of these summands.

The next theorem combines results of [**P153**] and [**DW90**].

THEOREM 5.19. *Let L be a link in a homology 3-sphere Σ and let χ be a character of πL of order k. Then*

(1) *if k is odd $H_1(M_\chi(L); \mathbb{Z}[\frac{1}{k}])$ is a direct double;*

(2) *if k is even the kernel of the natural homomorphism from $H_1(M_\chi(L); \mathbb{Z}[\frac{1}{k}])$ to $H_1(M_2(L); \mathbb{Z}[\frac{1}{k}])$ is a direct double.*

If L is a knot the corresponding assertions hold with coefficients $\mathbb{Z}$.

PROOF. The projection from $M_\chi(L)$ to $M_{\chi^r}(L)$ is surjective on homology with coefficients $\mathbb{Z}[\frac{1}{r}]$, by the transfer formula. Since $\varphi(n)$ is even if $n > 2$ it then follows from Corollary 5.18.1 that in each case the rank is even.

Let T be the $\mathbb{Z}$-torsion subgroup of $H_1(M_\chi(L); \mathbb{Z})$, and let $lk : T \times T \to \mathbb{Q}/\mathbb{Z}$ be the torsion linking pairing of $M_\chi(L)$. Then lk is non-singular and symmetric, and $t \in \Lambda$ acts isometrically. The pairing S defined by $S(x, y) = lk((t - t^{-1})x, y)$ for $x, y \in T$ is alternate, i.e., $S(x, x) = 0$ for all $x \in T$, and has radical $\mathrm{Ker}((t^2 - 1)id_T)$. As $H_1(M_\chi(L); \mathbb{Z})$ is annihilated by $\nu_k(t)$ and $\nu_k(1) = k$ and $\nu_k(-1) = (1 - (-1)^k)/2$, it follows that if k is odd then $\mathbb{Z}[\frac{1}{k}] \otimes (t^2 - 1)$ is injective. Thus $\mathbb{Z}[\frac{1}{k}] \otimes T$ is a direct double, since it supports a non-singular alternate pairing. If K is a knot then $t - 1$ is invertible and it is not necessary to invert k.

A similar argument applies in case (2), since the kernel in question is annihilated by $\nu_{k/2}(t^2)$. $\square$

Note that it is not assumed that the cover is the total linking number cover. As the lens spaces $L(k, q)$ may be obtained as k-fold cyclic branched covers of S^3, branched over the Hopf link, the result does not extend to k-primary torsion in general.

The rank of $H_1(M_2(L); \mathbb{Z})$ is $\eta(L) - 1$, and the Goeritz matrix G associated to any diagram for L is a presentation matrix for this group. If $\eta(L) = 1$ then the linking pairing on the finite group

$H_1(M_2(L);\mathbb{Z})$ is determined (up to sign) by the matrix G^{-1}. (See [**GL78**] for modern proofs of these observations, due originally to Seifert for knots and Kyle for links.) The Murasugi signature $\sigma(L)$ is the signature of the intersection pairing on $H_2(W;\mathbb{Q})$, where W is the 2-fold cyclic branched cover of D^4, branched over a properly embedded surface with boundary L [**Vi73, KT76**].

If R is a Dedekind domain and N is a finitely generated R-module then the kernel of any epimorphism $f : R^n \to N$ is projective, and $[N] = \rho(\mathrm{Ker}(f))^{-1}$ is a well defined element of the ideal class group $Cl(R)$, by Schanuel's Lemma. (If N is R-torsion free then it is projective, and $[N] = \rho(N)$.) We shall use an overline to denote the involution on $Cl(\mathbb{Z}[\zeta_n, \frac{1}{n}])$ induced by complex conjugation. If $n = p^r$ then $\mathbb{Z}[\zeta_n, \frac{1}{n}] = \mathbb{Z}[\zeta_n][(1-\zeta_n)^{-1}]$ and the localization induces an isomorphism of ideal class groups.

LEMMA 5.20. *Let L be a μ-component link in a homology 3-sphere Σ, and let $\chi : \pi L \to Z/nZ$ be an epimorphism. Let T_χ be the $\mathbb{Z}$-torsion submodule of $H_\chi = R_\chi \otimes H_1(M_\chi(L);\mathbb{Z})$ and let $P_\chi = H_\chi/T_\chi$. Then*

(1) *$T_\chi \cong Hom_{R_\chi}(T, \mathbb{Q}(\zeta_n)/R_\chi)$;*
(2) *$[T_\chi]\rho(P_\chi)\overline{\rho(P_\chi^*)} = 1$ in $Cl(R_\chi)$.*

PROOF. Let C_* be the cellular chain complex for $M_\chi(L)$, considered as a complex of $\mathbb{Z}[Z/nZ]$-modules, and let $D_* = R_\chi \otimes C_*$. The first assertion follows from Poincaré duality, which also gives $R_\chi \otimes H_2(M_\chi(L);\mathbb{Z}) \cong \overline{H_\chi^*} = \overline{P_\chi^*}$.

Since R_χ is a direct summand of a localization of $\mathbb{Z}[Z/nZ]$ it is a flat $\mathbb{Z}[Z/nZ]$-module. Moreover $R_\chi \otimes \mathbb{Z} = 0$. Hence there are exact sequences

$$0 \to Z_2 \to D_2 \oplus D_0 \to D_1 \to H_\chi = T_\chi \oplus P_\chi \to 0$$

and $0 \to D_3 \to Z_2 \to P_\chi^* \to 0$, from which the second assertion follows. $\square$

If $n = p$ is prime these necessary conditions are sufficient *mod* p-torsion, and provide a complete criterion in the knot-theoretic case $\mu = 1$ [**Da95**].

5.9. Families of coverings

Let F be a field, and let $F(\rho)$ be the 1-dimensional $F[\pi]$-module, determined by a homomorphism $\rho : \pi \to F^\times$. Let $H_1(\pi; \rho) = H_1(\pi; F(\rho))$ and $\beta_1(\pi; \rho) = dim_F H_1(\pi; \rho)$. The action ρ corresponds to a ring epimorphism $\rho : F\Lambda_\mu \to F$, i.e., to an F-rational closed point of the algebraic torus $Spec(F\Lambda_\mu)$. In particular, the trivial 1-dimensional representation ε_F corresponds to the augmentation $\varepsilon_F = id_F \otimes \varepsilon : F\Lambda_\mu \to F$. The ith *characteristic variety* of L is $V_i(L) = V(E_i(L)) = \{\wp \in Spec(F\Lambda_\mu) \mid E_i(L) \leq \wp\}$. Then $V_{=+1}(L) \subseteq V_i(L)$, and $V_i(L)$ is the support of $\wedge_i(H_1(X(L); F\Lambda_\mu))$, by the argument of Theorem 3.1.

THEOREM 5.21. *Let L be a μ-component 1-link and F a field. Then*

    (1) $V_1(L) = V(\Delta_1(L))$ *if $\mu \leq 2$ and*
           $V_1(L) = V(\Delta_1(L)) \cup \{\mathrm{Ker}(\varepsilon)\}$ *if $\mu > 2$.*
    (2) $\beta_1(\pi; \varepsilon_F) = \mu$ *and $\beta_1(\pi; \rho) = \max\{i \mid \rho \in V_i(L)\} = \max\{i \mid \rho(E_i(\pi)) = 0\}$ otherwise.*

PROOF. The first assertion follows immediately from Theorems 3.1 and 4.12.

If $\rho = \varepsilon_F$ then $\beta_1(\pi; \rho) = dim_F H_1(\pi; F) = \mu$. If ρ is nontrivial then $H_0(\pi; \rho) = 0$, and tensoring the augmentation sequence for π with $F(\rho)$ gives a short exact sequence

$$0 \to H_1(\pi; \rho) \to F(\rho) \otimes_{\mathbb{Z}[\pi]} A(\pi) = F(\rho) \otimes_{\mathbb{Z}[\pi]} I(\pi) \to F(\rho) \to 0,$$

and so $\beta_1(\pi; \rho) = \max\{i \mid \rho \in V_i(L)\} = \max\{i \mid \rho(E_i(\pi)) = 0\}$. $\square$

The support is all of $Spec(F\Lambda_\mu)$ if $\Delta_1(L) = 0$ and has dimension $\mu - 1$ otherwise, except if $F \otimes (\pi'/\pi'') = 0$, when it is empty. (This can only happen if $\mu \leq 2$.) The first Betti number jumps along the subvarieties of $Spec(F\Lambda_\mu)$ determined by the elementary ideals. In particular, $\rho \in V_1(\pi)$ if and only if either $\rho = \varepsilon$ and $\mu \geq 2$ or $\rho \neq \varepsilon$ and $H_1(\pi; \rho) \neq 0$.

An epimorphism $\sigma : \pi \to \mathbb{Z}^k$ induces an embedding $Spec(\sigma)$ of $Spec(F\Lambda_k)$ as a subtorus of $Spec(F\Lambda_\mu)$. A translate of a subtorus is a subvariety determined by equations of the form $\Pi t_i^{n(i)} = f$, with $n(i) \in \mathbb{Z}$ for $1 \leq i \leq \mu$ and $f \in F$.

THEOREM 5.22. [**Sa90, Li02**] *Let L be a μ-component iterated torus link and F a field. Then $V_i(L)$ is the union of finitely many translates of subtori, for all $i \geq 1$.*

PROOF. (Sketch.) We may assume that L may be obtained from a simpler iterated torus link $\mathcal{L}$ (with $m \leq \mu$ components) by composing the knot $\mathcal{L}_1$ with an m'-component link $\mathcal{L}'$ in $S^1 \times D^2$ which is either a torus knot or the union of such a knot and the core $S^1 \times \{0\}$ [**SW77**]. Let $\mathcal{L}'^+$ denote the corresponding 2- or 3-component link in S^3, and let $T = \partial X(\mathcal{L}_1)$. As $X(L) = X(\mathcal{L}) \cup_T X(\mathcal{L}'^+)$, and all the linking numbers are nonzero we find that

$$H_1(X(L); \Lambda_\mu) \cong (\Lambda_\mu \otimes H_1(X(\mathcal{L}); \Lambda_m)) \oplus (\Lambda_\mu \otimes H_1(X(\mathcal{L}'^+); \Lambda_{m'+1})).$$

(The tensor products are taken over the appropriate subrings for each summand.) Thus we may argue by induction, once the claim is established for the model links $\mathcal{L}'^+$. We refer to [**Li02**] for these. $\square$

Let $Aff(F)$ be the group of *affine transformations* $\alpha : F \to F$, such that $\alpha(z) = rz + s$ for some $r, s \in F$ with $r \neq 0$ and for all $z \in F$. Let $x_1, \ldots, x_\mu$ be a meridianal basis for π, and suppose that $f_0 : \pi \to Aff(F)$ is a homomorphism. Let $Aff(\pi; f_0)$ be the set of homomorphisms $f : \pi \to Aff(F)$ such that $f(x_i) = f_0(x_i)$ for all i. Then there is a representation $\rho : \pi \to F^\times$ such that $f(g)(z) - f(g)(0) = \rho(g)z$ for all $g \in \pi$, $z \in F$ and $f \in Aff(\pi; f_0)$. If $f \in Aff(\pi; f_0)$ let $\tilde{f}(g\pi'') = f(g)(0)$ for all $g \in \pi'$. Then $\tilde{f}(gh\pi'') = \tilde{f}(g\pi'') + \tilde{f}(h\pi'')$ and $\tilde{f}(kgk^{-1}\pi'') = \rho(k)\tilde{f}(g\pi'')$ for all $g, h \in \pi'$ and $k \in \pi$, so $\tilde{f} : \pi'/\pi'' \to F(\rho)$ is a homomorphism. The function from $Aff(\pi; f_0)$ to $Hom_{\mathbb{Z}[\pi]}(\pi'/\pi'', F(\rho))$ which sends f to $\tilde{f}$ is 1-1. It is easily seen to be bijective in the knot-theoretic case [**deR67**]. However, this map is no longer onto if $\mu \geq 2$.

Coverings of the exterior X of a μ-component link with covering group isomorphic to $\mathbb{Z}^k$ are parametrized by k-planes in $H^1(X; \mathbb{Q})$. The variation of the homology of such coverings of finite complexes as the k-plane moves in the Grassmannian $G_k(H^1(X; \mathbb{Q}))$ is studied in [**DF87, Fr87**]. We shall specialize the argument of [**DF87**] to the case of classical links. (The further subtleties of [**Fr87**] are irrelevant here, as $V_1(L)$ has dimension $\geq \mu - 1$ or is empty.)

Let $\sigma : \pi \to \mathbb{Z}^k$ be an epimorphism. Since $F\Lambda_k \otimes_\sigma A(L)$ is an extension of the maximal ideal I_k of $F\Lambda_k$ by $H_1(X; \sigma^*F\Lambda_k)$ and $E_1(L) = \Delta_1(L)I_\mu$ it is easily seen that the support of $H_1(X; \sigma^*F\Lambda_k)$ contains $Spec(\sigma)^{-1}(V(\Delta_1(L)))$ and is contained in $Spec(\sigma)^{-1}(V_1(L))$. As a finitely generated $F\Lambda_k$-module is finite dimensional over F if and only if its support is finite, $H_1(X; \sigma^*F\Lambda_k)$ is finite dimensional if and only if $V_1(L) \cap \mathrm{Im}(Spec(\sigma))$ is finite [**DF87**].

This is trivially the case if $k = 0$ or if $\mu \leq 2$ and $\Delta_1(L) = 1$. Otherwise $V_1(L)$ has codimension ≤ 1, and so $V_1(L) \cap \mathrm{Im}(Spec(\sigma))$ can only be finite if $k = 1$ and $\Delta_1(L) \neq 0$. An epimorphism $\sigma : \pi \to \mathbb{Z}$ sends each t_i to some power t^{a_i}, for $1 \leq i \leq \mu$. As $F\Lambda_1$ is a PID $H_1(X; \sigma^*F\Lambda_1)$ is infinite-dimensional if and only if $\sigma(\Delta_1(L))$ is identically 0. This imposes finitely many linear constraints on the exponents a_i, and so the set of infinite cyclic covers of X with finite dimensional homology corresponds to the $\mathbb{Q}$-rational points of a (possibly empty) Zariski open subset of the projective space $\mathcal{P}(H^1(X; \mathbb{Q}))$.

Let $\phi_n : \pi L \to (Z/nZ)^\mu$ be the epimorphism sending meridians to the standard generators. Then there is $q \in \mathbb{N}$ and polynomials $p_i(t)$ for $1 \leq i \leq q$ such that $\beta_1(M_{\phi_n}(L); \mathbb{C}) = p_i(n)$ if $i \equiv n \ mod \ (q)$ [**ASR94**]. The Betti numbers are said to be *polynomially periodic*. As observed in §7 above, when $\mu = 1$ the Betti numbers are in fact periodic. The argument depends on the fact that the algebraic group $(\mathbb{G}_m)^\mu = Spec(\Lambda_\mu)$ has the strong approximation property and on the proof of a conjecture of Lang about torsion points in an algebraic subvariety of $Spec(\Lambda_\mu)$. In general, may we always assume that the polynomials $p_i(t)$ have degree $< \mu$?

CHAPTER 6

Twisted Polynomial Invariants

The Alexander invariants of a 1-link L are invariants of π/π'', where $\pi = \pi L$. They are readily computable, in terms of determinants of matrices over a commutative ring. However, they are of little use if, for instance, π' is perfect. Representations into finite groups are computable, but do not dig very deeply into the group. Taken in conjunction with a homomorphism onto a free abelian group they give a manageable class of representations which penetrate below the commutator group. Twisted Alexander invariants combine features of representability into well-understood non-solvable groups (such as linear groups) with computability of invariants defined in terms of matrices over a commutative ring.

The twisted Alexander polynomials of a knot associated to a linear representation of the knot group were defined by Lin, in terms of the free differential calculus, applied to the presentation of the knot group derived from a "free" Seifert surface for the knot. (His paper [**Li01**] first circulated in 1990, as a Columbia University preprint.) This was extended to arbitrary finite presentations of groups with an epimorphism to a free abelian group in [**Wa94**]. It was given a more homological formulation, as a Reidemeister-Franz torsion, first for knots in [**Ki96**] and then for infinite cyclic covers of finite complexes in [**KL99**], where it was presented in terms of local coefficients. We shall take the latter point of view.

6.1. Definition in terms of local coefficients

Let X be a based finite complex with basepoint $*$, fundamental group π and universal covering space $\widetilde{X}$. Then $C_* = C_*(\widetilde{X})$ is a finitely generated free left $\mathbb{Z}[\pi]$-chain complex. Let $\alpha : \pi \to \mathbb{Z}^n$

be an epimorphism, R a noetherian factorial domain and $\rho : \pi \to$ $\mathrm{GL}(m, R)$ a representation, and let V be the associated *right* $R[\pi]$-module. (We shall view the R-module R^m underlying V as a space of *row* vectors.)

Let $V[\mathbb{Z}^n] = R\Lambda_n \otimes_R V$ have the $(R\Lambda_n, R[\pi])$-bimodule structure given by

$$r\lambda(\lambda' \otimes v)sg = \lambda\lambda'\alpha(g) \otimes vrsg,$$

for all $r, s \in R$, $\lambda, \lambda' \in \mathbb{Z}^n$, $v \in V$ and $g, h \in \pi$. (Thus left and right muliplication by elements of R agree.) The module $V[\mathbb{Z}^n]$ corresponds to the representation $\rho \otimes \alpha : \pi \to \mathrm{GL}(m, R\Lambda_n)$ given by $\rho \otimes \alpha(g) = \alpha(g)\rho(g)$, for all $g \in \pi$. (Note that $\mathrm{GL}(m, \varepsilon) \circ (\rho \otimes \alpha) = \rho$. Thus $\rho \otimes \alpha$ determines ρ, and hence α.)

The homology modules $H_*(X; \rho \otimes \alpha) = H_*(X; V[\mathbb{Z}^n])$ are finitely generated $R\Lambda_n$-modules, and the elementary ideals of these modules are the *twisted Alexander invariants* of X (with respect to $\rho \otimes \alpha$). Two representations ρ_1 and ρ_2 are conjugate if and only if their representation modules V_1 and V_2 are isomorphic. In that case $V_1[\mathbb{Z}^n] \cong V_2[\mathbb{Z}^n]$, and the associated twisted invariants agree.

We may identify $H_*(X; \rho \otimes \alpha)$ in terms somewhat closer to those of the original definitions as follows. Let $N = \mathrm{Ker}(\alpha)$. If M is a left $R[\pi]$-module the diagonal action of π on $V \otimes_R M$ (determined by $g(v \otimes m) = vg^{-1} \otimes gm$ for all $g \in \pi$, $v \in V$ and $m \in M$) induces a left action of $\pi/N = \mathbb{Z}^n$ on $V \otimes_{R[N]} M$. Let $V|_N$ be V considered as a right $R[N]$-module by restriction. Then $H_*(X_K; V|_N) = H_*(V \otimes_{R[N]} C_*(\widetilde{X}))$ has the additional data of the $R\Lambda_n$-module structure.

THEOREM 6.1. *Let X be a finite complex with $\pi_1(X) = \pi$ and $\rho : \pi \to \mathrm{GL}(m, R)$ a representation with representation module V. Let $\alpha : \pi \to \mathbb{Z}^n$ be an epimorphism with kernel N. Then $H_*(X_N; V|_N)$ and $H_*(X; V[\mathbb{Z}^n])$ are canonically isomorphic as $R\Lambda_n$-modules.*

PROOF. Define $\theta : V \otimes_{R[N]} R[\pi] \to V[\mathbb{Z}^n]$ by $\theta(v \otimes g) = \alpha(g) \otimes vg$, for all $v \in V$ and $g \in \pi$, and fix a section $\sigma : \mathbb{Z}^n \to \pi$. Then

$$\theta(v \otimes kg) = \alpha(g) \otimes vkg = \theta(vk \otimes g),$$

$$\theta(\lambda(v \otimes g)) = \theta(v\sigma(\lambda)^{-1} \otimes \sigma(\lambda)g) = \lambda\alpha(g) \otimes vg = \lambda\theta(v \otimes g)$$

and

$$\theta((v \otimes g)h) = \theta(v \otimes gh) = \alpha(g)\alpha(h) \otimes vgh = (\theta(v \otimes g))h,$$

for all $v \in V$, $k \in N$, $g, h \in \pi$ and $\lambda \in \mathbb{Z}^n$. Then θ is an $(R\Lambda_n, R[\pi])$-bimodule isomorphism, with inverse given by

$$\theta^{-1}(\lambda \otimes v) = v\sigma(\lambda)^{-1} \otimes \sigma(\lambda),$$

for all $\lambda \in Z^n$ and $v \in W$. This induces a canonical isomorphism of chain complexes $V[\mathbb{Z}^n] \otimes_{R[\pi]} C_*(\widetilde{X}) \cong V \otimes_{R[N]} C_*(\widetilde{X})$, and so $H_*(X; V[\mathbb{Z}^n])$ and $H_*(X; \alpha, V)$ are canonically isomorphic as $R\Lambda_n$-modules. $\qquad\square$

When $n = 1$ we may assume that σ is a homomorphism. However, if $X = X(L)$ for some μ-component 1-link L and $n = \mu > 1$ the abelianization homomorphism from πL to $\mathbb{Z}^\mu$ usually does not split. (See Corollary 4.12.1.)

COROLLARY 6.1.1. *If $n > 0$ then $H_0(X; \rho \otimes \alpha)$ is a torsion $R\Lambda_n$-module. If, moreover, either $n > 1$ or there is a $k \in N = \mathrm{Ker}(\alpha)$ such that $\det(\rho(k) - I) \neq 0$ then $H_0(X; \rho \otimes \alpha)$ is pseudonull.*

PROOF. Since $H_0(X; \rho \otimes \alpha) \cong V \otimes_{\mathbb{Z}[N]} \mathbb{Z}$, it is finitely generated as an R-module, and if $\det(\rho(k) - I) \neq 0$ for some $k \in N$ then it is a torsion R-module. $\qquad\square$

Let L be a link, $X = X(L)$ and $\pi = \pi L$. If $V = R$ with the trivial π-action then $H_*(X; \rho \otimes \alpha) = H_*(X; R\Lambda_n)$, and the associated twisted Alexander invariants are specializations of the ordinary Alexander invariants of L. If α is trivial ($n = 0$), R is a field and V is 1-dimensional the notation agrees with that of §9 of Chapter 5. The invariants of greatest interest for us are $\Delta(\pi; \rho \otimes \alpha) = \Delta_0(H_1(\pi; \rho \otimes \alpha))$ and, in particular, $\Delta(L; \rho \otimes \alpha) = \Delta(\pi L; \rho \otimes \alpha)$.

6.2. Presentations

Applying the functor $V[\mathbb{Z}^n] \otimes_{\mathbb{Z}[\pi]} -$ to the augmentation sequence

$$0 \to I(\pi) \to \mathbb{Z}[\pi] \to \mathbb{Z} \to 0$$

gives a "Crowell sequence"

$$0 \to H_1(\pi; \rho \otimes \alpha) \to V[\mathbb{Z}^n] \otimes_{\mathbb{Z}[\pi]} I(\pi) \to V[\mathbb{Z}^n] \to H_0(\pi; \rho \otimes \alpha) \to 0.$$

Since $V[\mathbb{Z}^n]$ is a free $R\Lambda_n$-module it follows that

$$TH_1(\pi; \rho \otimes \alpha) = T(V[\mathbb{Z}^n] \otimes_{\mathbb{Z}[\pi]} I(\pi)).$$

If $n = 1$ the kernel of the epimorphism from $V[\mathbb{Z}^n]$ to $V[\mathbb{Z}^n] \otimes_{\mathbb{Z}[\pi]} \mathbb{Z} = H_0(\pi; \rho \otimes \alpha)$ is Λ_1^m, while if $n > 1$ then $H_0(\pi; \rho \otimes \alpha)$ is pseudonull, by Corollary 6.1.1. In either case,

$$\Delta(\pi; \rho \otimes \alpha) \doteq \Delta_m(V[\mathbb{Z}^n] \otimes_{\mathbb{Z}[\pi]} I(\pi)).$$

Suppose that π has a g generator, r relator presentation

$$\langle x_1, \ldots, x_g \mid w_1, \ldots, w_r \rangle.$$

Then the free differential calculus gives an $r \times g$ presentation matrix P for $I(\pi)$ as a right $\mathbb{Z}[\pi]$-module. (See §6 of Chapter 10.) Let $\Phi(P) \in M_{mr,mg}(\mathbb{Z}\Lambda_n)$ be the matrix obtained by replacing each entry of P by its image under $\rho \otimes \alpha$. Then $\Phi(P)$ is a presentation matrix for $V[\mathbb{Z}^n] \otimes_{\mathbb{Z}[\pi]} I(\pi)$.

Suppose for simplicity that the presentation has deficiency $g - r = 1$. Let P_j be the square matrix obtained from P by deleting the jth column, and let $\Phi(P_j)$ be the corresponding $m(n-1) \times m(n-1)$ matrix, for $1 \leq j \leq n$. Let $Q_j = \det \Phi(P_j)$. Applying Φ to the expansions $w_j - 1 = \Sigma_{1 \leq i \leq n}(x_i - 1)\frac{\partial w_j}{\partial x_i}$ gives identities

$$\Sigma_{1 \leq i \leq n} \Phi(\frac{\partial w_j}{\partial x_i}) \Phi(x_i - 1) = 0,$$

for all $1 \leq j \leq n$. Elementary row operations then imply that

$$Q_j \det(\rho \otimes \alpha(x_i) - I) = \pm Q_i \det(\rho \otimes \alpha(x_j) - I),$$

for all $1 \leq i < j \leq n$. If $\det(\rho \otimes \alpha(x_i) - I) \neq 0$ for some $1 \leq i \leq n$ then Tietze arguments show that the *Wada* invariant

$$W_{\rho \otimes \alpha} = Q_i / \det(\rho \otimes \alpha(x_i) - I)$$

is well-defined, up to multiplication by units of Λ_n. A similar construction involving *h.c.f.*s of minors applies for presentations of deficiency ≤ 0 [**Wa94**]. If $n = 1$ and $H_1(\pi; \rho \otimes \alpha)$ is a torsion Λ-module then

$$\Delta(\pi; \rho \otimes \alpha) = W_{\rho \otimes \alpha} \Delta_0(H_0(\pi; \rho \otimes \alpha)),$$

and so $\Delta(\pi; \rho \otimes \alpha) = W_{\rho \otimes \alpha}$ if $H_0(\pi; \rho \otimes \alpha)$ is pseudonull [**KL99**].

6.3. Reidemeister-Franz torsion

Let F and $F(t)$ be the fields of fractions of R and $R\Lambda_n$, respectively. Involutions of R and $R\Lambda_n$ extend canonically to these fields. Let $V(t) = F(t) \otimes_{R\Lambda_n} V[\mathbb{Z}^n]$, and assume that the complex $C_*(X; V(t)) = V(t) \otimes_{\mathbb{Z}[\pi]} C_*(\widetilde{X})$ is acyclic. A basis for V and lifts of the cells of X together determine a class of bases for $C_*(X; V[\mathbb{Z}^n])$, and the torsion of this based complex is the image in $F(t)^{\times}/R\Lambda^{\times}$ of

$$\tau(X; \rho \otimes \alpha) = \Pi \Delta_0(H_i(X; \rho \otimes \alpha))^{e(i)}$$

(where $e(i) = (-1)^{i+1}$, for all $i \geq 0$), as in Theorem 3.15. If X is aspherical we may also write $\tau(\pi; \rho \otimes \alpha) = \tau(X; \rho \otimes \alpha)$.

LEMMA 6.2. *If* $\Delta(L; \rho \otimes \alpha) \neq 0$ *then* $\tau(X(L); \rho \otimes \alpha)$ *is defined, and if also either* $n > 1$ *or* $\det(\rho(k) - I) \neq 0$ *for some* $k \in \mathrm{Ker}(\alpha)$ *then* $\tau(X(L); \rho \otimes \alpha) \doteq \Delta(L; \rho \otimes \alpha)$.

PROOF. Up to homotopy, we may assume that X is a finite 2-complex, with one 0-cell, $r + 1$ 1-cells and r 2-cells, for some $r \geq 0$. We then have an exact sequence

$$0 \to H_2(X; V[\mathbb{Z}^n]) \to V[\mathbb{Z}^n]^r \to V[\mathbb{Z}^n]^{r+1} \to H_1(X, *; V[\mathbb{Z}^n]) \to 0,$$

and a Crowell sequence

$$0 \to H_1(X; V[\mathbb{Z}^n]) \to H_1(X, *; V[\mathbb{Z}^n]) \to V[\mathbb{Z}^n] \to V[\mathbb{Z}^n] \otimes_{\mathbb{Z}[\pi]} \mathbb{Z} \to 0.$$

If $\Delta(L; \rho \otimes \alpha) \neq 0$ then $V(t)$ and $H_1(X, *; V(t))$ each have $R\Lambda_{\mu}$-rank n, by the Crowell sequence, and so $H_2(X; V[\mathbb{Z}^n]) = 0$, by the previous sequence. Since $H_0(X; V[\mathbb{Z}^n])$ is a torsion $R\Lambda_n$-module $C_*(X; V(t))$ is acyclic. Therefore $\tau(X(L); \rho \otimes \alpha)$ is defined, and $\tau(X(L); \rho \otimes \alpha) = \Delta(L; \rho \otimes \alpha)/\Delta_0(H_0(X; \rho \otimes \alpha))$.

The denominator is 1 if $n > 1$ or if $\det(\rho(k) - I) \neq 0$ for some $k \in \mathrm{Ker}(\alpha)$, by Corollary 6.1.1, and then $\tau(X(L); \rho \otimes \alpha) \doteq \Delta(L; \rho \otimes \alpha)$. $\square$

There is also a sign-determined torsion [**Ki07**].

6.4. Duals and pairings

We shall allow the ring R to have an involution (possibly trivial), which we extend to $R\Lambda_n$ by inversion of the variables. Transposition of matrices composed with involution of the entries defines an anti-involution $\dagger$ on $GL(m, R)$. The *dual* representation ρ^D is then given by $\rho^D(g) = \rho(g^{-1})^\dagger$.

Let $V^D = \overline{Hom_R(V, R)}$, with $(r\phi)(v) = \bar{r}\phi(v)$ and $(\phi g)(v) = \phi(vg^{-1})$, for all $r \in R$, $\phi \in W$, $v \in V$ and $g \in \pi$.

LEMMA 6.3. *The module associated to ρ^D is V^D.*

PROOF. Let $\{v_i | 1 \leq i \leq m\}$ be the standard basis for $V = R^m$ and let $\{w_j | 1 \leq j \leq m\}$ be the dual basis for V^D, so that $w_j(v_i) = 1$ if $i = j$ and 0 otherwise. Then $v_i g = \sum_{i=1}^{m} \rho(g)_{ij} v_j$. Hence

$$(w_k \rho^D(g))(v_l) = (\sum_{i=1}^{m} \rho^D_{ki} w_i)(v_l) = \sum_{i=1}^{m} \overline{\rho^D(g)_{ki}}(w_i(v_l))$$

$$= \overline{\rho^D(g)_{kl}} = \rho(g^{-1})_{lk}$$

$$= w_k(v_l g^{-1}) = (w_k g)(v_l),$$

for all $1 \leq k, l \leq m$. Thus $w_k g = w_k \rho^D(g)$ for all $1 \leq k \leq m$. □

Evaluation defines a sesquilinear pairing $ev : V \times V^D \to R$ by $ev(v, \phi) = \phi(v)$. This is easily seen to be non-singular and π-invariant.

LEMMA 6.4. *A representation $\rho : \pi \to GL(n, R)$ is conjugate to its dual if and only if there exists a non-singular sesquilinear pairing $(v, w) \mapsto \{v, w\} \in R$ on V such that $\{vg, wg\} = \{v, w\}$, for all $v, w \in V$ and $g \in \pi$.*

PROOF. Assume that ρ^D is conjugate to ρ. Then there exists a matrix $A \in GL(n, R)$ such that $A^{-1}\rho(g)A = \rho(g^{-1})^\dagger$, for all $g \in \pi$. Define a sesquilinear pairing on V by $\{v, w\} = vA\bar{w}^{tr}$, for all $v, w \in V$. Since A is invertible, this pairing is non-singular. It is easy to check that $\{v \cdot g, w \cdot g\} = \{v, w\}$, for all $v, w \in V$ and $g \in \pi$.

Conversely, assume that $\{-, -\}$ is a non-singular sesquilinear pairing which is invariant under ρ. Then there is an invertible matrix

$A \in GL(n, R)$ such that $\{v, w\} = vA\bar{w}^{tr}$, for all $v, w \in V$. Since ρ preserves the pairing,

$$v\rho(g)A\rho(g)^{\dagger}\bar{w}^{tr} = \{v \cdot g, w \cdot g\} = \{v, w\} = vA\bar{w}^{tr},$$

for all $v, w \in V$ and $g \in \pi$. It follows that $\rho(g)A\rho(g)^{\dagger} = A$, and so $A^{-1}\rho(g)A = \rho(g^{-1})^{\dagger}$, for all $g \in \pi$. Hence ρ^D is conjugate to ρ. $\square$

If $R = \mathbb{R}$ then ρ is *orthogonal* if $\rho(\pi) < O(m)$, while if $R = \mathbb{C}$ then ρ is *unitary* if $\rho(\pi) < U(m)$. The corresponding sesquilinear pairings are then the standard inner products on R^m. Conversely, if $R = \mathbb{R}$, and the form is positive-definite, then by considering a basis for V that is orthonormal with respect to the pairing, we see that A is the identity matrix. If P is the change-of-basis matrix, and $\rho'(g) = P\rho(g)P^{-1}$ for $g \in \pi$, then $\rho'(g) = \rho'(g^{-1})^{tr}$ for all $g \in G$. Hence ρ is conjugate to an orthogonal representation. Similarly, if $R = \mathbb{C}$ and the pairing is hermitian and positive-definite then ρ is conjugate to a unitary representation.

We may extend ev to a π-invariant sesquilinear pairing

$$Ev : V[\mathbb{Z}^n] \times V^D[\mathbb{Z}^n] \to R[\mathbb{Z}^n],$$

by $Ev(\lambda \otimes \phi, \mu \otimes v) = \lambda\bar{\mu}\phi(v)$, for all $\lambda, \mu \in R\Lambda_n$, $\phi \in V^D$ and $v \in V$. If $\{v_i\}$ and $\{w_j\}$ are dual bases for the free R-modules V and V^D, as in Lemma 6.2, then $\{1 \otimes v_i\}$ and $\{1 \otimes v_j^*\}$ are dual bases for the free $R\Lambda_n$-modules $V[\mathbb{Z}^n]$ and $V^D[\mathbb{Z}^n]$. Hence Ev is non-singular. The group $Hom_{R\Lambda_n}(V[\mathbb{Z}^n], R\Lambda_n)$ is a $(R[\pi], R\Lambda_n)$-bimodule. Let $\overline{V^D[\mathbb{Z}^n]}$ be the conjugate $(R[\pi], R\Lambda_n)$-bimodule, with

$$rg(\lambda \otimes \phi)\lambda' = \overline{\alpha(g)r\lambda'}\lambda \otimes \phi g^{-1},$$

for all $r \in R$, $\lambda, \lambda' \in R\Lambda_n$, $\phi \in V^D$ and $g \in \pi$. Then $Ad(Ev)$ induces a $(R[\pi], R\Lambda_n)$-bimodule homomorphism

$$f : \overline{V^D[\mathbb{Z}^n]} \to Hom_{R\Lambda_n}(V[\mathbb{Z}^n], R\Lambda_n),$$

which is an isomorphism, since Ev is non-singular. If M is a left $R[\pi]$-module we may extend f to a homomorphism

$$f_M : Hom_{R[\pi]}(M, \overline{V^D[\mathbb{Z}^n]}) \to Hom_{R\Lambda_n}(V[\mathbb{Z}^n] \otimes_{R[\pi]} M, R\Lambda_n)$$

of (right) $R\Lambda_n$-modules, by $f_M(\Psi)(\lambda \otimes v \otimes m) = \Psi(m)(\lambda \otimes v)$, for all $\Psi \in Hom_{R[\pi]}(M, \overline{V^D[\mathbb{Z}^n]})$, $\lambda \in R\Lambda_n$, $v \in V$ and $m \in M$. These define a natural transformation of left exact contravariant functors in the variable M. Since $f_{R[\pi]} = f$ is an isomorphism then so is f_M, for all finitely generated left $R\Lambda_n$-modules M.

6.5. Reciprocity

We shall assume henceforth that $X = X(L)$, where L is a μ-component 1-link, with $\mu \geq 1$, and that X has a finite triangulation.

Untwisted Alexander polynomials are reciprocal, in the sense that $\overline{\Delta_i(L)} \doteq \Delta_i(L)$ for all $i \geq 0$. This is a consequence of Poincaré duality and the Universal Coefficient Theorem, applied to suitable localizations of Λ_μ. (See Theorem 5.1.)

In general, twisted Alexander invariants need not be reciprocal. If U is the trivial knot, $\alpha : \pi U \cong \mathbb{Z}$ and $\rho : \pi U \to \mathbb{Q}^\times = GL(1, \mathbb{Q})$ has image $\{2^n | n \in \mathbb{Z}\}$ then $\tau(X; \rho \otimes \alpha) = (2t - 1)^{-1}$, which is not reciprocal. We shall give an example with non-reciprocal twisted Alexander polynomial below.

However, twisted Alexander polynomials are reciprocal if the defining representation is conjugate to its dual, and if the contribution from the boundary is well-behaved.

THEOREM 6.5. *Suppose that $\mu \geq 2$ and all linking numbers ℓ_{ij} are nonzero. If $n \geq 2$ and $\rho : \pi \to GL(n, R)$ is conjugate to its dual then*

$$\Delta(\pi L; \rho \otimes \alpha) \doteq \overline{\Delta(\pi L; \rho \otimes \alpha)}.$$

PROOF. We may assume that $\Delta(\pi L; \rho \otimes \alpha) \neq 0$. Since ρ is conjugate to its dual we may fix an isomorphism $V \cong V^D$ of right $R[\pi]$-modules. Since $\overline{V^D[\mathbb{Z}^n]}$ is a left $R[\pi]$-module, we may define cohomological twisted polynomial invariants in terms of $H^q(X; \overline{V^D[\mathbb{Z}^n]})$.

Let $C_*^D(\widetilde{X}, \partial \widetilde{X})$ be the chain complex determined by the cell structure dual to the triangulation of X. Define chain complexes

$$C_* = V^D[\mathbb{Z}^n] \otimes C_*(\widetilde{X}) \quad \text{and} \quad D_* = V[\mathbb{Z}^n] \otimes C_*^D(\widetilde{X}, \partial \widetilde{X}).$$

These are finitely generated free $R\Lambda_n$-complexes. Evaluation defines a sesquilinear pairing $D_q \times C_{3-q} \to R\Lambda_n$ by

$$(6.1) \qquad \langle p \otimes v \otimes z_1, q \otimes w \otimes z_2 \rangle = \sum_{g \in \pi} (z_1 \cdot gz_2) p\bar{q}.w(vg),$$

where $z_1 \cdot gz_2$ is the algebraic intersection number in $\mathbb{Z}$ of cells z_1 and gz_2. Then this pairing induces a $R\Lambda_n$-module isomorphism

$$\overline{D_{3-q}} \to \mathrm{Hom}_{R\Lambda_n}(C_q, R\Lambda_n) \cong \mathrm{Hom}_{R[\pi]}(C_*(\widetilde{X}), \overline{V^D[\mathbb{Z}^n]}).$$

Consequently, there are isomorphisms

$$\overline{H_{3-q}(X, \partial X; V[\mathbb{Z}^n])} \cong H^q(X; \overline{V^D[\mathbb{Z}^n]}),$$

for all q.

The linking number condition implies that the restriction of α to each peripheral subgroup $\pi_1(\partial X(L_i))$ is a monomorphism. Let $\{m_i, \ell_i\}$ be a meridian-longitude pair for L_i. Then

$$H_0(\partial X(L_i); V[\mathbb{Z}^n]) = V[\mathbb{Z}^n]/(\rho(m_i)\alpha(m_i) - I, \rho(\ell_i)\alpha(\ell_i) - I)$$

is psuedonull, while $H_i(\partial X(L_i); V[\mathbb{Z}^n]) = 0$ for $i > 0$. Thus the natural map from $H_1(X, \partial X; V[\mathbb{Z}^n])$ to $H_1(X, \partial X; V[\mathbb{Z}^n])$ is a pseudo-isomorphism.

Applying the functors and natural isomorphisms

$$f_- : \mathrm{Hom}_{R[\pi]}(-, \overline{V^D[\mathbb{Z}^n]}) \Rightarrow \mathrm{Hom}_{R\Lambda_n}(V[\mathbb{Z}^n] \otimes_{R[\pi]} -, R\Lambda_n)$$

from §6.4 to the complex $C_*(\widetilde{X})$, we see that

$$H^q(X; \overline{V^D[\mathbb{Z}^n]}) \cong H^q(\mathrm{Hom}_{R\Lambda_n}(V[\mathbb{Z}^n] \otimes_{R[\pi]} C_*(\widetilde{X}; R), R\Lambda_n)),$$

for all q. We may approximate the right-hand side by means of the UCSS. On localizing at height 1 primes, the UCSS reduces to the Universal Coefficient Theorem for PIDs, and gives

$$TH^q(X; \overline{V^D[\mathbb{Z}^n]})_\wp \cong TH_{q-1}(X; V[\mathbb{Z}^n])_\wp,$$

for all height 1 primes $\wp$ and all q. Hence

$$\Delta_0(TH^q(X; \overline{V^D[\mathbb{Z}^n]})) \doteq \Delta_0(TH_{q-1}(X; V[\mathbb{Z}^n])),$$

for all q. The theorem now follows, since $\overline{TH_1(X; V[\mathbb{Z}^n])}$ is pseudo-isomorphic to $\overline{TH_1(X, \partial X; V[\mathbb{Z}^n])} \cong TH^2(X; \overline{V^D[\mathbb{Z}^n]})$. $\qquad\square$

When $n = 1$ the contributions from the boundary need not be pseudonull. We shall follow [**Ki96, KL99**] in using standard duality properties of torsion.

LEMMA 6.6. *Let* $\alpha : \mathbb{Z}^2 \to \mathbb{Z}$ *be nonzero, and* $\rho : \mathbb{Z}^2 \to GL(m, R)$. *Then* $\Delta_0(H_1(\mathbb{Z}^2; \rho \otimes \alpha)) \doteq \Delta_0(H_0(\mathbb{Z}^2; \rho \otimes \alpha))$, *and so* $\tau(\mathbb{Z}^2; \rho \otimes \alpha) \doteq 1$.

PROOF. We may assume that $\mathbb{Z}^2 = \langle x, y \rangle$, where $\alpha(x) = n > 0$ and $\alpha(y) = 0$. We then have $\Delta_0(H_0(\mathbb{Z}^2; \rho \otimes \alpha)) = \det(I - t^n \rho(x))$ and $\Delta_0(H_1(\mathbb{Z}^2; \rho \otimes \alpha)) = \det(t^n \rho(x) - I)$, while $H_i(\mathbb{Z}^2; \rho \otimes \alpha)) = 0$, for all $i \geq 2$. Hence $\tau(\mathbb{Z}^2; \rho \otimes \alpha) \doteq 1$. □

THEOREM 6.7. [**HSW10**] *Suppose that* $n = 1$ *and that* α *has nonzero restriction to each peripheral subgroup. If* ρ *is conjugate to its dual and* $\Delta(L; \rho \otimes \alpha) \neq 0$ *then* $\tau(X(L); \rho \otimes \alpha)$ *and* $\Delta(L; \rho \otimes \alpha)$ *are reciprocal.*

PROOF. Let $W = V^D$, and let $V(t) = F(t) \otimes V$ and $W(t) = F(t) \otimes W$, where $F(t)$ is the field of fractions of $R\Lambda$. The intersection pairing defined above induces non-singular pairings

$$H_q(X; V(t)) \times H_{3-q}(X, \partial X; W(t)) \to F(t).$$

Since the torsion $\tau(X, \rho \otimes \alpha)$ of C_* is defined, by our hypothesis, so is the torsion $\tau(X, \partial X; \rho^D \otimes \alpha)$ of D_*.

Choose a basis $\{v_i\}$ over F for V and lifts to $\widetilde{X}$ of simplices of X to get a preferred $F(t)$-basis for C_*. Basis elements have the form $1 \otimes v_i \otimes z_j$. The dual basis for W and the dual cells in $\widetilde{X}$ of the fixed lifts of simplices of X determine a basis for D_*. These bases are dual to the sesquilinear form $\langle -, - \rangle$ defined in Theorem 6.5.

Then $\tau(X; \rho \otimes \alpha) \doteq \tau(X, \partial X; \rho^D \otimes \bar{\alpha})$ by Theorem 1' of [**Mi62**]. Since the torsion is invariant under subdivision, we may also compute $\tau(X, \partial X; \rho^D \otimes \bar{\alpha})$ in terms of the given triangulation of $(X, \partial X)$. Then

$$\tau(X, \partial X; \rho^D \otimes \bar{\alpha}) = \tau(X, \partial X; \rho \otimes \bar{\alpha}) \quad \text{(since } \rho \text{ is conjugate to } \rho^D\text{)}$$
$$= \overline{\tau(X, \partial X; \rho \otimes \alpha)}$$
$$= \overline{\tau(X; \rho \otimes \alpha)},$$

using Lemma 6.6 and the multiplicativity of torsion in short exact sequences. Hence

$$\tau(X; \rho \otimes \alpha) \doteq \bar{\tau}(X; \rho \otimes \alpha).$$

The denominator of $\tau(X; \rho \otimes \alpha)$ is $\Delta_0 = \Delta_0(H_0(\pi; \rho \otimes \alpha))$, which is the $h.c.f.$ in $R\Lambda$ of $\{\det(t^{\alpha(g)}\rho(g) - I_m)|g \in \pi\}$. Since ρ is conjugate to its dual it follows that $\Delta_0 = \overline{\Delta_0}$. Hence $\Delta(L; \rho \otimes \alpha)$ is also reciprocal. $\square$

COROLLARY 6.7.1. [**Ki96, KL99**] *If ρ is orthogonal or unitary then $\Delta(L; \rho \otimes \alpha)$ is reciprocal.* $\square$

COROLLARY 6.7.2. *If $\rho : \pi \to \mathrm{Sp}_{2n}(\mathbb{C})$ is a symplectic representation, then $\Delta(L; \rho \otimes \alpha)$ is reciprocal.*

PROOF. The representation preserves the bilinear form given by

$$A = \begin{pmatrix} 0_n & I_n \\ -I_n & 0_n \end{pmatrix}.$$ $\square$

COROLLARY 6.7.3. *If the involution on R is trivial and $\rho : \pi \to \mathrm{SL}(2, R)$ then $\Delta(L; \rho \otimes \alpha)$ is reciprocal.*

PROOF. Since $V = R^2$ there is a natural skew-symmetric pairing $V \times V \to \wedge_2 V = R$. This is preserved by $\mathrm{SL}(2, R)$, and so Corollary 6.7.2 applies. $\square$

Corollary 6.7.3 shows that to find examples of representations into $\mathrm{SL}(n, R)$ with non-reciprocal invariants we must assume $n \geq 3$. The following example is from [**HSW10**]. Let $f(t) = t^2 - t - 1$ and let K be a knot with (untwisted) Alexander polynomial $\Delta(K) = f\bar{f}$. Let α be the abelianization, and let $\rho : \pi \to \mathrm{SL}(3, \mathbb{Z})$ be the representation with image generated by

$$C = \begin{pmatrix} 0 & 0 & -1 \\ 1 & 0 & 0 \\ 0 & 1 & 2 \end{pmatrix}.$$

(This is the companion matrix for $(t - 1)f(t)$.) Then

$$\Delta(\pi; \rho \otimes \alpha) = \tau(X; \rho \otimes \alpha) = (t - 1)(t + 1)^2(t^4 - 7t^2 + 1)(t^2 - t - 1),$$

which is not reciprocal.

6.6. Applications

Twisted Alexander polynomials associated to representations of knot groups in $SL(2, \mathbb{F}_7)$ were used to distinguish the two 11-crossing knots with Alexander polynomial 1 [**Wa94**]. They have since proven to be very powerful invariants in a wide range of contexts. As there is an excellent survey of such work in [**FV11**], we shall confine ourselves to brief outlines of some of these results.

It is noteworthy that a number of these applications use only representations ρ with finite image. Perhaps this is related to the fact that 3-manifold groups are residually finite. It should be noted, however, that the current proofs provide no bounds on the sizes of the representations that need to be considered.

6.6.1. Unknotting. If K is a nontrivial knot then πK admits a finite representation such that the corresponding twisted Alexander polynomial is not a unit [**SW06**]. The key topological fact is that some cyclic branched cover M_r has a finite regular cover which is not a homology sphere. This is clear if $\Delta_1(K)$ is not a unit. If $\delta_1(K) \doteq 1$ then either K is a satellite knot and $M_r(K)$ is Haken, for all $r \geq 2$, or K is hyperbolic (and not 4_1) and $M_r(K)$ is hyperbolic, for all $r \geq 3$ [**GL84**]. In all cases, $\pi_1(M_r(K))$ is infinite and residually finite, and so $M_r(K)$ has a regular covering of degree > 120. Since the only perfect group which acts freely on a homology 3-sphere is the binary icosahedral group, the covering space is not a homology 3-sphere. If Q is the covering group then the composite epimorphism $\pi' \to \pi_1(M_r) \to Q$ lifts to an epimorphism from π to $Q^r \rtimes Z/rZ$, and the associated permutation representation has the desired property. Friedl and Vidussi have extended this result to links with non-abelian group.

Distinct prime knots can be distinguished by their cyclic branched covers. (See [**BP08**].) Does it follow that any two distinct (prime?) knots can be distinguished by means of twisted Alexander polynomials?

6.6.2. Fibering and the Taubes conjecture. Let N be a 3-manifold such that ∂N is empty or a union of tori, and let α :

$\pi = \pi_1(N) \to \mathbb{Z}$ be an epimorphism. Then N fibres over S^1 with projection inducing α if and only if for each $\rho : \pi \to \mathrm{GL}(m, \mathbb{Z})$ with finite image $\Delta(\pi; \rho \otimes \alpha)$ is a monic polynomial and $m\|\alpha\|_T$ is the degree of $\tau(\pi; \rho \otimes \alpha)$. (Here $\|\alpha\|_T$ is the Thurston norm of α.) There is strong evidence suggesting that α is induced by a fibre bundle projection if $\Delta(\pi; \rho \otimes \alpha) \neq 0$, for all such ρ.

This leads to a proof of the Taubes Conjecture: if N is a closed orientable 3-manifold then $N \times S^1$ admits a symplectic structure if and only if N fibres over S^1 [**FV11'**]. Further work by Friedl and Vidussi provides some evidence suggesting that if M is a symplectic 4-manifold with a free S^1-action, then the orbit space is fibred.

6.6.3. Concordance. Standard duality properties of torsion imply that if $X = \partial W$ where W is an even-dimensional manifold, α and ρ each extend to $\pi_1(W)$ and ρ is conjugate to its dual then $\tau(X, \rho \otimes \alpha) = f\bar{f}$ for some $f \in F(t)$. This leads to a purely 3-dimensional definition of certain invariants of knot concordance. Let K be a 1-knot and let X_m and Y_m be the m-fold cyclic covering spaces of $X = X(K)$ and $Y = M(K)$, respectively. Let M_m be the m-fold cyclic branched covering space of S^3, branched over K. The composition of the inclusion of $\pi = \pi_1(Y_m)$ into $\pi_1(Y)$ with the Hurewicz homomorphism gives an epimorphism α from π to $m\mathbb{Z} \cong \mathbb{Z}$. The inclusion of X_m into Y_m induces an isomorphism on homology in degree ≤ 1, while the inclusion of X_m into M_m induces a surjection to $H_1(M_m; \mathbb{Z})$. Hence a character $\chi : H_1(M_m; \mathbb{Z}) \to Z/dZ$ determines a 1-dimensional representation V_χ of π in $\mathbb{Q}(\zeta_d)$. If K is a slice knot and $m = p^r$ and $n = q^s$, where p and q are primes, then $\Delta_1(Y_m, \alpha, V_\chi) = f(t)f(t^{-1})(t-1)^s$, for certain χ, while $\tau(Y_m, \alpha, V_\chi)$ is the determinant class of the Casson-Gordon Witt class defined in [**CG86**]. This work is used in [**KL99'**] to show that the knot 8_{17} is not concordant to its reverse -8_{17}.

6.6.4. Sublinks. Suppose that $L^+ = A \cup L$ is a $(\mu+1)$-component 1-link with a distinguished component A. Fix a basepoint $* \in X(+) = X(L^+)$, and let $X = X(L)$, $\pi = \pi L$ and $\pi(+) = \pi(L^+)$. Let $\alpha : \pi \to \pi/\pi'$ be the abelianization homomorphism, $f : \pi(+) \to \pi$

be the natural epimorphism and $\nu = f^{-1}(\pi')$. Then $\mathbb{Z}[\pi/\pi'] \cong \Lambda_\mu$ and $\mathbb{Z}[\pi(+)/\pi(+)'] \cong \Lambda_{\mu+1} = \Lambda_\mu[u, u^{-1}]$, where u is the image of a meridian m_A for the component A. The epimorphism f induces an epimorphism of Λ_μ-algebras $\psi : \Lambda_\mu[u, u^{-1}] \to \Lambda_\mu$ such that $\psi(u) = 1$.

Morifuji has used Wada's approach to establish a "Torres condition" relation between the twisted polynomials of L and L^+ associated to compatible unimodular representations. Let $\ell(i) = \mathrm{lk}(A, L_i)$, for $1 \le i \le \mu$, and let $\xi = \Pi_{i \le \mu} t_i^{\ell(i)}$.

THEOREM. [**Mo07**] *With the above notation, if $\rho : \pi \to \mathrm{GL}(m, R)$ then $\psi(\Delta(L^+; \rho f \otimes \alpha f)) = P(\xi)\Delta(L; \rho \otimes \alpha)$, where P is a polynomial of degree m. If $\rho(\pi) < \mathrm{SL}(m, R)$ and R is a field then we may assume that P is monic, and has constant term $(-1)^m$.*

He considers only the case $m = 2$ in detail, but is able to describe the coefficients of $\xi(L^+)$ for upper triangular representations. In this case the coefficients are sums of powers of products of eigenvalues of the images of a generating set for π.

6.6.5. Cyclically periodic links. The following two theorems are analogues of classical results of Murasugi ([**Mu71**]; see Theorem 9.9 and its corollary 9.9.1 below). In each case we assume that K is a knot which is invariant under a rotation of order q about a trivial knot A, $\lambda = \mathrm{lk}(A, K)$ and $\bar{L} = \bar{A} \cup \bar{K}$ is the orbit link.

THEOREM. [**HLN06**] *let $\bar{\rho} : \pi\bar{K} \to \mathrm{GL}(m, R)$, where $R = \mathbb{Z}$ or $\mathbb{Q}$. Let ρ be the lift of $\bar{\rho}$ to πK. Then there is a polynomial $F(t, u) \in R\Lambda_2$ such that*

$$\Delta_{K,\rho}(t) = \Delta_{\bar{K},\bar{\rho}}(t) \prod_{i=1}^{q-1} P(t, \zeta_q^i).$$

THEOREM. [**HLN06**] *Suppose that $q = p^r$, where p is a prime. Let $\bar{\rho} : \pi\bar{K} \to \mathrm{GL}(n, \mathbb{F}_p)$ and let ρ be the lift of $\bar{\rho}$ to πK. If $\Delta_{K,\rho}(t) \ne 0$, then*

$$\Delta_{K,\rho}(t) = \Delta_{\bar{K},\bar{\rho}}(t)^q (\delta_{L,\rho}(t)/D_{K,\rho}(t))^{q-1},$$

where $\delta_{L,\rho}(t) \in \mathbb{F}_p\Lambda$.

Here P is related to a twisted polynomial of $\bar{L}$. The term $\delta_{L,\rho}(t)$ is more complicated than in the classical case, but is often determined by data about K and ρ alone. The term $D_{K,\rho}(t)$ in the denominator is related to twisted homology in dimension 0 and is easily computed in examples. In the classical setting $D_{K,\rho}(t)$ is simply $(1-t)$ and $\delta_{L,\rho}(t) = 1 - t^\lambda$, so the classical δ_λ becomes our $\delta_{L,\rho}(t)/D_{K,\rho}(t)$.

Part 2

Applications: Special Cases and Symmetries

CHAPTER 7

Knot Modules

In this chapter and the next we shall consider the special cases $\mu = 1$ and $\mu = 2$, respectively. The knot theoretic case has been the most studied. In higher dimensions knot modules and the associated pairing provide complete invariants for significant classes of knots ([**Le70, Fa83**]), and the Witt class of the Blanchfield pairing on the middle dimensional knot module is a complete invariant of concordance for $(2q + 1)$-dimensional knots with $q \geq 1$ [**Le69, Ke75'**]. The question of which modules and pairings are realized by high-dimensional knots has been almost completely answered by Levine [**Le77**], and the algebraic study of such modules was pursued in [**Lev**]. We shall make some complementary observations on the structure of knot modules, and give some further detail on Blanchfield pairings and cyclic branched covers for classical knots. In the final section we use ribbon links to realize arbitrary Laurent polynomials with augmentation 1 as the first nonzero Alexander polynomials of the groups of higher-dimensional links.

7.1. Knot modules

Let $\Lambda = \Lambda_1 = \mathbb{Z}[t, t^{-1}]$ be the ring of integral Laurent polynomials in one variable, and let $R\Lambda = R \otimes_{\mathbb{Z}} \Lambda$, for any noetherian ring R. Then $R\Lambda$ is a PID if R is a field and is a factorial noetherian domain of global dimension 2 if R is a PID. Since Λ is finitely generated as a ring, its prime ideals are intersections of maximal ideals. Moreover, projective Λ-modules are free.

Let M be a finitely generated Λ-module and F a field. Then $FM = F \otimes M$ is a direct sum of cyclic modules, since $F\Lambda$ is a PID. In fact $FM \cong \oplus_{i \geq 1}(F\Lambda/\lambda_i(M))$, and so its structure is determined by the invariants $\lambda_i(M)$. However, Λ is not a PID, and the Λ-modules arising in knot theory need not be direct sums of cyclic modules.

Let zM denote the maximal finite Λ-submodule of M. Then zM is naturally isomorphic to $e^2 e^2 M$, by Theorem 3.22. Every finite Λ-module can be decomposed uniquely as a direct sum of m-primary modules, where the summation is over the maximal ideals m of Λ. By the Krull-Schmidt Theorem these summands are in turn direct sums of indecomposable modules in an essentially unique way. An m-primary finite Λ-module may be considered as a module over the m-adic completion $\hat{\Lambda}_m$. (The latter ring is isomorphic to a ring of power series $S[[X]]$ in one variable with coefficients in an unramified extension S of the p-adic integers, where $(p) = m \cap \mathbb{Z}$.)

If M is a torsion Λ-module then it is *pure* if $zM = 0$, i.e., if it has no nonzero finite submodule. A torsion module M is pure if and only if $E_0(M)$ is principal if and only if M has a square presentation matrix, by Theorems 7, 8, 21 and 22 of Chapter 3.

A *knot module* is a finitely generated Λ-module on which $t-1$ acts invertibly. The homology modules of the infinite cyclic covering of a knot complement are knot modules, by Theorem 2.1. A knot module M is a torsion Λ-module, since $t - 1$ acts invertibly on $\mathbb{Q} \otimes_{\mathbb{Z}} M$. Its $\mathbb{Z}$-torsion submodule is the direct sum of its p-primary components. Since $t - 1$ acts invertibly on these components, and since a finitely generated $\mathbb{F}_p \lambda$-module on which $t-1$ acts invertibly is finite induction on the exponent of the $\mathbb{Z}$-torsion shows that it is finite. Hence zM is the submodule of elements of finite additive order, and the pure torsion quotient M/zM is $\mathbb{Z}$-torsion free.

Knot modules are also modules over $L = \Lambda[(1-t)^{-1}]$, since $t - 1$ acts invertibly. Let $P = \mathbb{Z}[x]$ and embed P in L via $x \mapsto (1 - t)^{-1}$. Then $t = 1 - x^{-1}$, $\bar{x} = 1 - x$ and $L = P[(x\bar{x})^{-1}]$. A *lattice* in a pure knot module M is a P-submodule N which is finitely generated over $\mathbb{Z}$ and such that $M \cong L \otimes_P N$. (See also §4 of Chapter 11 below.)

THEOREM 7.1. *Every pure knot module M has lattices.*

PROOF. We may assume that M is annihilated by a polynomial $\lambda(t)$ such that $\lambda(1) = 1$. If λ has degree n then $\xi(x) = x^n \lambda(1 - x^{-1})$ is in P and annihilates M. This is a monic polynomial, and so any finitely generated P-module annihilated by ξ is finitely generated over $\mathbb{Z}$. In particular, if $B = \{m_1, \ldots, m_g\}$ generates M as a λ-modukle and N is the P-submodule generated by B then N is finitely generated over $\mathbb{Z}$ and $M \cong L \otimes_P N$. $\qquad\qquad\square$

7.2. A Dedekind criterion

If θ is an irreducible element of Λ then (θ) is a prime ideal, and $R_\theta = \Lambda/(\theta)$ is a 1-dimensional noetherian domain. It is then a subring of the algebraic number field $K_\theta = \mathbb{Q}\Lambda/(\theta)$. If R_θ is integrally closed (and so a localization of the ring of integers in K_θ) we shall call θ a *Dedekind element*.

THEOREM 7.2. *Let R be a Dedekind domain with infinitely many prime ideals. If $\mathfrak{m}$ is a maximal ideal of $R[Y]$ then $\mathfrak{m} = (\wp, g(Y))$ for some prime $\wp$ in R and some polynomial $g(Y)$ whose image in $R/\wp[Y]$ is irreducible.*

PROOF. Let $F = R_0$ be the field of fractions of R. If $\mathfrak{m} \cap R = 0$ then $\mathfrak{m}_0 = \mathfrak{m}F[Y]$ is a proper maximal ideal of $F[Y]$, and so is principal. Therefore after enlarging the coefficient ring R to A by localizing away from finitely many primes of R we may assume that $\mathfrak{m} = (f)$ for some non-constant polynomial f. Let $\mathfrak{q}$ be a nonzero prime of A and let p in A generate the maximal ideal of $A_\mathfrak{q}$. Then p maps to a nonzero element of the field $A[Y]/(f)$, so $pg - 1 = hf$ for some g, h in $A[Y]$. Therefore f maps to a unit in $(A_\mathfrak{q}/\mathfrak{q})[Y]$, and so one coefficient of f is a unit in $A_\mathfrak{q}/\mathfrak{q}$, while all the other coefficients are in $\mathfrak{q}$. But at least one of these coefficients is nonzero and so is divisible by only finitely many primes. Since A has infinitely many primes, this gives a contradiction. Hence $\mathfrak{m} \cap R$ is a nonzero prime ideal. The rest follows easily. $\qquad\square$

If R has only finitely many primes $\wp_1, \ldots, \wp_n$ then it is a PID. If p_i generates $\wp_i$, for $1 \leq i \leq n$ then $m = 1 - Y\Pi p_i$ generates a maximal ideal with quotient field $R[Y]/(m) \cong F$.

COROLLARY 7.2.1. *The maximal ideals in $\mathbb{Z}[t]$ or Λ have the form $(p, g(t))$, where p is a prime number and g is a polynomial with integral coefficients whose image in $\mathbb{F}_p[t]$ is irreducible.* $\qquad\square$

In particular, the localizations of $\mathbb{Z}[t]$ or Λ at maximal ideals are regular noetherian domains of dimension 2. (See also page 123 of [**AM**].)

An interesting consequence is that if S is the multiplicative system generated by $\Pi \cup \{t^n - 1 \mid (n, p) = 1 \; \forall p \in \Pi\}$, where Π is a set of rational primes, then $\mathbb{Z}[t]_S$ is a PID. For if d is the degree of $g(t)$ and $n = p^d - 1$ then $\mathfrak{m}$ must contain $t^n - 1$. (The case $\Pi = \emptyset$ inverts all cyclotomic polynomials, while at the other extreme we obtain $\mathbb{Q}[t]$.)

Taking instead $R = k[X]$, where k is an algebraically closed field, we obtain the "Weak Nullstellensatz". (The argument extends easily to the case of more than two variables.)

COROLLARY 7.2.2. *If k is an algebraically closed field and $\mathfrak{m}$ is a maximal ideal of $k[X, Y]$ then $\mathfrak{m} = (X - \alpha, Y - \beta)$ for some $\alpha, \beta \in k$.* □

The next result corresponds to an algebraic criterion for smoothness of a point on a plane curve. (See Chapter 10.)

THEOREM 7.3. *Let θ be an irreducible element of Λ. Then θ is Dedekind if and only if it is not contained in $\mathfrak{m}^2$, for any maximal ideal $\mathfrak{m} = (p, g(t))$ such that p divides the resultant $R(\theta, \theta')$ and the image of $g(t)$ divides the image of θ in $\mathbb{F}_p \Lambda$.*

PROOF. The maximal ideals $\mathfrak{m}$ of Λ which contain θ correspond bijectively to the maximal ideals $\mathfrak{n}$ of R_θ under the canonical epimorphism. It shall suffice to show that for such an $\mathfrak{n}$ the localization $R_{\theta, \mathfrak{n}}$ is a discrete valuation ring if and only if θ is not in $\mathfrak{m}^2$. Let $p = \mathfrak{m} \cap \mathbb{Z}$ and $S = \Lambda_\mathfrak{m}$. Then S is a local ring with maximal ideal generated by p and $g(t)$, for some $g(t)$ representing an irreducible factor of the image of $\theta \bmod p$. Since $0 < (\theta) < \mathfrak{m}S$ is a chain of distinct prime ideals, $\mathfrak{m}S$ cannot be principal, and so $\mathfrak{m}S/(\mathfrak{m}S)^2$ has dimension 2 as a vector space over the field $S/\mathfrak{m}S$, by Nakayama's Lemma. The maximal ideal of $R_{\theta, \mathfrak{n}}$ is $\mathfrak{m}S/(\theta)$ and so is principal if and only if there is some s in S such that $\mathfrak{m}S = (\theta, s)$. In this case the images of θ and s in $\mathfrak{m}S/(\mathfrak{m}S)^2$ would form a basis, so θ is not in $\mathfrak{m}^2$. Conversely if θ is not in $\mathfrak{m}^2$ then there is some s in S such that the images of θ and s generate $\mathfrak{m}S/(\mathfrak{m}S)^2$, and hence $\mathfrak{m}S = (\theta, s)$, by Nakayama's Lemma again.

If θ is in $\mathfrak{m}^2$ then the derivative θ' is in $\mathfrak{m}$, so the images of θ and θ' in $\mathbb{F}_p\Lambda$ have a common root in an extension of $\mathbb{F}_p$. Thus p must divide the resultant $R(\theta, \theta')$, and the image of $g(t)$ must divide the image of θ in $\mathbb{F}_p\Lambda$. $\square$

This provides a quick method for showing that rings of cyclotomic integers are integrally closed, without first computing the discriminant. Let $\Phi_n \in \mathbb{Z}[Y]$ be the nth cyclotomic polynomial, and ζ_n a primitive n^{th} root of unity. Since $X^n - 1$ (and hence Φ_n) has distinct roots over any field of characteristic prime to n, the only primes dividing $disc(\Phi_n)$ are divisors of n. If $n = mq$ with $q = p^r$ and $(m, p) = 1$ then $\Phi_n(X) = \Phi_m(X^q)/\Phi_m(X^{q/p})$ so $\Phi_n \equiv \Phi_m^{\varphi(q)}$ $mod\ (p)$. Moreover, $\Phi_n(X)$ divides $\Phi_p(X^{n/p})$ and so $\Phi_n(\zeta_m)$ divides $\Phi_p(1) = p$. Therefore Φ_n is not in $(p, g)^2$ for any g representing an irreducible factor of Φ_m in $\mathbb{F}_p[Y]$, and so $\mathbb{Z}[\zeta_n] = \mathbb{Z}[Y]/(\Phi_n)$ is the full ring of integers in $\mathbb{Q}(\zeta_n)$.

In general an irreducible element need not be Dedekind but it is at least true that every maximal ideal of R_θ can be generated by at most two elements (since this is so for Λ) and so every finitely generated torsion free R_θ-module is reflexive. (See Theorem 41 of [**Mat**].) Moreover, R_θ is Gorenstein, since it has a finite injective resolution

$$0 \to R_\theta \to K_\theta \to K_\theta/R_\theta \to 0$$

over itself. (See Theorem 29 of [**Mat**].) This may be of relevance in studying the duality pairings arising in knot theory.

7.3. Cyclic modules

Since Λ is factorial the conditions $E_0(M) = 0$ and $E_1(M) = \Lambda$ characterize free cyclic modules, by Corollary 3.7.1.

If M is a cyclic torsion Λ-module then clearly $E_1(M) = \Lambda$ and for each factor θ of $\Delta_0(M)$ the quotient module $M/\theta M$ is pure cyclic. In particular, if θ is irreducible the SFS invariants of $M/\theta M$ (considered as a module over $R = \Lambda/(\theta)$) are both 1. If

$\theta = \phi\psi$ then $M/\theta M$ may be considered as an extension of $\Lambda/(\psi)$ by $\Lambda/(\phi)$. Such extension modules have a presentation matrix $\begin{pmatrix} \phi & 0 \\ \eta & \psi \end{pmatrix}$ for some η and are determined up to isomorphism by the class of η in the quotient of $Ext^1_\Lambda(\Lambda/(\psi), \Lambda/(\phi)) \cong \Lambda/(\phi, \psi)$ by the action of $Aut(\Lambda/(\psi)) \times Aut(\Lambda/(\phi)) \cong (\Lambda/(\psi))^\times (\Lambda/(\phi))^\times$.

If ϕ and ψ have no common factor this group is finite. If also $E_1(N) = \Lambda$ the class of η in $\Lambda/(\phi, \psi)$ is a unit, and so N is determined by a unit $u_{\phi,\psi(N)}$ in $U(\phi, \psi) = (\Lambda/(\phi, \psi))^\times / (\Lambda/(\psi))^\times (\Lambda/(\phi))^\times$. (If $(\phi, \psi) = \Lambda$ we set $U(\phi, \psi) = 1$.) For the module N to be cyclic we must have $u(N) = 1$. Modules P which are extensions of a pure cyclic module $\Lambda/(\Delta)$ by a finite cyclic module Λ/J ($= zP$) are similarly classified by an invariant $u_{\Delta,J}(P)$ in the quotient of the unit group $(\Lambda/(\Delta, J))^\times$ by its subgroup $(\Lambda/(\Delta))^\times (\Lambda/J)^\times$. In this section we shall show that these necessary conditions for a Λ-module to be a cyclic pure torsion module are also sufficient. (However, we do not know how to compute $u(N)$ from an arbitrary presentation matrix for N if $U(\phi, \psi) \neq 1$.)

LEMMA 7.4. *Let M be a pure torsion Λ-module with $E_1(M) = \Lambda$. Let $\Delta_0(M) = \Pi_{i=1}^{i=n}\delta_i^{e(i)}$ be the factorization of $\Delta_0(M)$ into powers of distinct irreducibles, and suppose that $M/\delta_i M \cong \Lambda/(\delta_i)$, for all $1 \leq i \leq n$. Then M has a composition series $\{M_i \mid 0 \leq i \leq n\}$ such that $M_i/M_{i-1} \cong \Lambda/(\delta_i^{e(i)})$, for all $1 \leq i \leq n$.*

PROOF. We shall induct on n, the number of distinct irreducible factors of $\Delta_0(M)$. Let $M_n = M$ and $N = M/\delta_n^{e(n)}M$. Then $P = N/TN$ is a pure δ_n-primary Λ-torsion module such that $\Delta_0(P) = \delta_n^{e(n)}$ and $P/\delta_n P \cong M/\delta_n M \cong \Lambda/(\delta_n)$, and so $P \cong \Lambda/(\delta_n^{e(n)})$, by Nakayama's Lemma. Let M_{n-1} be the kernel of the projection of M onto P. It shall suffice to show that M_{n-1} satisfies assumptions similar to those on M. Certainly $\Delta_0(M_{n-1}) = \Pi_{i=1}^{i=n-1}\delta_i^{e(i)}$. Since M_n/M_{n-1} is cyclic, consideration of presentation matrices as in Lemma 3.12 shows that $E_1(M_n) \leq E_1(M_{n-1})$. Hence $E_1(M_{n-1}) = \Lambda$. Finally if for each $i < n$ we apply the Snake Lemma to the endomorphism of the short exact sequence

$$0 \to M_{n-1} \to M_n \to \Lambda/(\delta_n^{e(n)}) \to 0$$

given by multiplication by δ_i we obtain an exact sequence of cokernels

$$0 \to M_{n-1}/\delta_i M_{n-1} \to M_n/\delta_i M_n \cong \Lambda/(\delta_i) \to \Lambda/(\delta_i, \delta_n^{e(n)}) \to 0,$$

from which it follows that $M_{n-1}/\delta_i M_{n-1} \cong \Lambda/(\delta_i)$, for $i < n$. $\square$

An easy induction now gives $p.d._\Lambda M \leq 1$, i.e., $e^2 M = 0$.

THEOREM 7.5. *A Λ-torsion module M is cyclic if and only if $E_1(M) = \Lambda$, $\rho(M/\delta M) = 1$ for all irreducible factors δ of $\Delta = \Delta_0(M)$, $u_{\phi,\psi}(M/\psi M) = 1$ for any two factors ϕ and ψ of Δ such that $\phi\psi$ is square free, and $u_{\Delta,J}(M) = 1$, where $J = \Delta^{-1} E_0(M)$.*

PROOF. The conditions are clearly necessary. Suppose that they hold. If P is an $m \times n$ presentation matrix for M and θ divides $\Delta_0(M)$ then $\binom{P}{\theta I_n}$ is a presentation matrix for $M/\theta M$. Therefore $E_0(M/\theta M) = (E_0(M), \theta E_1(M), \ldots) = (\theta)$, since $E_1(M) = \Lambda$, so $M/\theta M$ is a pure module. Let θ be the product of the distinct irreducible factors of $\Delta_0(M)$. Since the SFS invariants are trivial, the lemma implies that $M/\theta M$ has a composition series with pure cyclic factors. By a finite induction on the number of factors of $\Delta_0(M)$, and using the conditions $u_{\phi,\psi}(M/\phi\psi M) = 1$ repeatedly, it follows that $M/\theta M$ is cyclic. Nakayama's Lemma then implies that $M/\Delta M \cong \Lambda/(\Delta_0(M))$.

Consideration of presentation matrices (as in Lemma 3.12) gives $E_1(M) = (\Delta E_1(zM), E_0(zM)) \leq E_1(zM)$, so $E_1(zM) = \Lambda$. The finite module zM may be regarded as a module over the ring $R = \Lambda/Ann(zM)$. Since this ring is artinian, $R/rad\,R \cong \Pi(\Lambda/m_i)$ is a finite product of fields. Since $E_0(zM) \leq Ann(zM)$, by part (1) of Theorem 3.1, and $E_1(zM) = \Lambda$ we have $zM/m_i zM \cong \Lambda/m_i$. Hence $zM/\sqrt{Ann(zM)}zM \cong R/rad\,R$. Nakayama's Lemma now implies that zM is cyclic. If $zM \cong \Lambda/J$ we must have $E_0(M) = \Delta J$.

The theorem now follows on using the condition $u_{\Delta,J}(M) = 1$. $\square$

Let $\delta = 13t^2 - 25t + 13$ and J be the ideal generated by the image of $(3, t+1)$ in $R = \Lambda/(\delta) \cong \mathbb{Z}[\frac{1}{13}, \frac{1+\sqrt{-51}}{2}]$. As a Λ-module J has a presentation matrix $P_J = \begin{pmatrix} 3 & t+1 \\ 13(t+1) & 17 \end{pmatrix}$. It is a pure torsion module and $E_0(J) = (\delta)$ and $E_1(J) = \Lambda$, but $\rho(J) = [J]$ is nontrivial and J is not cyclic.

The module M with presentation matrix δP_J is pure, δ-primary, has all elementary ideals principal, and $M/\delta M \cong (\Lambda/(\delta))^2$, but $\delta M \cong J$ and M does not have a composition series with pure cyclic factors. For otherwise it would have a 3×3 presentation matrix of the form

$$\begin{pmatrix} \delta & 0 & 0 \\ \alpha & \delta & 0 \\ \beta & \gamma & \delta \end{pmatrix}.$$

Then $\alpha\gamma - \beta\delta \in E_1(M)$ would imply that either α or γ is divisible by δ, and so may be assumed 0. In either case it would follow easily that $\delta M \cong \Lambda/(\delta)$.

Let $\psi = 5t^2 - 9t + 5$ and $\theta = t^2 - t + 1$. Then ψ and θ are Dedekind, and $\Lambda/(\psi)$ and $\Lambda/(\theta)$ are PIDs. The units of $\Lambda/(\psi)$ are represented by $\{\pm(5t - 4)^n \mid n \in Z\}$, while those of $\Lambda/(\theta)$ are represented by $\{\pm 1, \pm t, \pm(t - 1)\}$. The element $\sigma = 1 + 2t$ represents a unit of $\Lambda/(\psi, \theta)$ ($\cong \mathbb{Z}/4\mathbb{Z}[\tau]$ where $\tau^2 = \tau - 1$) which is not in the subgroup generated by the images of $(\Lambda/(\psi))^\times$ and $(\Lambda/(\theta))^\times$. Therefore the module N presented by $\begin{pmatrix} \psi & 0 \\ \sigma & \theta \end{pmatrix}$ is not cyclic, although $E_0(N) = (\psi\theta)$, $E_1(N) = \Lambda$, $N/\psi N \cong \Lambda/(\psi)$ and $N/\theta N \cong \Lambda/(\theta)$ (and thus the SFS invariants of these quotients are all 1).

Let $\delta = t^2 - t + 1$. Then the module with presentation matrix $\begin{pmatrix} \delta^2 & 0 \\ 2\delta & \delta^2 \end{pmatrix}$ is pure, δ-primary and $M/\delta M \cong (\Lambda/(\delta))^2$, but $E_1(M)$ is not principal and so M is not a direct sum of cyclic modules.

Let $\delta = 13t^2 - 25t + 13$ and M_n be the module presented by $\delta^n \begin{pmatrix} 3 & t+1 \\ 13(t+1) & 17 \end{pmatrix}$ over Λ. Then $E_0(M_n) = (\delta^{2n+1})$, $E_1(M_n) = (\delta^n)$ and $M_n/\delta^n M_n \cong (\Lambda/\delta^m)^2$, but $\delta^n M_n \cong J$ and so M_n is not a direct sum of cyclic modules.

7.4. Recovering the module from the polynomial

In this section we shall show that the Alexander polynomial $\Delta_0(M)$ of a pure knot module M determines the module up to finite ambiguity if and only if $\lambda_1(M)$ has no repeated factors.

If M is a finitely generated Λ-module and $\delta \in \Lambda$ is irreducible the (δ)-primary submodule of M is $M(\delta) = \{m \in M \mid \delta^n m = 0 \; \forall n \gg 0\}$. If M is pure and $(\delta) \neq (\delta')$ then $M(\delta) \cap M(\delta') = 0$, and so there is a monomorphism $\oplus M(\delta) \to M$. However, M need not be the direct sum of its primary submodules. It is at least an iterated extension of pure primary modules.

LEMMA 7.6. *Let M be a pure knot module and let $\Delta = \Pi \delta_i^{e(i)}$ be a factorization of $\Delta = \Delta_0(M)$ into irreducibles. Then M has a filtration $0 = M_0 < \cdots < M_n = M$ such that M_i/M_{i-1} is a pure knot module with $E_0(M_i/M_{i-1}) = (\delta_i)^{(e(i))}$, for $1 \leq i \leq n$. Conversely if M admits such a filtration it is a pure knot module, and $E_0(M)$ is the product of the 0th elementary ideals of the subquotients.*

PROOF. Let $M_0 = 0$. If M_j has been determined for $j < i$ and M/M_{i-1} is pure, let M_i be the preimage in M of $(M/M_{i-1})(\delta_i)$. Then M_i is finitely generated, since Λ is noetherian, and M_i/M_{i-1} is a finitely generated Λ-torsion module annihilated by a power of δ_i. It is easily seen by induction on i that M/M_i is pure, and $M/M_i(\delta_j) = 0$ for $j \leq i$. Therefore $M_n = M$, since M/M_n is annihilated by Δ. Moreover $E_0(M_n) = \Pi_{i=1}^{i=n} E_0(M_i/M_{i-1})$. Hence $E_0(M_i/M_{i-1}) = (\delta_i)^{e(i)}$, for each $1 \leq i \leq n$. The converse is straightforward. $\square$

THEOREM 7.7. *Let $\Delta \in \Lambda$ have augmentation $\varepsilon(\Delta) = 1$. There are only finitely many knot modules M with $E_0(M) = (\Delta)$ and $Ann(M) = \sqrt{(\Delta)}$.*

PROOF. Let $\Delta = \Pi \delta_i^{e(i)}$ be the factorization into irreducibles. We may assume that the factors δ_i are polynomials of degree $d(i)$, with nonzero constant term. If $E_0(M) = (\Delta)$ then M is a pure knot module. Suppose that $Ann(M) = \sqrt{(\Delta)} = (\Pi \delta_i)$. Let $M = \cup M_i$ be a filtration as in the Lemma, and for each $1 \leq j \leq n$ let $N_j = M_j/M_{j-1}$. Then $E_0(N_j) = (\delta_j)^{e(j)}$ and so N_j is a pure knot module also. Moreover $Ann(N_j) = (\delta_j)$, and so we may consider N_j as a torsion free R_j-module, of rank $e(j)$, where $R_j = \Lambda/(\delta_j)$. Let α_j be a root of δ_j, and let $\beta_j = (1 - \alpha_j)^{-1}$. Then β_j is an algebraic integer, since it is a root of $t^{d(j)} \delta_j (1 - t^{-1})$. Since $(t - 1)$ acts invertibly on N_j we may view N_j as a finitely generated torsion free S_j-module, where $S_j = \mathbb{Z}[\beta_j, \beta_j^{-1}, (\beta_j - 1)^{-1}] = \mathbb{Z}[\alpha_j, \alpha_j^{-1}, (1 - \alpha_j)^{-1}]$. Since S_j is a localization of an order in the number field $\mathbb{Q}\Lambda/(\delta_j)$ the Jordan-Zassenhaus Theorem holds, and so N_j is determined up to a finite ambiguity by its rank.

If P and Q are finitely generated Λ-modules then $Ext_\Lambda^1(P, Q)$ is finitely generated and is annihilated by $(Ann(P), Ann(Q))$. Therefore if P and Q are pure knot modules such that $E_0(P)$ and $E_0(Q)$ have no common factors then $Ext_\Lambda^1(P, Q)$ is finite. The theorem now follows by a finite induction. □

In particular, if Δ has no repeated factors there are only finitely many pure Λ-torsion modules with Alexander polynomial Δ.

On the other hand, if Δ has repeated factors, then there are infinitely many isomorphism classes of finitely generated pure Λ-torsion modules M with $E_0(M) = (\Delta)$. To see this it suffices to assume that Δ is a power of a single irreducible element, $\Delta = \delta^r$, say, with $r \geq 2$. Let $0 \leq s \leq r/2$. For each $u \in \Lambda$ let M_u be the Λ-module with generators e, f and g and relations $\delta^{s-r} e = \delta^{r-1} f = \delta g + uf = 0$. Then $E_0(M_u) = (\Delta)$ and $E_1(M_u) = \delta^s(\delta, u)$. Thus $Ann(M_u) = (\delta^r)$ if u is not divisible by δ, while if u and v have distinct residue classes $mod\ (\delta)$ then M_u and M_v are not isomorphic. (However, the modules $\mathbb{Q} \otimes M_u$ are all isomorphic.)

Although the orders λ and ξ of a knot module M and of a lattice in M are related by a change of variables, they can have different algebraic behaviour. In general, $P/(\xi) \leq \Lambda/(\lambda) = P/(\xi)[(x\bar{x})^{-1}]$. For instance, $\lambda(t) = 9t^4 - 3t^3 - 11t^2 - 3t + 9$ is a knot polynomial, and is irreducible since it is irreducible $mod\ (2)$. The ring $\Lambda/(\lambda)$ is a Dedekind domain. However, $\xi(x) = x^4\lambda(1 - x^{-1})$ is in $(3, x - 2)^2$, and so $P/(\xi)$ is not Dedekind.

7.5. Homogeneity and realizing π-primary sequences

A pure torsion module M is *homogeneous* if for each irreducible factor δ of $\Delta_0(M)$ there is a $k \geq 1$ such that if δ^d is the power of δ dividing $\Delta_1(M)$ then δ^d divides $\lambda_i(M)$ for $1 \leq i \leq k$, while δ does not divide $\lambda_{k+1}(M)$. (The pair (d, k) may vary with the factor δ.) It is of type λ if $\Lambda_k(M) = \lambda$ or 1 for all $k \geq 1$ are either λ or 1.

THEOREM 7.8. *A homogeneous Λ-module M of type π^e, with π irreducible, is free as a module over $\Lambda/(\pi^e)$ if and only if all of its elementary ideals are principal and $\rho(M/\pi M) = 1$.*

PROOF. The conditions are clearly necessary. Conversely, they imply that $M/\pi M$ satisfies the hypotheses of Corollary 3.7.1 for some r and so is free over $\Lambda/(\pi)$. By the homogeneity of M and Nakayama's Lemma there is a surjection $(\Lambda/(\pi^e))^r \to M$, which must be an isomorphism since the kernel is a pure Λ-module, and its 0th elementary ideal is Λ, by Lemma 3.12.   □

This extends Lemma 15.1 of [**Lev**] which assumes that πM is the kernel of multiplication by π^{e-1} and that $M/\pi M$ is free. Let $\delta = t^2 - t + 1$. Then the module with presentation matrix $\left(\begin{smallmatrix} \delta & 0 \\ 2 & \delta \end{smallmatrix} \right)$ is pure, δ-primary and homogeneous, but does not satisfy this assumption.

By similar arguments we may relax the hypothesis of Levine's π-primary sequence realization theorem (§10-16 of [**Lev**]) from "$\Lambda/(\pi)$ is Dedekind" to "*the lower derivative $\Lambda/(\pi)$-modules involved are projective*". The key constructive step is Lemma 14.1 of [**Lev**], which we may restate as follows.

LEMMA 7.9. *Let π be an irreducible element of Λ and M a finitely generated projective $\Lambda/(\pi)$-module, and let $R = \Lambda/(\pi^e)$. Then there is a projective R-module $\tilde{M}$ such that $\tilde{M}/\pi\tilde{M} \cong M$.*

PROOF. We may assume that $M \cong (\Lambda/(\pi))^{r-1} \oplus I$ for some ideal I of $\Lambda/(\pi)$, by the Stable Range Theorem [**Ba64**]. Clearly we may assume in fact that $M = I$. Since I can be generated by two elements, by Corollary 3.20.1, the rest of Levine's construction applies.   □

We may similarly improve Lemma 15.2 of [**Lev**] to a characterization of projective R-modules. If M is homogeneous of type π^e and $M/\pi M$ is projective of rank r over $\Lambda/(\pi)$ then $E_{r-1}(M) \leq (\pi^e)$ while $(E_r(M), \pi) = \Lambda$, and so also $(E_r(M), \pi^e) = \Lambda$. Thus the image in R of $E_{r-1}(M)$ is 0, while that of $E_r(M)$ is R, and so M is projective as an R-module. The hypotheses of Lemmas 15.3 and 15.5

of [**Lev**] may be altered accordingly. (Lemma 15.5 follows from 15.3 by Nakayama's Lemma, as in Theorem 7.8 above.)

The proof of the realization theorem may now be completed as in Section 16 of [**Lev**].

We may give another partial answer to our general question.

THEOREM 7.10. *Let M be a homogeneous pure Λ-torsion module such that*

    (1) *all of its elementary ideals are principal;*

    (2) $\rho(M/\delta M) = 1$ *for each irreducible factor δ of $\Delta_0(M)$;*

    (3) *if δ and δ' are distinct irreducible factors of $\Delta_0(M)$ then* $(\delta, \delta') = \Lambda$.

Then M is a direct sum of cyclic modules and satisfies the Elementary Divisor Theorem.

PROOF. Condition (3) implies that M is the direct sum of its primary submodules. Conditions (1) and (2) with Theorem 7.8 then imply that these are direct sums of cyclic modules. Finally the Elementary Divisor Theorem follows on using (3) again. $\square$

Conditions (1) and (2) of the Theorem are evidently necessary and are easily checked. Condition (3) simplifies our task in two ways. On the one hand we thereby avoid extension problems. On the other hand, it enables us to identify $M(\delta)/\delta M(\delta)$ with $M/\delta M$ and so to determine $M(\delta)$ (by Theorem 7.5) directly from a presentation for M. In general it seems difficult to use a presentation for a module to determine the structure of its primary submodules (apart from their polynomial invariants). Even if the elementary ideals and SFS invariants of a primary module are principal, it need not be a sum of cyclic modules. However, if so it satisfies the Elementary Divisor Theorem.

7.6. The Blanchfield pairing

We shall describe the Blanchfield pairing of a classical knot K from a surgical point of view. The archetype of all subsequent applications of constructive surgery to knot theory is the following result of Levine [**Le65**].

THEOREM 7.11. *Let $\lambda \in \Lambda$ be such that $\bar{\lambda} = \lambda$ and $\varepsilon(\lambda) = 1$. Then there is a 1-knot K such that $B(K) \cong \Lambda/(\lambda)$.*

PROOF. Let $X_o = X(U) \cong S^1 \times D^2$, where U is a trivial knot. Suppose that $\lambda = \Sigma_{0 \leq i \leq s} a_i(t^i + t^{-i})$. Let $K_0, \ldots, K_s$ be $s + 1$ disjoint trivial knots contained in a 3-disc $D \subset X_o$, and such that $K_1 \amalg \cdots \amalg K_s$ is a trivial s-component link while $\mathrm{lk}(K_0, K_i) = a_i$ for $1 \leq i \leq s$. Form a new knot C by taking the sum of the knots $K_0, \ldots, K_s$ along bands B_i running from K_{i-1} to K_i (for $1 \leq i \leq s$) and which each pass once around the solid torus X_o in the positive direction (relative to the meridian of U). We may choose the twisting of the bands so that C is unknotted in $S^3 = X_o \cup D^2 \times S^1$. Note also that $\mathrm{lk}(C, U) = 0$. Let N be a regular neighbourhood of C in X_o. Let h be a self-homeomorphism of the solid torus $X(C) = S^3 \setminus int\, N$ which twists it once positively. (In other words, if $g : X(C) \to S^1 \times D^2$ is a homeomorphism which preserves the ambient orientation and the meridian then $ghg^{-1}(z, d) = (z, zd)$ for all $(z, d) \in S^1 \times D^2$.) Since the image of U lies in $X(C)$ we may define a new knot K by $K = h \circ U$.

Since U is trivial $X'_o \cong \mathbb{R} \times D^2$. Let $Z = X(U \cup C) = X_o \setminus int\, N$ and let Z' be the infinite cyclic covering space induced by the inclusion $Z \subset X_o$. Since $\mathrm{lk}(C, U) = 0$ the inclusion of N into X_o lifts to X'_o. Let d be a meridianal disc for N with oriented boundary m, and which is contained in D. (Thus m is a meridian for C.) Let $\tilde{N}$ be a fixed lift of N to X'_o and let $\tilde{c}$ and $\tilde{d}$ be the induced lifts of C and d. Let α and $\tilde{m}$ be the lifts of m and $h(m)$ (respectively) to $Z' \subset X'_o$ which meet $\tilde{d}$. Let $\tilde{D}$ be the lift of D which meets $\tilde{N}$, and let $\tilde{K}_0, \ldots, \tilde{K}_s$ be the lifts of $K_0, \ldots, K_s$ to $\tilde{D}$. An easy excision argument shows that $H_2(X_o, Z; \Lambda)$ is freely generated by the relative 2-cycle corresponding to $\tilde{d}$. Moreover $H_2(X_o, Z; \Lambda) \cong H_1(Z; \Lambda)$, since X'_o is contractible, and $H_1(Z; \Lambda)$ is generated by α.

We may recover $X(K)'$ from Z' by adjoining copies of lifts of N along lifts of $h|_{\partial N}$. Thus $B(K) \cong H_1(Z; \Lambda)/\Lambda\tilde{m}$. Since $int\, X'_o \cong \mathbb{R}^3$ we may apply linking number arguments to identify $\tilde{m}$. Clearly $\mathrm{lk}(\alpha, t^i \tilde{c}) = 1$ if $i = 0$ and is 0 otherwise, and so $\tilde{m} = \Sigma \mathrm{lk}(\tilde{m}, t^i \tilde{c}) t^i \alpha$. Now $\tilde{c}$ is the sum of the knots $\tilde{K}_0, t\tilde{K}_1, \ldots, t^s \tilde{K}_s$ along lifts of the bands B_i. If $i \neq 0$ then $\tilde{m}$ is homotopic to $\tilde{c}$ in the complement of $t^i\tilde{c}$, and so $\mathrm{lk}(\tilde{m}, t^i\tilde{c}) = \mathrm{lk}(\tilde{c}, t^i\tilde{c}) = a_{|i|}$. Hence $\tilde{m} = \lambda'\alpha$, where $\lambda' = b_0 + \Sigma_{1 \leq i \leq s} a_i(t^i + t^{-i})$. If $\tilde{\beta}, \tilde{\gamma}$ are loops in $\mathbb{R}^3 = int\, X'_o$ which project to disjoint loops β, γ in $S^1 \times \mathbb{R}^2 \subset \mathbb{R}^3$ then $\mathrm{lk}(\beta, \gamma) = \Sigma\mathrm{lk}(\tilde{\beta}, t^k\tilde{\gamma})$. Therefore $1 = \mathrm{lk}(m, c) = \Sigma\mathrm{lk}(\tilde{m}, t^i\tilde{c})$ and so $b_0 + 2\Sigma_{1 \leq i \leq s} a_i = 1 = a_0 + 2\Sigma_{1 \leq i \leq s} a_i$. Hence $\lambda' = \lambda$ and $B(K) \cong \Lambda/(\lambda)$. $\square$

Surgery constructions have been applied to a variety of other situations. For instance, every Laurent polynomial $\lambda \in \Lambda$ such that $\overline{\lambda} = \lambda$ and $\varepsilon(\lambda) = 1$ is realized by a strongly invertible knot with unknotting number 1 [**Sa83**]. If λ is also a monic polynomial then it is realized by a fibred 1-knot [**Bu66**], and we may assume further that it has unkotting number 1 [**Qua**]. The unknotting number is an upper bound for the number of surgery tori needed to undo the knot; are these numbers equal? (Knots with unknotting number 1 are prime [**Sc85**].) Analogous constructions may be used in higher dimensions to show that the properties of the modules $H_q(X(K); \Lambda)$ given in Theorem 2.1 essentially characterize such modules [**Le77**]. (There remains a question related to torsion in $H_1(X(K); \Lambda)$.)

An elaboration of this technique leads to a presentation for the knot module $B(K)$ of a classical knot, and hence to a description of the Blanchfield pairing b_K. If M is a matrix with entries in Λ let $M^\dagger$ be the transposed conjugate matrix, with $M^\dagger_{ij} = \overline{M_{ji}}$.

THEOREM 7.12. *A Λ-module is isomorphic to $B(K)$ for some 1-knot K if and only if it has a presentation matrix $\mathbb{D}$ such that $\mathbb{D}^\dagger = \mathbb{D}$ and $\varepsilon(\mathbb{D}) = diag[\pm 1, \ldots, \pm 1]$. Moreover if $\theta : \Lambda^n \to B(K)$ is the associated epimorphism and $\{\tilde{\alpha}_1, \ldots, \tilde{\alpha}_n\}$ is the standard basis of Λ^n the Blanchfield pairing of K is given by*

$$\langle \theta(\Sigma u_i \tilde{\alpha}_i), \theta(\Sigma v_j \tilde{\alpha}_j) \rangle = -v^\dagger \mathbb{D}^{-1} \varepsilon(\mathbb{D}) u \quad mod \quad \Lambda,$$

where (u_i) and (v_j) are regarded as column vectors in Λ^m.

PROOF. By the Addendum to Lemma 1.11 there is an embedding T of $nS^1 \times D^2$ in $X_o = S^1 \times D^2$ (the exterior of the unknot U) with core $T|_{nS^1 \times \{0\}}$ a trivial link such that $\mathrm{lk}(T_i, U) = 0$ for each $1 \leq i \leq n$ and a self homeomorphism h of $X(T) = S^3 \setminus \cup T$, where $\cup T$ is the interior of $T(nS^1 \times D^2)$, such that $h(T_i(1, s)) = T_i(s^{e(i)}, s)$, for some $e(i) = \pm 1$ and for all $s \in S^1$ and $1 \leq i \leq n$ and $h \circ U = K$ (as knots in $S^3 = X(T) \cup T(nS^1 \times D^2)$). We may recover $X(K)$ from $Z = X_o \backslash \cup T$ by attaching a 2-cell and a 3-cell for each surgery torus. Each surgery torus lifts to X_o', since it has linking number 0 with U, and so $X(K)'$ may be obtained from the lift $\tilde{Z}$ of Z in X_o' by adjoining 2- and 3-cells. Hence there is an epimorphism $\theta : H_1(Z; \Lambda) \to B(K) = H_1(X; \Lambda)$ with kernel the submodule generated by the attaching maps for the 2-cells. Moreover the natural map from $B(K)$ to $H_1(X, \partial X; \Lambda)$ is an isomorphism, since the longitude of K lifts to a loop in X' which is null homologous there.

Since $X_o \backslash Z$ is a disjoint union of open solid tori which are nullhomologous in X_o, the module $H_2(X_o, Z; \Lambda)$ is freely generated by relative 2-cycles corresponding to meridianal discs of the components of $X_o \backslash Z$, while $H_1(X_o, Z; \Lambda) = 0$. Therefore $H_1(Z; \Lambda) \cong H_2(X_o, Z; \Lambda)$. Let c_i and m_i be the centreline and meridian of T_i and let $\tilde{c}_i$ and $\tilde{m}_i$ be fixed lifts of c_i to X_o' and $h(m_i)$ to $Z' \subset X_o'$, respectively. Then $H_1(Z; \Lambda)$ is free of rank n, with basis $\{\tilde{\alpha}_1, \ldots, \tilde{\alpha}_n\}$, where $\tilde{\alpha}_i$ is Alexander dual to $\tilde{c}_i$ in $S^3 = (int\, X_o') \cup \{\infty\}$ (i.e., $\mathrm{lk}(\tilde{\alpha}_i, t^p \tilde{c}_j) = 1$ if $i = j$ and $p = 0$, and is 0 otherwise). The relator obtained by sewing in a meridianal disc to $h(m_i)$ (for $1 \leq i \leq n$) is $R_i = \Sigma_{p,j} r_{ipj} t^p \tilde{\alpha}_j$, where $r_{ipj} = \mathrm{lk}(\tilde{m}_i, t^p \tilde{c}_j)$.

Thus the $n \times n$-matrix $\mathbb{D}$ with (i, j)-entry $\mathbb{D}_{ij} = \Sigma_p r_{ipj} t^p$ is a presentation matrix for $B(K)$. If $i \neq j$ or $p \neq 0$ then as $\tilde{m}_i$ is homologous to $\tilde{c}_i$ in X_o' we have

$$\mathrm{lk}(\tilde{m}_i, t^p \tilde{c}_j) = \mathrm{lk}(\tilde{c}_i, t^p \tilde{c}_j) = \mathrm{lk}(t^{-p} \tilde{c}_i, \tilde{c}_j) = \mathrm{lk}(\tilde{c}_j, t^{-p} \tilde{c}_i) = \mathrm{lk}(\tilde{m}_j, t^{-p} \tilde{c}_i).$$

Hence $\mathbb{D}$ is hermitean, i.e., $\mathbb{D}_{ij} = \overline{\mathbb{D}_{ji}}$. Moreover, on projecting into X_o we see that $\varepsilon(\mathbb{D}) = diag[e(1), \ldots, e(n)]$, where $e(i) = \pm 1$ for $1 \leq i \leq n$. Since $B(K)$ is a torsion module $\Delta = \det(\mathbb{D}) \neq 0$ and since $\mathbb{D}$ is hermitean $\bar{\Delta} = \Delta$.

Let z be a 1-cycle on Z' and S a 2-chain on X' such that $\partial S \subseteq Z'$ and which is transverse to z in X'. The intersection number of z and S in X' is $I_{X'}(z, S) = I_{Z'}(z, S \cap Z') = I_{S^3}(z, S \cap Z')$, which in turn equals $\mathrm{lk}(z, \partial(S \cap Z')) = \mathrm{lk}(z, \partial S) + \mathrm{lk}(z, S \cap \partial Z')$. Since $\Delta B(K) = 0$

there is a 2-chain S_j on X' such that $\partial S_j = \Delta \tilde{\alpha}_j$. Since $\tilde{m}_i$ is homologous to $\Sigma_{j=1}^{j=m} \mathbb{D}_{ij} \tilde{\alpha}_j$ and $S_j \cap \partial Z'$ and ∂S_j together bound $S \cap Z'$ in Z', $S_j \cap \partial Z'$ is homologous to $-\Sigma_{k=1}^{k=m} \Delta [\mathbb{D}^{-1}]_{kj} \tilde{m}_k$. Let $\mathbb{E} = -\Delta \mathbb{D}^{-1}$, for brevity in the following equations. Then

$$\langle \theta(\tilde{\alpha}_i), \theta(\tilde{\alpha}_j) \rangle \equiv (\Delta)^{-1} \sum_{n \in \mathbb{Z}} I_{X'}(\tilde{\alpha}_i, t^n S_j) t^n$$

$$\equiv \sum_{n \in \mathbb{Z}} \Delta^{-1} (\mathrm{lk}(\tilde{\alpha}_i, t^n \Delta \tilde{\alpha}_j) + \mathrm{lk}(\tilde{\alpha}_i, t^n \sum_{k=1}^{k=m} \mathbb{E}_{kj} \tilde{m}_k) t^n)$$

$$\equiv \sum_{n \in \mathbb{Z}} \sum_{k=1}^{k=m} \Delta^{-1} \mathbb{E}_{kj} \mathrm{lk}(\tilde{\alpha}_i, t^n \tilde{m}_k) t^n$$

$$\equiv - \sum_{n \in \mathbb{Z}} \sum_{k=1}^{k=m} [\mathbb{D}^{-1}]_{kj} \delta_{ik} \delta_{n0} e(i) t^n$$

$$\equiv - [\mathbb{D}^{-1}]_{ij} e(i)$$

(where the congruences are *mod* Λ).

Conversely, every such presentation matrix $\mathbb{D}$ may be realized by some classical knot. The 1×1 case corresponds to Theorem 7.11. We shall follow (in outline) the exposition of [**Bai**]. Let $D = [0, n] \times [0, 1]^2$ and $D_i = [i - 1, i] \times [0, 1]^2$, for $1 \le i \le n$. Let $j : D \to S^1 \times D^2$ be an embedding and T_i an unknotted solid torus in D_i, for $1 \le i \le n$. Modify T_1 as in Theorem 7.11 to give $\mathbb{D}_{11}$, the coefficient of $\tilde{\alpha}_1$ in R_1. Since $\varepsilon(\mathbb{D}_{21}) = 0$ the coefficient of $\tilde{\alpha}_1$ in R_2 has the form $\Sigma_{k \ne 0}(t^k - 1)c_k$. Let N be a small neighbourhood of T_1 in $j(D_1)$. Let S_k^+ and S_k^- be small circles in N bounding disjoint discs in X_o and such that $\mathrm{lk}(S_k^+, T_1) = -\mathrm{lk}(S_k^-, T_1) = c_k$. Let a_k' be an arc from S_k^+ to S_k^- in N and let u_k represent t^k in $H_1(X_o) \cong Z$. Let a_k be the connected sum of a_k' and u_k along an arc between them, and let S_k be the boundary connected sum of S_k^+ and S_k^- along an untwisted band with core a_k (so S_k and T_i are unlinked). A boundary connected sum of T_2 with the S_k then gives the required torus for the relator R_2, while retaining the correct coefficient for $\tilde{\alpha}_1$, and leaving T_1 and T_2 still unlinked in S^3. Continuing in this way we obtain the required family of surgery tori. $\square$

In practice one can effectively calculate the entries of $\mathbb{D}$ for a given knot. (See [**Rol**].) A derivation of the Blanchfield pairing for a 1-knot via the free differential calculus is given in [**Ke79**].

Every $f \in \Lambda_\mu$ such that $\varepsilon(f) = 1$ and $\overline{f} = f$ is $\Delta_\mu(\pi)$ for some μ-component boundary 1-link [**Fr06**]. However the characterization of $\Delta_1(L)$ for more general multi-component 1-links L remains an open problem. In Chapter 8 we shall use surgery to characterize presentation matrices for the modules $B(L)$ in the 2-component case.

7.7. Blanchfield pairings and Seifert matrices

One drawback of Theorem 7.12 is that it is not obvious that every $(+1)$-Blanchfield pairing has such a presentation. Kearton used high-dimensional handlebody theory to show that every $(-1)^q$-Blanchfield pairing is realized by some simple $(2q + 1)$-knot, for all $q \geq 2$ [**Ke75, Le77**]. If q is even the Seifert matrix associated to a Seifert hypersurface of this knot can be realized by a 1-knot [**Ke65, Le69**], which thus has the same Blanchfield pairing. We shall sketch Trotter's more algebraic approach to the connection between Blanchfield pairings and Seifert matrices.

Let M be a knot module. Then $\mathbb{Q}M = \mathbb{Q} \otimes_{\mathbb{Z}} M$ is a $\mathbb{Q}\Lambda$-module, of finite-dimension over $\mathbb{Q}$. Let $E(\mathbb{Q}M) = Hom_{\mathbb{Q}\Lambda}(\mathbb{Q}M, \mathbb{Q}(t)/\mathbb{Q}\Lambda)$ and $F(\mathbb{Q}M) = Hom_{\mathbb{Q}}(\mathbb{Q}M, \mathbb{Q})$, and let $(t\phi)(v) = \phi(tv)$, for all $v \in \mathbb{Q}M$ and $\phi \in F(\mathbb{Q}M)$. Since $\mathbb{Q}\Lambda$ is a PID, it is easy to see that these modules are non-canonically isomorphic to $\mathbb{Q}M$. We shall give an explicit isomorphism $E(\mathbb{Q}M) \cong F(\mathbb{Q}M)$ (which defines an equivalence of functors), using a "generalized trace" due to Trotter [**Tr73**]. (See also [**Le77, Sto**].)

Recall that a proper fraction is a rational function $\frac{p}{q}$ in which the degree of the numerator is strictly less than that of the denominator: $\deg(p) < \deg(q)$. We may write $\mathbb{Q}(t) \cong \mathbb{Q}L \oplus Pr$, where Pr is the subspace of proper fractions with denominators q prime to t and $1 - t$, by the elementary theory of proper fractions. Let $\chi(f) = 0$ for $f \in \mathbb{Q}L$ and $\chi(f) = f'(1)$ for $f \in Pr$. An easy calculation shows that the latter formula holds also if $f = \frac{p}{q}$ with $\deg(p) = \deg(q)$. In particular, $\chi((t - 1)f) = f(1)$, for all $f \in Pr$. The chain rule gives

$\chi(\bar{f}) = -\chi(f)$, for all $f \in Pr$, and so for all $f \in \mathbb{Q}(t)$. (Note that Pr is not closed under the involution.) Thus we obtain a nonzero additive function from $\chi : \mathbb{Q}(t)/\mathbb{Q}L \to \mathbb{Q}$.

LEMMA 7.13. *Let U be a finite-dimensional $\mathbb{Q}L$-module. Then the homomorphism $\chi_U : E(U) \to F(U)$ induced by composition with χ is an isomorphism.*

PROOF. If $f \in \mathbb{Q}L$ then $(t-1)^n f \in \mathbb{Q}L$, and so $\chi((t-1)^n f) = 0$, for all $n \in \mathbb{Z}$. Conversely, if $f = \frac{p}{q} \in Pr$ and $\chi((t-1)f) = 0$, then p is divisible by $(t-1)$, since $0 = \chi((t-1)f) = \frac{p(1)}{q(1)}$. A finite induction now shows that if $\chi((t-1)^{-k}f) = 0$, for $1 \le k \le \deg(p)$, then $p = 0$. Thus $\chi((t-1)^n f) = 0$ for all $n \in \mathbb{Z} \Leftrightarrow f \in \mathbb{Q}L$.

It follows immediately that χ_U is a monomorphism. Since $E(U)$ and $F(U)$ are torsion $\mathbb{Q}\Lambda$-modules of the same finite length, χ_U is an isomorphism for all such U. $\square$

Suppose now that $b : M \times M \to \mathbb{Q}(t)/\Lambda$ is a ε-Blanchfield pairing on a pure knot module M, and let $b_\mathbb{Q} : \mathbb{Q}M \times \mathbb{Q}M \to \mathbb{Q}(t)/\mathbb{Q}\Lambda$ be the natural extension. Let $[v, w] = s(b_\mathbb{Q}(v, w))$, for all $v, w \in \mathbb{Q}M$. Then $[-, -]$ is a non-singular $-\varepsilon$-symmetric bilinear pairing, and t acts isometrically: $[tv, tw] = [v, w]$, for all $v, w \in \mathbb{Q}M$.

Let N be a lattice in M. Then the values of $[n, n']$ have bounded denominators, and so after rescaling we may assume that $[N, N] \le \mathbb{Z}$. Let $B = \{b_1, \dots, b_k\}$ be a basis for N, and define integer matrices $S_B = [s_{ij}]$ and $\Gamma_B = [\gamma_{ij}]$ by $s_{ij} = [b_j, b_i]$ and $xb_j = \sum \gamma_{ij}b_i$. Let $N^{\#} = \{v \in V | [v, n] \in \mathbb{Z} \; \forall \; n \in N\}$. Then $N \le N^{\#}$ and $[N^{\#} : N]$ is finite. Hence the localizations at primes agree, for almost all primes. Trotter showed that one could modify the choices to obtain a self-dual lattice, with $N = N^{\#}$ [**Tr77**]. Then S_B is invertible and $V = \Gamma_B S_B^{-1}$ is an integral matrix such that $\det(V - \varepsilon V^{tr}) = \pm 1$. (See §2 of [**Tr73**].) In particular, if $\varepsilon = +1$ then k is even, since $V - V^{tr}$ is skew-symmetric. The matrix V is the Seifert matrix associated to an embedding $j : F \to S^3$ of a once-punctured surface F of genus $g = \frac{k}{2}$, and $K = j|_{\partial F}$ is a 1-knot with $b_K = b$.

7.8. Branched covers

Let K be a 1-knot and X_q and M_q be the unbranched and branched q-fold cyclic covering spaces corresponding to an epimorphism $\phi : \pi = \pi K \to Z/qZ$. Then $H_1(X_q; \mathbb{Z}) \cong H_1(M_q; \mathbb{Z}) \oplus \mathbb{Z}$. If q is odd then $H_1(M_q; \mathbb{Z})$ is a direct double, as is the kernel of the natural homomorphism from $H_1(M_{2k}; \mathbb{Z})$ to $H_1(M_2; \mathbb{Z})$, by Theorem 5.19. If we assume moreover that $q = p$ is a prime then $H_1(M_q; \mathbb{Z})$ is a finite $\mathbb{Z}[\zeta_p]$-module of order prime to p. When $p = 2$ it is easy to see that any finite group of odd order may be realized. (Cyclic groups of odd order may be realized by knots with cyclic knot module; the general case follows on taking sums.) The conditions of Lemma 5.20 characterize such homology modules when p is odd [**Da95**].

If $H_1(M_q; \mathbb{Z})$ is finite its order is determined by $\Delta_1(K)$. However, the Alexander polynomials need not determine the structure of this group. For example, $E_1(6_1) = E_1(9_{46}) = (2t^2 - 5t + 2)$, $E_2(6_1) = \Lambda$, $E_2(9_{46}) = (3, t - 1)$ and $E_j(6_1) = E_j(9_{46}) = \Lambda$ for all $j > 2$. Hence these knots each have $\Delta_1 = 2t^2 - 5t + 2$ and $\Delta_j = 1$ for all $j > 1$, but $H_1(M_2(6_1); \mathbb{Z}) \cong Z/9Z$ and $H_1(M_2(9_{46}); \mathbb{Z}) \cong (Z/3Z)^2$.

Moreover, the Alexander polynomials alone do not even determine the prime divisors of the order of the torsion of $H_1(M_q; \mathbb{Z}[\frac{1}{q}])$, if this group is infinite [**We80**]. Let $K = 3_1 \sharp 9_5$. Then $\Delta_1(K) = \delta\eta$, where $\delta = t^2 - t + 1$ and $\eta = 6t^2 - 11t + 6$, and $H_1(M_6(K); \mathbb{Z})$ has nontrivial 5-torsion. On the other hand, there is a 1-knot K' with $B(K') \cong \Lambda/\delta\eta$, by Theorem 7.11, and $H_1(M_6(K'); \mathbb{Z}) \cong \mathbb{Z}^2$ for any such knot. It remains open to what extent the Alexander ideals determine the torsion.

LEMMA 7.14. *Let P be a finitely generated $\mathbb{Z}[u]$-torsion module such that $(u^n - 1)P = (u - 1)P$ for all $n \geq 1$. Then $(u - 1)P = 0$.*

PROOF. Let $S = \{(u - 1)^n \mid n \geq 1\}$. If m is a maximal ideal in $R = \mathbb{Z}[u]_S$ then m contains some $u^n - 1$, by Theorem 7.2, and so $mP_S = P_S$. Since P_S is finitely generated this implies that $P_S = 0$. Thus $(u - 1)^k P = 0$ for some $k \geq 1$.

Let $Q = (u - 1)P$. Then Q is a finitely generated abelian group and $Q = \nu_n(u)Q$ for all $n \geq 1$. If p is a prime then $\nu_p(u) \equiv (u-1)^{p-1}$

mod (p) and so $(u-1)$ acts invertibly on Q/pQ. Therefore $Q = pQ$ for all primes p and so $Q = 0$. $\square$

The following criteria for the homology of the branched cyclic covers of a 1-knot to be periodic are due to Gordon [**Go72**].

THEOREM 7.15. *Let K be a knot in a homology 3-sphere. Then the following are equivalent*

(1) $H_1(M_{km}; \mathbb{Z}) \cong H_1(M_m; \mathbb{Z})$ *for all $k \geq 1$;*
(2) $H_1(M_k; \mathbb{Z}) \cong H_1(M_{(k,m)}; \mathbb{Z})$ *for all $k \geq 1$;*
(3) $\lambda_1(K)$ *divides $t^m - 1$.*

PROOF. Let $H = H_1(X; \Lambda)$. Then $H_1(M_k; \mathbb{Z}) \cong H/(t^k - 1)H$. It follows easily that (3) implies (2) and (2) implies (1). Suppose that (1) holds. Then $(t^{km} - 1)H = (t^m - 1)H$ for all $k \geq 1$. Let $u = t^m$. Considering H as a $\mathbb{Z}[u]$-module and applying Lemma 7.13 we see that $(u - 1)H = 0$. $\square$

Let $b_m = |TH_1(M_m; \mathbb{Z})|$ be the order of the torsion subgroup. Then $b_m = |Res(\Delta_1(K), t^m - 1)|$, when this is nonzero (equivalently, when $H_1(M_m; \mathbb{Z})$ is finite). Such resultants satisfy a linear recursion formula of length 3^d over $\mathbb{Z}$, where d is the degree of $\Delta_1(K)$ [**Le33, St00**].

Silver and Williams have applied ideas from symbolic dynamics to study the asymptotic behaviour of the homology of abelian branched covers of links. The statement of their result uses the notion of Mahler measure $\mathcal{M}(\Delta)$ of a multivariable polynomial Δ. (If $p = C \prod_{i=1}^n (t - \xi_i)$ is a polynomial in one variable then $\mathcal{M}(p) = |C| \prod_{i=1}^n \max\{1, |\xi_i|\}$.) They identify the torsion subgroup with the set of connected components of periodic points in an associated algebraic dynamical system, and show that

THEOREM. [**SW02**] *Let L be a μ-component 1-link such that $\Delta_1(L) \neq 0$. Then*

$$\overline{\lim} |A|^{-1} \log |TH_1(M_\phi(L); \mathbb{Z})| = \log \mathcal{M}(\Delta_1(L)). \quad \square$$

Here the limit is taken over epimorphisms $\phi : \pi L \to A \cong \oplus_{i=1}^{i=\mu} (\mathbb{Z}/\lambda_i(A)\mathbb{Z})$, where $|\lambda_\mu(A)| \to \infty$. If $\Delta_{\alpha(L)}(L)$ is used instead of $\Delta_1(L)$ on the right, then the result extends to all 1-links [**Le10**].

When $\mu = 1$ we may replace $\overline{\lim}$ by ordinary limit, as $|A| \to \infty$. Thus

$$\lim_{m \to \infty} b_m^{\frac{1}{m}} = \mathcal{M}(\Delta_1(K)).$$

Hence if K is a knot such that $\Delta_1(K)$ is not a product of cyclotomic polynomials then b_m grows exponentially, and so only finitely many branched cyclic covers M_m are homology 3-spheres. Such m are bounded by a function of d [**KW92**]. Moreover, either the b_m are periodic or for any $N > 0$ there is an m such that b_m has at least N prime factors [**SW02'**].

If we use coefficients $\mathbb{Z}/p^r\mathbb{Z}$ then the homology is always periodic.

THEOREM 7.16. *Let K be a 1-knot, p a prime integer and $r \geq 1$. Then there is an integer $n \geq 1$ such that $H_1(M_{k+n}; \mathbb{Z}/p^r\mathbb{Z}) \cong H_1(M_k; \mathbb{Z}/p^r\mathbb{Z})$ as Λ-modules, for all $k \geq 0$.*

PROOF. Let $H = H_1(X; \Lambda)$, and let C_* be the singular chain complex for $X(K)'$ with coefficients $\mathbb{Z}/p^r\mathbb{Z}$. Then $H(r) = H_1(C_*) \cong H/p^r H$ is a finitely generated Λ-module annihilated by $(\lambda_1(K), p^r)$, and so is finite, Hence the automorphism of H induced by t has finite order, n say. Since $H_1(M_k; \mathbb{Z}/p^r\mathbb{Z}) \cong H(r)/(t^k - 1)H(r)$ the result is immediate.   □

7.9. Alexander polynomials of ribbon links

In this section we shall show that if $n \geq 2$ and L is a μ-component n-link then the condition $\varepsilon(\Delta_\mu(\pi)) = 1$ is the only constraint on such link polynomials. Every Laurent polynomial $f(t) \in \Lambda$ with $f(1) = 1$ is the Alexander polynomial of some ribbon 2-knot [**Ki61**]. (See also [**AS88, AY81**], for the fibred case.) We shall give a slightly stronger result, which extends to higher-dimensional links.

The case $n = 2$ is of particular interest, for then $H_1(X'; \mathbb{Z})$ and duality determine the other homology modules. (When $n = 1$ and L is a boundary link we must also have $\overline{\Delta} = \Delta$; there is as yet no such characterization for other classical links.)

An n-knot is a ribbon knot of 1-fusion if it is the fusion of the components of a 2-component trivial link along a single band.

THEOREM 7.17. [**Yj69**] *Let* $f = f(t) \in \Lambda$ *be such that* $f(1) = 1$, *and let* $n > 1$. *Then there is an n-knot* K *which is a ribbon knot of 1-fusion such that* $\pi'/\pi'' \cong \Lambda/(f)$, *where* $\pi = \pi K$.

PROOF. We may assume that $f \in \mathbb{Z}[t]$ and $f(0) \neq 0$. Let d be the degree of f and let $g = g(t) = (f(t) - 1)/(t - 1)$. Then $g(t) = \Sigma_{i=0}^{i=d-1} g_i t^i \in \mathbb{Z}[t]$. Let

$$\pi = \langle a, t \mid a = wtw^{-1}t^{-1} \rangle,$$

where $w = w(a,t) = \Pi_{i=0}^{i=d-1} t^i a^{g_i} t^{-i} = a^{g_0} \cdots t^{d-1} a^{g_{d-1}} t^{1-d} t^{-g(1)}$. (The final term ensures that $w(1,t) = 1$ in $\langle t \rangle$.) Clearly $\pi/\pi' \cong \mathbb{Z}$, and it is easily seen that $\pi'/\pi'' \cong \Lambda/(f)$, since $f = (t - 1)g + 1$.

Let $x = at$ and let $w_k \ldots w_1$ be a word of length k in the alphabet $\{t, t^{-1}, x, x^{-1}\}$ representing $w(xt^{-1}, t)$. Then π also has the deficiency-1 Wirtinger presentation

$$\langle t, x_0, \ldots, x_k, x \mid t = x_0, \ x = x_k, \ x_i = w_i x_{i-1} w_i^{-1} \ \forall i, \ 1 \leq i \leq k \rangle.$$

The elementary construction of §1.8 gives an n-ribbon $R : D^{n+1} \rightarrow S^{n+2}$ with π as its ribbon group, for any $n \geq 1$. This has k parallel throughcuts, and the corresponding slits are in the two extreme components of the complement of the throughcuts. Hence $K = R|_{\partial D^{n+1}}$ is the fusion of a 2-component trivial link along a band, and so is a ribbon knot of 1-fusion. If $n \geq 2$ then $\pi K \cong \pi$. $\square$

The group π is a 1-relator group. (This is so for the group of any ribbon knot of 1-fusion.) Since $\pi/\pi' \cong \mathbb{Z}$ the relator $atwt^{-1}w^{-1}$ is not a proper power. Therefore $c.d.\pi \leq 2$ [**Ly50**]. The conditions $c.d.\pi = 1$, $\pi \cong \mathbb{Z}$ and $f(t) = 1$ are clearly equivalent for groups with such presentations.

When $n = 1$ the knot $K = R|_{\partial D^2}$ provided by the construction of §1.8 bounds a disc knot $D^2 \subset D^4$ with group π, obtained by desingularizing the ribbon immersion R, and so K has Alexander polynomial $f\overline{f}$. Ribbon knots realizing such polynomials were first constructed in [**Te59**]. However the ribbon immersion R realizing π is not uniquely determined, and we do not know whether we can arrange that πK be a 1-relator group.

ADDENDUM. *The knot constructed in the above theorem is fibred if and only if the extreme coefficients of f are ± 1.*

PROOF. If the extreme coefficients of f are ± 1 then π' is free with basis represented by $\{t^i a t^{-i} \mid 0 \leq i < d\}$. Since $n > 1$ and K is a ribbon knot of 1-fusion it is fibred, by a theorem of Yoshikawa [**AS88, Yo81**].

The converse is clear. $\qquad\square$

We may realize arbitrary finite sequences δ_i with δ_{i+1} dividing δ_i in Λ as the higher polynomial invariants $\Delta_i(A(\pi))$, by taking sums of knots. If the summands are all fibred so is their sum.

We turn next to links. Let $A = \{a_1, \ldots, a_\mu\}$ and $T = \{t_1, \ldots, t_\mu\}$, and let $\frac{\partial}{\partial a_i}$ be the free derivation of $\mathbb{Z}F(A \cup T)$ with respect to the generator a_i. (See §6 of Chapter 12.) Let $\partial_i : \mathbb{Z}F(A \cup T) \to \Lambda_\mu$ be the composite of $\frac{\partial}{\partial a_i}$ with the retraction onto $\mathbb{Z}[F(T)]$ which sends a_j to 1 and t_j to t_j for $j \leq \mu$.

LEMMA 7.18. *Given $f_i = \Sigma_{m \in F(T)} f_{im} m \in \mathbb{Z}[F(T)]$, for $1 \leq i \leq \mu$, there is a word $W \in F(A \cup T)$ with trivial image in $F(T)$ and such that $\partial_i(W) = f_i$, for all $i \leq \mu$.*

PROOF. Let $v_i = \Pi_{m \in F(T)} m a_i^{f_{im}} m^{-1}$, where the factors are taken in some fixed order, and let $W = \Pi v_i = v_1 \cdots v_\mu$. Then the v_i and W have trivial image in $F(T)$. Moreover, $\partial_i(v_i) = \Sigma_{m \in F(T)} f_{im} m = f_i$, and $\partial_j(v_i) = 0$, if $j \neq i$, so $\partial_i(W) = \partial_i(v_i) = f_i$, for all $i \leq \mu$. $\qquad\square$

The order on $F(T)$ used in this lemma is not important.

THEOREM 7.19. *Let $f \in \Lambda_\mu$ be such that $\varepsilon(f) = 1$, and let $n > 1$. Then there is a μ-component ribbon boundary n-link L with $\Delta_\mu(\pi L) = f$.*

PROOF. We may write $f = 1 - \Sigma(t_i - 1)f_i$. Choose $F_i \in \mathbb{Z}[F(T)]$ with image $f_i \in \Lambda_\mu$. There are words $W_i \in F(A \cup T)$ with trivial image in $F(T)$ and such that $\partial_j W_i = F_i$ for all $j \leq \mu$, by the lemma. Let π be the group with presentation

$$\langle a_i, t_i, 1 \leq i \leq \mu \mid a_i = t_i W_i t_i^{-1} W_i^{-1}, \ \forall \ i \rangle.$$

The free differential calculus gives a presentation matrix $[I_\mu - D, 0_\mu]$ for $A(\pi)$, where D is a $\mu \times \mu$ matrix with $D_{ij} = (t_i - 1)f_i$ for all $i, j \le \mu$ and 0_μ is a null $\mu \times \mu$ matrix. As the columns of D are all equal, it is easy to see that $\Delta(\pi) = \det(I_\mu - D) = 1 - \Sigma(t_i - 1)f_i = f$.

As in Theorem 7.17, the group π has an equivalent Wirtinger presentation of deficiency μ, and the elementary construction of §1.8 gives a μ-component ribbon n-link L with group $\pi L \cong \pi$ and meridians corresponding to the generators t_i. Since the projection of π onto $\pi/\langle\langle a_1, \ldots, a_\mu \rangle\rangle \cong F(T)$ carries the meridians to a free basis, L is a boundary link. $\qquad\square$

Let Y be the finite 2-complex corresponding to the above presentation, and let Z be the complex obtained by adjoining 2-cells along maps corresponding to the generators t_i. Then Z is 1-connected and $\chi(Z) = 1$, and so it is a finite contractible 2-complex. Thus if the Whitehead Conjecture is true Y is aspherical, and so $c.d.\pi L \le 2$.

When $n = 1$ this construction gives a ribbon boundary link L with $\Delta(\pi) = f\overline{f}$. This condition is satisfied by the first nonzero Alexander polynomial of every classical slice link, by Corollary 4.17.2, and every such product is the Alexander polynomial of a ribbon 1-link [**Na78**]. However it is not clear whether the earlier construction always gives boundary links.

CHAPTER 8

Links with Two Components

Linking as distinct from knotting is first apparent when $\mu = 2$. In this case there are also some simplifications and special results, both in the algebra and the topology. Bailey has characterized the Λ_2-modules which arise as $B(L)$ for some 2-component link L as the modules having certain presentation matrices. Link-homotopy is determined by linking number alone. On the other hand, the classification of links up to I-equivalence is poorly understood, even in the 2-component case.

Throughout this chapter "link" shall mean "1-link", and $\delta \in \Lambda_\mu$ is a (μ-component) *link polynomial* if $\delta = \Delta_1(L)$ for some μ-component link L. To simplify the notation we shall use x and y instead of t_1 and t_2 for our Laurent polynomial variables, and $\ell = \ell_{12}$ for the linking number. If M is a matrix with entries in Λ_2 let $M^\dagger$ be the transposed conjugate matrix, with $M^\dagger_{ij} = \overline{M_{ji}}$. Let $\nu_k(t) = (t^k - 1)/(t - 1)$, for $k \in \mathbb{Z}$.

8.1. Bailey's Theorem

Let L be a 2-component link with exterior X and group $\pi = \pi L$. The equivariant chain complex of X' is homotopy equivalent to a finite free complex D_* with $D_0 = \Lambda_2$, $D_1 = (\Lambda_2)^{n+1}$ and $D_2 = (\Lambda_2)^n$. Let $Z_D \leq D_1$ be the submodule of 1-cycles. Then there are exact sequences

$$(\Lambda_2)^n \to Z_D \to B(L) = \pi'/\pi'' \to 0$$

and

$$0 \to Z_D \to (\Lambda_2)^{n+1} \to \Lambda_2 \to \mathbb{Z} \to 0.$$

Hence Z_D has rank n, and is projective, by Schanuel's Lemma. Therefore $B(L)$ has a square presentation matrix, as projective Λ_μ-modules are free. Moreover $p.d._\Lambda A(L) \leq 2$, by Lemma 4.11, and so $B(L)$ has no nontrivial finite Λ_2-submodule. (See the remark following Theorem 3.22.)

THEOREM 8.1. [**Bai**] *A Λ_2-module is isomorphic to $B(L)$ for some 2-component link L with $\mathrm{lk}(L_1, L_2) = \ell \geq 0$ if and only if it has a square presentation matrix of the form $\begin{pmatrix} \nu & \beta\gamma \\ \gamma^\dagger & A \end{pmatrix}$ where $\nu = \nu_\ell(xy)$, $\beta = -(x-1)(y-1)\nu_{\ell-1}(xy)$, γ is a row vector, $A = A^\dagger$ and $\varepsilon(A) = diag[\pm 1, \ldots, \pm 1]$. Moreover $A(t,1)$ and $A(1,t)$ are presentation matrices for $B(L_1)$ and $B(L_2)$, respectively.*

PROOF. We shall simplify L by surgery, in two stages. The link L is link homotopic to the $(2, 2\ell)$-torus link $L^* = L(2, 2\ell)$, since their linking numbers are equal. By the addendum to Lemma 1.11 there is an embedding T of $nS^1 \times D^2$ in $X(L^*)$ with core $T|_{nS^1 \times \{0\}}$ a trivial link such that $\mathrm{lk}(T_i, L_1^*) = \mathrm{lk}(T_i, L_2^*) = 0$ for each $1 \leq i \leq n$ and a self homeomorphism h of $X(T) = S^3 \setminus int\, T(nS^1 \times D^2)$ such that $h(T_i(1, s)) = T_i(s^{e(i)}, s)$, for some $e(i) = \pm 1$ and for all $s \in S^1$ and $1 \leq i \leq n$, and $h \circ L^* = L$. The map h carries meridians of L^* to meridians of L and so induces an isomorphism $H_1(X(L^*); \mathbb{Z}) \cong H_1(X(L); \mathbb{Z})$. (Note also that $B(L^*) \cong \Lambda_2/(\nu_\ell(xy))$.)

Suppose first that $\ell = 1$. Our argument in this case is closely related to that of Theorem 7.12. The model is now the Hopf link $(L^* = Ho)$, with group $\pi Ho \cong \mathbb{Z}^2$ and exterior $X_o = X(Ho) \cong S^1 \times S^1 \times [0, 1]$ (rather than the unknot, $\mathbb{Z}$ and $S^1 \times D^2$), and we may again apply linking number arguments, since $int\, X_o' \cong \mathbb{R}^3$. Let $\cup T$ denote the interior of $T(nS^1 \times D^2)$ and let $Z = X_o \setminus \cup T$. We may recover $X(L)$ from Z by attaching a 2-cell and a 3-cell for each surgery torus. Hence $B(L) = H_1(X; \Lambda_2)$ is the quotient of $H_1(Z; \Lambda_2)$ by the submodule determined by the attaching maps for the 2-cells.

Let c_i and m_i be the centreline and meridian of T_i and let $\tilde{c}_i$ and $\tilde{m}_i$ be fixed lifts of c_i to X_o' and $h(m_i)$ to $Z' \subset X_o'$, respectively. Then $H_1(Z; \Lambda_2)$ is free of rank n, with basis $\{\tilde{\alpha}_1, \ldots, \tilde{\alpha}_n\}$, where $\tilde{\alpha}_i$ is Alexander dual to $\tilde{c}_i$ in $S^3 = (int\, X_o') \cup \{\infty\}$ (i.e., $\mathrm{lk}(\tilde{\alpha}_i, x^p y^q \tilde{c}_j) = 1$

if $i = j$ and $p = q = 0$, and is 0 otherwise). The relator obtained by sewing in a meridianal disc to $h(m_i)$ (for $1 \leq i \leq n$) is $R_i = \Sigma_{p,q,j} r_{ipqj} x^p y^q \tilde{\alpha}_j$, where $r_{ipqj} = \mathrm{lk}(\tilde{m}_i, x^p y^q \tilde{c}_j)$. Thus the $n \times n$-matrix $A = A(x, y)$ with (i, j)-entry $A_{ij} = \Sigma_{p,q} r_{ipqj} x^p y^q$ is a presentation matrix for $B(L)$. As in the knot theoretic case A is hermitean ($A_{ij} = \overline{A_{ji}}$) and so $\det(A) = \overline{\det(A)}$.

There is an exact sequence

$$0 \to H_2(X(L_1), X(L); \Lambda_1) \xrightarrow{\ \delta\ } H_1(X(L); \Lambda_1) \to B(L_1) \to 0,$$

since $H_2(X(L_1); \Lambda_1) = 0$. The Wang sequence for ψ gives an exact sequence

$$0 \to \psi B(L) \to H_1(X(L); \Lambda_1) \xrightarrow{\ \tau\ } H_0(X(L); \Lambda_2) \to 0.$$

Since $\ell = 1$ we have $H_2(X(L_1), X(L); \Lambda_1) \cong \mathbb{Z}$, by excision, and the composite $\tau \delta : H_2(X(L_1), X(L); \Lambda_1) \to H_0(X(L); \Lambda_2)$ is an isomorphism. It follows that $B(L_1) \cong \psi B(L)$. Therefore $A(t, 1)$ is a presentation matrix for $B(L_1)$, and similarly $A(1, t)$ is a presentation matrix for $B(L_2)$. Moreover, if $\tilde{\alpha}, \tilde{\beta}$ are closed loops in $\mathbb{R}^3$ which project to disjoint loops α, β in $S^1 \times \mathbb{R}^2 \subset \mathbb{R}^3$ then $\mathrm{lk}(\alpha, \beta) = \Sigma \mathrm{lk}(\tilde{\alpha}, t^k \tilde{\beta})$. Applying this observation twice we find that $A(1, 1) = diag[\pm 1, \ldots, \pm 1]$. The bordered matrix $\begin{pmatrix} 1 & 0 \\ \gamma^\dagger & A \end{pmatrix}$ is another presentation matrix for $B(L)$ which is as in the enunciation (with $\nu = \nu_1(xy) = 1$ and γ arbitrary).

In general, we may reduce L^* to the Hopf link Ho by untwisting it $\ell - 1$ times. Let $B = D^2 \times I$ be a 3-ball in S^3 which meets Ho in two parallel, unknotted arcs $\{d_\pm\} \times I$, where $d_\pm = (\pm\frac{1}{2}, 0) \in D^2$, with one arc on each component of Ho and oriented compatibly with the natural orientation of $I = [0, 1]$. Let $g(z, t) = (e^{2\pi it} z, t)$ for $(z, t) \in B$. Then g is a self homeomorphism of B which twists B once around its axis $\{0\} \times I$. Let $Y = X_o \setminus int\, B$. Then $X_o = Y \cup (B \setminus N)$ and $X(L^*) = Y \cup_h (B \setminus N)$, where N is an open regular neighbourhood of Ho and $h = g^{\ell - 1}|_{(\partial B) \setminus N}$. We may clearly arrange that these changes take place in the complement of the surgery tori.

The space $Y \simeq S^1 \vee S^1$ is a handlebody and $\pi_1(Y) \cong F(2)$ is freely generated by loops ξ and ζ corresponding to meridians of L^*. Let $E = \{0\} \times D^1 \times I$ be the "vertical" 2-disc separating the arcs

$\{d_{\pm}\} \times I$ in B. The loop ∂E represents the conjugacy class of the commutator $[\xi, \zeta]$ in $\pi_1(Y)$, and lifts to a generator of $H_1(Y; \Lambda_2) \cong \Lambda_2$. Moreover $X_o \simeq Y \cup E$ and $X(L^*) \simeq Y \cup_{h|E} E$. (In other words, we may recover L from Ho by performing a Dehn surgery on ∂E. Note that as the surgery torus has linking number 1 with each component of Ho it does not lift to nontrivial abelian covers.)

We now let $Z = Y \setminus \cup T$ and let $\tilde{Z}$ be the preimage of Z in $Y' \subset X'_o = R^3$. Let $\tilde{\alpha}_0$ be a lift of ∂E to a loop in $\tilde{Z}$. Then $H_1(Z; \Lambda_2)$ is free of rank $n+1$, with basis $\{\tilde{\alpha}_0, \ldots, \tilde{\alpha}_n\}$. Since the disc E lifts to the complement of the surgery tori in X'_o we have $\mathrm{lk}(\tilde{\alpha}_0, x^p y^q \tilde{c}_j) = 0$ for all p, q and $1 \leq j \leq n$. The abelian covering space $(B \setminus Ho)'$ is $\mathbb{Z}^2$-equivariantly homeomorphic to a regular neighbourhood of the planar lattice $\mathbb{R} \times \mathbb{Z} \cup \mathbb{Z} \times \mathbb{R}$ in $\mathbb{R}^3 = \mathbb{R}^2 \times \mathbb{R}$. Let $\tilde{c}_0$ be the closure in $\mathbb{R}^3 \cup \{\infty\}$ of the union of the rays $(-\infty, 0] \times \{0\}$ and $\{0\} \times [0, \infty)$. Then $\mathrm{lk}(\tilde{\alpha}_i, x^p y^q \tilde{c}_0) = 0$ for all p, q and $1 \leq i \leq n$, while $\mathrm{lk}(\tilde{\alpha}_0, x^p y^q \tilde{c}_0) = 1$ if $p = q = 0$ and is 0 otherwise. Thus the $\tilde{\alpha}_i$s are Alexander dual to the $\tilde{c}_j$s, for $0 \leq i, j \leq n$.

Let $\tilde{m}_0$ be a lift of $h(\partial E)$ to $\tilde{Z} \subseteq Y'$. Since $h(\partial E)$ is freely homotopic to $(\xi\zeta)^\ell (\zeta\xi)^{-\ell}$ in Y, the image of $\tilde{m}_0$ in $H_1(Y; \Lambda_2)$ is $\nu_\ell(xy)\tilde{\alpha}_0$. Therefore the coefficient of $\tilde{\alpha}_0$ in $\tilde{m}_0$ is $\nu_\ell(xy)$.

The cycle $(1 - x)(1 - y)\tilde{c}_0$ is homologous in $S^3 \setminus Y'$ (the one-point compactification of $(B \setminus Ho)'$) to a loop running once around the unit square, i.e., to the lift of the commutator of the meridians to a loop at $(0, 0, 0)$ in $(B \setminus Ho)'$. Since $h(\partial E)$ is freely homotopic to $(xy)^{\ell-1}(yx)^{1-\ell}$ in $B \setminus Ho$, $\tilde{m}_0$ is homologous in $S^3 \setminus Y'$ to $-(1 - x)(1 - y)\nu_{\ell-1}(xy)\tilde{c}_0$. Hence the coefficient of $\tilde{\alpha}_m$ in R_0 is $-\Sigma_{j,k}(1-x)(1-y)\nu_{\ell-1}(xy)r_{0jkm}x^j y^k$, where $r_{0jkm} = \mathrm{lk}(\tilde{c}_0, x^j y^k \tilde{c}_m)$.

We thus obtain an $(n+1) \times (n+1)$-presentation matrix for $B(L)$, whose (i, j)-entry is the coefficient of $\tilde{\alpha}_j$ in R_i, for $0 \leq i, j \leq n$. Let $A(x, y)$ be the submatrix obtained by deleting the row and column corresponding to α_0 and R_0, and $\beta(x^{-1}, y^{-1})^\dagger$ be the column matrix corresponding to α_0, omitting the R_0 entry. Symmetry of the linking invariants implies that $A(x, y) = A(x^{-1}, y^{-1})^\dagger$ and that the top row is $[\nu_\ell(xy), -(1 - x)(1 - y)\nu_{\ell-1}(xy)\beta(x, y)]$. (This row corresponds to R_0.)

Let L'' be the link obtained from Ho by using the surgeries on T that lead from L^* to L (but *not* twisting along D). Then $L_1'' \cong L_1$, $L_2'' \cong L_2$ and $\mathrm{lk}(L_1'', L_2'') = 1$, and it is clear from the above argument that $A(x, y)$ is a presentation matrix for $B(L'')$. Now $B(L_1'') \cong \psi B(L'')$, since the linking number is 1, so $A(t, 1)$ and $A(1, t)$ are presentation matrices for $B(L_1) \cong B(L_1'')$, and $B(L_2)$, respectively, and $A(1, 1) = diag[\pm 1, \ldots, \pm 1]$, as before.

The argument for the converse is very similar to that for the knot theoretic case considered in Theorem 7.12, which follows [**Le67**] closely. We embed unknotted and unlinked surgery tori T_i in Y together with surgery coefficients corresponding to the relators R_i for $1 \le i \le n$. The main novelty is due to the twisting. Let $D = [0, n+1] \times [0, 1]^2$ and $D_i = [i-1, i] \times [0, 1]^2$, for $1 \le i \le n+1$. Let T_i be an unknotted solid torus in D_i, for $1 \le i \le n$ and let $j : D \to X_o$ be an embedding where $\alpha_0 \subset j(int\, D) \cap B \subset j(D_{n+1})$. (Thus j maps $D \setminus D_{n+1}$ into Y.) If the coefficient of $\tilde{\alpha}_0$ in R_k is $\Sigma_{(i,j) \in I} c_{ij} x^i y^j$, then place disjoint 2-discs D_{ij} in $j(D_{n+1})$ so that $\partial D_{ij} = S_{ij}$ is homotopic to $(\tilde{\alpha}_0)^{c_{ij}}$ in $j(D_{n+1}) \setminus B$ and use these for further modifications of T_k as in [**Le67**]. □

This theorem is due to Bailey [**Bai**]. As the constructive part of Bailey's Theorem is based on [**Le67**], it does not provide new proofs of the following special cases found by Levine.

COROLLARY 8.1.1. *If Δ is a 2-component link polynomial and $\lambda \in \Lambda_2$ is such that $\lambda = \bar{\lambda}$ and $\varepsilon(\lambda) = 1$ then $\lambda \Delta$ is a 2-component link polynomial.*

PROOF. If $\Delta = \det(M)$, where $M = \begin{pmatrix} \nu & \beta\gamma \\ \gamma^\dagger & A \end{pmatrix}$, then $M_1 = \begin{pmatrix} M & 0 \\ 0 & \lambda \end{pmatrix}$ also satisfies Bailey's criteria, and $\det(M_1) = \lambda \Delta$. □

There is a similar result for μ-component link polynomials.

COROLLARY 8.1.2. *A polynomial $\Delta \in \Lambda_2$ is the first Alexander polynomial of a 2-component link L with $\ell = \pm 1$ if and only if $\overline{\Delta} \doteq \Delta$ and $|\varepsilon(\Delta)| = 1$.*

PROOF. This follows from Corollary 8.1.1, since $\Delta(Ho) = 1$. □

An alternative presentation matrix for $B(L)$ was given by Cooper [**Cp82**]. (See also [**Le82**] for the case $\ell = 0$.) Presentation matrices of 3-component links with all linking numbers 0, and of links resulting from surgery on the trivial μ-component link (for μ arbitrary) have been characterized by Turaev [**Tu86**] and Platt [**Pl88**], respectively.

8.2. Consequences of Bailey's Theorem

The Torres conditions characterize link polynomials with $\ell = 0$ or 1, by Corollary 8.1.2 above and Corollary 8.4.1 below. We shall assume henceforth that $\ell > 1$.

Bailey's Theorem implies that a 2-component link polynomial has the form $\nu f + \beta g$, where $f = \bar{f}$, $\varepsilon(f) = \pm 1$ and $g = \bar{g}$. (We may take $f = \det(A)$ and $g = \det(B)$, where $B = \begin{pmatrix} 0 & \gamma \\ \gamma^\dagger & A \end{pmatrix}$.) He showed moreover that a polynomial δ in Λ_2 has this form if and only if it satisfies the Torres conditions. The representation of such a polynomial is not unique. If f, g and f', g' both give rise to δ then there is some $h \in \Lambda_2$ such that $f - f' = \beta h$ and $g - g' = \nu h$, so we may get rid of the ambiguity by passing to a quotient ring in which ν and β each map to 0. If $\wp$ is a prime ideal of Λ_2 containing ν and β then it must contain either $(x - 1)$ or $(y - 1)$, since $\nu_\ell(xy)$ and $\nu_{\ell-1}(xy)$ together generate Λ_2. If $y - 1$ is in $\wp$ then $\nu_\ell(x)$ is in $\wp$, and so $\wp$ must contain the d-cyclotomic polynomial $\phi_d(x)$ for some $d > 1$ which divides ℓ. (Similarly if $x - 1 \in \wp$.) The ideal $(\phi_d(x), y - 1)$ is a height 2 prime ideal of Λ_2. Any prime ideal properly containing it is maximal, and has the form $(p, k(x), y - 1)$, for some prime integer p and polynomial $k(x)$ representing an irreducible factor of $\phi_d(x)$ in $\mathbb{F}_p[x]$. (See Lemma 7.2.) Thus if we wish our quotient ring to be an integral domain we may as well fix a primitive d^{th} root of unity ζ_d and consider the homomorphism $F : \Lambda_2 \to \mathbb{Z}[\zeta_d]$ mapping x to ζ_d and y to 1. Notice that the involution on Λ_2 induces complex conjugation on $\mathbb{Z}[\zeta_d]$. We may also consider the images in the finite field $\mathbb{Z}[\zeta_d]/(p, k(\zeta_d)) = \mathbb{F}_p[x]/(k(x)) = \mathbb{F}_p\Lambda_1/(k(x))$.

As F factors through the projection of Λ_2 onto $\Lambda_2/(y - 1)$ we see that $F(f) = \zeta_d(\zeta_d - 1)F(\partial\delta/\partial x)/\ell$ (the value of $\delta(x, 1)/\nu_\ell(x)$ at

$x = \zeta_d$). Hence $F(g) = \zeta_d(\zeta_d - 1)^{-1}(F(\partial\delta/\partial y) - \zeta_d F(\partial\delta/\partial x))$. Since $f = \bar{f}$ and $g = \bar{g}$ these values are real. Moreover whether $F(f)$ is 0 is independant of the choice of ζ_d, since δ has coefficients in $\mathbb{Z}$.

THEOREM 8.2. *Let L be a 2-component link with linking number $\ell > 1$ and let $\Delta = \Delta_1(L)$.*

(1) *Suppose that $\Delta(x, 1)/\nu_\ell(x) = \phi_d(x)h(x)$, where $d > 1$ divides ℓ and $h(\zeta_d) \neq 0$. Then the ideal in $\mathbb{Z}[\zeta_d]$ generated by $h(\zeta_d)^{-1}(\partial\Delta/\partial y)(\zeta_d, 1)$ is of the form $J\bar{J}$ for some ideal J.*

(2) *Suppose that $\Delta(x, 1)/\nu_\ell(x) = ph(x) + q(x)k(x)$, for some prime $p \in \mathbb{Z}$ and $k(x) \in \Lambda_1$ representing an irreducible factor of $\phi_d(x)$ in $\mathbb{F}_p[x]$, and that $h(x)$ has nonzero image in $\mathbb{F}_q = \mathbb{F}_p\Lambda_1/(k(x))$. Then $x(x - 1)^{-1}(\partial\Delta/\partial y)(x, 1)$ has image $\pm b\bar{b}h$ in $\mathbb{F}_q$, for some $b \in \mathbb{F}_q$.*

PROOF. There are hermitean square matrices A and $B = \begin{pmatrix} 0 & \gamma \\ \gamma^\dagger & A \end{pmatrix}$ such that $\Delta = \nu_\ell(xy)\det(A) - (x - 1)(y - 1)\nu_{\ell-1}(xy)\det(B)$, by Bailey's Theorem.

(1). Suppose that $\phi_d(x)$ divides $\Delta(x, 1)/\nu_\ell(x) = \det(A(x, 1))$. Then $\zeta_d(\zeta_d - 1)F(\partial\delta/\partial x)/\ell = \det(F(A)) = F(\det(A)) = 0$, and so $F(\det(B)) = \zeta_d(\zeta_d - 1)^{-1}F(\partial\Delta/\partial y)$. Since $\phi_d(x)$ is a knot polynomial $\phi_d(1) = 1$, and so d cannot be a prime power. Ideals in $\mathbb{Z}[\zeta_d]$ are uniquely factorizable as products of powers of prime ideals, since this ring is a Dedekind domain. Let $\wp$ be a prime ideal of $\mathbb{Z}[\zeta_d]$, $R = \mathbb{Z}[\zeta_d]_\wp$ and $S = F^{-1}(\mathbb{Z}[\zeta_d] \setminus \wp)$. Then F extends to an epimorphism from Λ_{2S} to R. Since $\det(F(A)) = 0$ and R is a local domain one of the rows of $F(A)$ is a linear combination of the others. Therefore we may find a Λ_{2S}-matrix P with determinant 1 such that $PAP^\dagger = \begin{pmatrix} e & u \\ u^\dagger & C \end{pmatrix}$, where $e = a\phi_d(x) + b(y - 1)$ and $u = \phi_d(x)v + (y-1)w$, for some $a, b \in \Lambda_{2S}$ and row vectors v, w. Thus $F(PAP^\dagger)$ has first row and column 0. (We perform the conjugate column operations also so as to preserve the hermitean character of the matrices.) Let $Q = [1] \oplus P$.

Since $\phi_d(x)$ divides the first row of $PA(x, 1)P^\dagger$ we have

$$F(a)F(\det(C)) = F(\phi_d(x)^{-1}\det(PA(x, 1)P^\dagger)) =$$

$$F(\phi_d(x)^{-1}\det(A(x,1))) = 1,$$

so $F(\det(C))$ is a unit. Moreover

$$F(\det(B)) = F(\det(QBQ^\dagger)) = -F(\rho)\overline{F(\rho)}F(\det(C)),$$

where ρ is the $(1,2)$-entry of $QBQ^\dagger$. Hence $((\partial\Delta/\partial y)(\zeta_d,1)) = (F(\det(B))) = (F(\rho))\overline{(F(\rho))}$, since $\zeta_d(\zeta_d - 1)$ is also a unit in R. Let $v(\wp)$ be the $\wp$-adic valuation of $F(\det(B))$. If $\wp = \overline{\wp}$ then $v(\wp)$ is even, while if $\wp \neq \overline{\wp}$ then $v(\wp) = v(\overline{\wp})$. Let U be the set of prime ideals of $\mathbb{Z}[\zeta_d]$ obtained by choosing one representative from each pair $\wp \neq \overline{\wp}$. Let $J = (\Pi_{q\in U}q^{v(q)})(\Pi_{r=\overline{r}}r^{v(r)/2})$. Then $((\partial\Delta/\partial y)(\zeta_d,1)) = J\overline{J}$.

(2). The argument is similar, but simpler. Since $\mathbb{F}_p[x]/(k(x))$ is a field, the only elementary matrices needed are those corresponding to adding multiples of one row to another, and these always lift to elementary Λ_2-matrices. $\square$

Although this theorem follows almost inevitably from Bailey's Theorem, its meaning is rather obscure. The partial derivative is an invariant of homotopies of the second component of the link, by Theorem 5.7, while the cyclotomic polynomials suggest that the homology of a cyclic cover of X is involved. As the proof of Bailey's Theorem involves using surgeries to change crossings of the components of the link with themselves, in other words to carry out a link homotopy, we might expect a deeper connection between these ideas.

COROLLARY 8.2.1. *Let $f = 4x - 7 + 4x^{-1}$, $g = 1$ and $\ell = 2m$, for some $m > 0$. Then $D(x,y) = \nu_{2m}(xy)f - (x-1)(y-1)\nu_{2m-1}(xy)$ is not a 2-component link polynomial.*

PROOF. Let $p = 5$ and $k(x) = x + 1$. Then we may take $h(x) = -3$. Now $x(x-1)^{-1}(\partial\Delta/\partial y)(x,1) \equiv -x\nu_{2m-1}(x) \equiv 1 \bmod (5, x+1)$. Since $1 = \pm b^2.3$ has no solutions $\bmod (5)$, part (2) of the theorem fails for D. $\square$

The first counterexample found was the polynomial $D(x,y) = \nu_6(xy)x^{-1}\phi_6(x) - 2(x-1)(y-1)\nu_5(xy)$. The ideal generated by $((\partial D/\partial y)(\zeta_6,1)) = (2)$ in $\mathbb{Z}[\zeta_6]$ is (2), which is not of the form $J\overline{J}$ in $\mathbb{Z}[\zeta_6]$. Platt found examples for all composite linking numbers.

THEOREM 8.3. [**P186**] *If d is divisible by at least two primes then there is an integer q such that the ideal $q\mathbb{Z}[\zeta_d]$ is not of the form $J\bar{J}$.*

PROOF. Let q be a prime which is congruent to $-1 \bmod (d)$. (There are infinitely many such primes, by a well-known theorem of Dirichlet.) Then q is unramified in the extension $\mathbb{Q}(\zeta_d)/\mathbb{Q}$, since the only ramified primes are the divisors of d, and the decomposition group of (q) is the subgroup of $Gal(\mathbb{Q}(\zeta_d)/\mathbb{Q}) \cong (Z/dZ)^{\times}$ generated by the image of q, and so contains complex conjugation. Hence $q\mathbb{Z}[\zeta_d]$ is a product of distinct primes which are invariant under conjugation. This proves the theorem.   □

It is easy to find explicit counterexamples for any such q. In particular, $D(x,y) = \nu_d(xy)x^{-\varphi(d)/2}\phi_d(x) - (x-1)(y-1)\nu_{d-1}(xy).q$ is not a 2-component link polynomial, Platt has also extended the argument of Theorem 8.2 to the cases when $\ell = p^r$ is a prime power greater than 2. (She assumes that $f(x,1)$ is an irreducible quadratic $a(x - 2 + x^{-1}) + 1$, and considers the image of g in $\mathbb{Z}[\zeta_d]/(f(\zeta_d,1))$.) The construction of counter-examples is rather more delicate, but involves similar number-theoretic ideas. Thus for any $\ell \geq 2$ there is a polynomial $\Delta(x,y)$ with $\varepsilon(\Delta) = \ell$ and which satisfies the Torres conditions, but which is not a 2-component link polynomial.

The Torres conditions characterize the Alexander polynomials of the localizations of the modules of 2-component links, after localization with respect to the multiplicative system $S = 1 + I_2$, by the following result of [**Le88**].

THEOREM 8.4. *Let $\Delta = \nu f + \beta g$, where $f = \bar{f}$, $\varepsilon(f) = \pm 1$ and $g = \bar{g}$. Then there is a 2-component link L such that $\Delta_1(L) = f^n\Delta$ for some $n \geq 0$.*

PROOF. Since $g = \bar{g}$ we may write g as a sum of terms of the form $\pm h\bar{h}$, where $h(x,y) = x^k y^l + 1$ or 1. Suppose $g = \Sigma_{i=1}^{i=k}\epsilon_i h_i \bar{h}_i$. Let A be the diagonal matrix $diag[\epsilon_1 f, \ldots, \epsilon_k f]$ and let $B = \begin{pmatrix} 0 & \gamma \\ \gamma^{\dagger} & A \end{pmatrix}$, where $\gamma = (h_1, \ldots, h_k)$. Then $\det(A) = (\Pi\epsilon_i)f^k$ and $f.\det(B) = -g.\det(A)$, and so $f^{k-1}\Delta$ is a link polynomial.   □

COROLLARY 8.4.1. *There is a 2-component link L with linking number $\ell = 0$ and $\Delta_1(L) = \Delta$ if and only if $\overline{\Delta} \doteq \Delta$ and Δ is divisible by $(x-1)(y-1)$.*

PROOF. Since $\nu = 0$ if $\ell = 0$ we may take $f = 1$ above.   □

The graded module associated to the completion $\widehat{B}(L)$ is a cyclic module over $Gr(\widehat{\Lambda_2}) \cong \mathbb{Z}[X,Y]$, and $Ann(Gr(\widehat{B}(L)))$ is generated by the "initial form" of the image of $\Delta_1(L)$ in $\widehat{\Lambda}_2$, i.e., the homogeneous polynomial consisting of the nonzero terms of lowest degree in this element. If $\ell \neq 0$ the initial form is just ℓ, while if $\ell = 0$ the Torres conditions imply that the initial form has even degree in X and Y, and if the total degree is d the coefficients of X^d and Y^d are both 0. These conditions characterize such initial forms, by Theorem 8.4.

If $\Delta_1(L) \neq 0$ then π'/π'' has nontrivial p-torsion for an integral prime p if and only if p divides $\Delta_1(L)$, in which case it divides the linking number. For $Ann(\pi'/\pi'')$ is generated by $\lambda_1(L)$, which is divisible by each of the prime factors of $\Delta_1(L)$. Given any $\lambda \in \Lambda_2$ such that $\overline{\lambda} = \lambda$ there is a 2-component link L with $l = 0$ such that $\Delta_1(L) = \lambda(x-1)(y-1)$, by the Corollary. Hence on taking $\lambda = p$ we see that π'/π'' need not be $\mathbb{Z}$-torsion free. If $\Delta_1(L) = 0$ and $E_2(L)$ is principal then π'/π'' is $\mathbb{Z}$-torsion free, by part (6) of Theorem 4.12.

If $\ell = 0$ we may assume that $\Delta_1(L) = (1 - x^{-1})(1 - y^{-1})g$ where $\overline{g} = g$. Jin has used Bailey's Theorem to show that the I-equivalence invariants η_i^{KY} introduced by Kojima and Yamasaki are given by $\eta_1^{KY}(x) = (x-1)^2 g(x,1)/x\Delta_1(L_1) = (x-1)(\partial/\partial y|_{y=1}\Delta_1(L))/\Delta_1(L_1)$ and $\eta_2^{KY}(y) = (y-1)^2 g(1,y)/y\Delta_1(L_2)$. (See [**Ji87, KY79, MM83**].)

8.3. The Blanchfield pairing

Let L be a 2-component link. If $\ell = \mathrm{lk}(L_1, L_2) = 1$ the inclusions of $\partial X(L_1)$ and $\partial X(L_2)$ into $X = X(L)$ induce isomorphisms on homology, and so $\partial X'$ is the union of two copies of $\mathbb{R}^2$. Hence the Blanchfield pairing on $B(L)$ is primitive. The arguments of Theorem 7.12 carry over to show that this pairing is determined by the matrix A of Bailey's Theorem. Is every $(+1)$-linking pairing on a torsion Λ_2-module M such that $\mathbb{Z} \otimes_\Lambda M = 0$ realized by such a link?

In general the Blanchfield pairing is not perfect, but we may remedy that by localization, as suggested in Chapter 2. Let Σ be the multiplicative system in Λ_2 generated by all nonzero 1-variable polynomials, $\Sigma = \{p(x)q(y)\}$. Then $R = \Lambda_{2\Sigma}$ is the subring of $\mathbb{Q}(x, y)$ generated by $\mathbb{Q}(x) \cup \mathbb{Q}(y)$, and is a PID containing $\mathbb{Q}$, since it is a localization of $\mathbb{Q}(x)[y]$. A prime ideal $\wp$ in R is generated by a polynomial $p(x, y)$ which is irreducible in $\mathbb{Q}[x, y]$ and is of positive degree in each variable. The residue field $R/\wp$ is an algebraic extension of $\mathbb{Q}(x)$, $K = \mathbb{Q}(x)[z]$, say (where $p(x, z) = 0$). Prime ideals invariant under the involution of R correspond to such p for which also $p(x^{-1}, y^{-1}) = a(x)b(y)p(x, y)$ for some 1-variable rational functions $a(x)$ and $b(y)$. Since p is irreducible we must have $p(x^{-1}, y^{-1}) = \pm x^r y^s p(x, y)$ for some sign and exponents r, s. If $\overline{\wp} = \wp$ the involution of R induces a nontrivial involution on $K = R/\wp$, and $K = K_+[\alpha]$, where K_+ is the fixed field of the involution and $\alpha = x - x^{-1}$. (Note that $\bar{\alpha} = -\alpha$.) If b is a ε-hermitean form on the K-vector space V the form αb is $(-\varepsilon)$-hermitean. Hence in discussing Witt groups it suffices to consider the case $\varepsilon = +1$, which is anyway the case relevant for classical links.

The Blanchfield pairing on $\hat{t}H_1(X; \Lambda_2)_\Sigma = TH_1(X; R)$ is perfect, since R is a PID. If L is split $TH_1(X; R) = 0$, so the Witt class $B_\Sigma(L)$ may be used to detect links not concordant to split links. If $\Delta_1(L) = 0$ then $TH_1(X; R)$ is annihilated by $\lambda_2(L)$, which has integral coefficients and augmentation $\varepsilon(\Lambda_2(L)) = \pm 1$. Any factorization of such a polynomial into irreducibles in R must come from a factorization in Λ_2, by the Gauss Content Lemma. Thus if $R/\wp^e$ is a direct summand of $TH_1(X; R)$ then $\wp$ is generated by an integral polynomial which divides $\lambda_2(L)$. On the other hand, if $q \in \Lambda_2$ is such that $\bar{q} = q$ and $\varepsilon(q) = \pm 1$ then there is a 2-component link L such that $H_1(X; \Lambda_2) \cong \Lambda_2 \oplus (\Lambda_2)/(q)$, by Bailey's Theorem. (We may assume L is a boundary link [**GL02**].) In particular, if q is irreducible (and not a unit in R) then $TH_1(X; R)$ has length 1 and so $b_\Sigma(L)$ is not neutral. For instance, the links obtained by Whitehead doubling each component of the Hopf link are boundary links with such modules, and so are not split concordant. More generally, the image of the set of such links in $W_{+1}(R_0, R, -)$ is not finitely generated.

This invariant may also be applied to links with nonzero first Alexander polynomial. For instance, the link $L = K \amalg \partial\Delta$ of Figure 1.2 is I-equivalent to the Hopf link, by Theorem 1.9. Since $\Delta_1(L) = (xy)^2 - x^2y - xy^2 + xy - x - y + 1$ is irreducible the localized module $TH_1(X(L); R)$ has length 1. Thus the localized Blanchfield pairing cannot be Witt equivalent to the trivial pairing, and so L cannot be concordant to the Hopf link.

Closer study of the UCSS over Λ_{2S} when $S = \{((x-1)(y-1))^n \mid n \geq 0\}$ shows that $b_S(L)$ is primitive, and is perfect if $\Delta_1(L) \neq 0$ or if L is a boundary link, and $\mathbb{Q} \otimes b_S(L)$ is always perfect. If $\Delta_1(L) = 0$ the kernel and cokernel of the adjoint map of the unlocalized Blanchfield pairing are determined by the rank 1 Λ_2-module $I = B(\pi)/TB(\pi)$ and the longitude-annihilating polynomials $b_1(y), b_2(x)$ [Le82].

8.4. Links with Alexander polynomial 0

Throughout this section we shall assume that $\Delta_1(L) = 0$.

THEOREM 8.5. *Let L be a 2-component link with group $\pi = \pi L$ and such that $E_1(L) = 0$. Let P be the submodule of $A(L)$ generated by the longitudes. Then $P \cong (\Lambda_2/(b_1(y), x-1)) \oplus (\Lambda_2/(b_2(x), y-1))$ for some $b_1(y), b_2(x) \in \Lambda_2$ such that $b_1(1) = b_2(1) = 1$. Hence $Ann(P) = E_0(P) = (b_1(y) + b_2(x) - 1, (x-1)b_2(x), (y-1)b_1(y))$ and P is $\mathbb{Z}$-torsion free.*

PROOF. Let P_1 and P_2 be the cyclic submodules of P generated by ℓ_1 and ℓ_2, respectively. Then $Ann(P_1) = (x - 1, b_1(y))$ and $Ann(P_2) = (y - 1, b_2(x))$, for some polynomials b_1 and b_2, by Theorem 5.4. Since the longitudes are in π' and so are nullhomologous in X, b_1 and b_2 must augment to a generator ± 1 of $\mathbb{Z}$. Therefore $P_1 \cong \Lambda_2/(b_1(y), x-1)$ and $P_2 \cong \Lambda_2/(b_2(x), y-1)$ are $\mathbb{Z}$-torsion free, by the argument of part (6) of Theorem 4.6.

We may assume that $b_1(1) = b_2(1) = +1$. Suppose that $a_1(x, y)$ and $a_2(x, y)$ in Λ_2 are such that $a_1(x, y)\ell_1 + a_2(x, y)\ell_2 = 0$. Then

$a_1(x,y)\ell_1 = a_1(x,y)b_2(x)\ell_1$ (since $b_2(x) \equiv 1 \bmod (x-1)) = 0$. Therefore $a_1(x,y)$ is in $Ann(P_1)$, and similarly $a_2(x,y)$ is in $Ann(P_2)$, and so $P \cong P_1 \oplus P_2$. In particular, P is $\mathbb{Z}$-torsion free.

Let $p = b_1(y) + b_2(x) - 1$, $q = (x-1)b_2(x)$, $r = (y-1)b_1(y)$, $s = (x-1)(y-1)$, $u = b_1(y)b_2(x)$, $b_1' = (b_1(y) - 1)/(y-1)$ and $b_2' = (b_2(x) - 1)/(x-1)$. Then $E_0(P) = (q,r,s,u) = (p,q,r)$, since $p = u - sb_1'b_2'$, $u = b_2(y)p - b_2'q$ and $s = -sp + (y-1)q + (x-1)r$. Clearly also $E_0(P) \leq Ann(P) = Ann(P_1) \cap Ann(P_2)$. Suppose that $a(x,y)$ is in $Ann(P)$. Then $a(x,y) = m(x,y)(x-1) + n(x,y)b_1(y)$ (since it is in $Ann(P_1)$). Now $m(x,y) \equiv \tilde{m}(x) = m(x,1)$ and $n(x,y) \equiv \tilde{n}(x) = n(x,1) \bmod (y-1)$, so $a(x,y) \equiv \tilde{m}(x)(x-1) + \tilde{n}(x)b_1(y) \bmod (r,s)$. Hence $a(x,y) \equiv \tilde{a}(x) = \tilde{m}(x)(x-1) + \tilde{n}(x)(1 - b_2(x)) \bmod (p,r,s)$. We also have $\tilde{a}(x) = (x-1)a'(x)$ for some $a'(x)$, since $b_2(1) = 1$. The term $\tilde{a}(x)$ is also in $Ann(P)$, and hence in $Ann(P_2)$, since $(p,q,r) \leq Ann(P)$. Therefore $a'(x)$ is $Ann(P_2)$, since multiplication by $(x-1)$ is injective on P_2. Thus $\tilde{a}(x)$ is in $(x-1)Ann(P_2) = (q,s)$ and so $a(x,y)$ is in (p,q,r,s). Hence

$$Ann(P) = E_0(P) = (b_1(y) + b_2(x) - 1, (x-1)b_2(x), (y-1)b_1(y)).$$

$\square$

This result was first announced by Crowell and Brown (in 1976 – not published) for L a homology boundary link.

The quotient of π'/π'' by its Λ_2-torsion submodule is isomorphic to an ideal I in Λ_2, since it is torsion free and of rank 1. The ideal I may be uniquely specified by requiring that $\tilde{I} = \Lambda_2$, and then $b_1(y) + b_2(x) - 1 \in I$ and $\varepsilon(I) = \mathbb{Z}$. Bailey's Theorem may be used to show that any such triple b_1, b_2, I is realized by some 2-component link L with $\Delta_1(L) = 0$ [**Le82**]. In [**Le87**] it is shown that $\Delta_1(L_1) \doteq b_2(x)b_2(x^{-1})\Delta_2(L)(x,1)$ and conversely that for any pair $(b_2(x), \delta(x))$ with δ a knot polynomial divisible by $b_2(x)b_2(x^{-1})$ there is a 2-component link L such that

$$\Delta_1(L_1) = \delta(x) = b_2(x)b_2(x^{-1})\Delta_2(L)(x,1).$$

COROLLARY 8.5.1. *If π maps onto $F(2)/F(2)''$ then π/π'' is torsion free and $\tau(E_2(L)) = (\tau(\Delta_2(L)))$.*

PROOF. Both assertions follow from Corollary 4.15.1, which establishes that $A(L)/P$ is $\mathbb{Z}$-torsion free and $E_2(L) = \Delta_2(L)E_0(P)$. It is easily seen that $\tau(E_0(P)) = \Lambda$, and so $\tau(E_2(L)) = (\tau(\Delta_2(L)))$. $\square$

The conditions of the next theorem imply that π maps onto $F(2)/F(2)''$, by Corollary 4.15.2.

THEOREM 8.6. *Let L be a 2-component link with group $\pi = \pi L$ and such that $E_1(L) = 0$. Then the following are equivalent*

(1) $E_2(L)$ *is principal;*

(2) $E_1(\pi'/\pi'')$ *is principal;*

(3) $p.d._\Lambda A(L) \leq 1;$

(4) $p.d._\Lambda \pi'/\pi'' \leq 1.$

PROOF. (1) $\Rightarrow$ (2, 3, 4). If $E_2(L)$ is principal then $A(L) \cong (\Lambda_2)^2 \oplus TA(L)$ and $TA(L)$ has a square presentation matrix, by Theorem 3.7, whence (3) holds, and $\pi'/\pi'' \cong \Lambda_2 \oplus TA(L)$ by Theorem 4.15, so (2) and (4) hold.

(2) $\Rightarrow$ (1, 3, 4). This is similar.

(3) $\Leftrightarrow$ (4). This follows from Lemma 4.5.

(4) $\Rightarrow$ (2). If $p.d._\Lambda \pi'/\pi'' \leq 1$ the longitudes of L must be in π'', by Theorems 3.8 and 4.14. Let Y be the closed 3-manifold obtained by surgery on the longitudes. As in Theorem 4.15 Poincaré duality and the UCSS give an exact sequence

$$H^2(Y; \Lambda_2) \to e^0 H_2(Y; \Lambda_2) \to e^2(\pi'/\pi'') = 0$$

and an isomorphism $H^2(Y; \Lambda_2) \cong \overline{H_1(Y; \Lambda_2)}$. Since $H_2(Y; \Lambda_2)$ has rank 1, we have $e^0 H_2(Y; \Lambda_2) \cong \Lambda_2$. Therefore $\pi'/\pi'' \cong \overline{H^2(Y; \Lambda_2)}$ maps onto $\overline{\Lambda_2} \cong \Lambda_2$, so $\pi'/\pi'' \cong \Lambda_2 \oplus T(\pi'/\pi'')$, $p.d._\Lambda T(\pi'/\pi'') \leq 1$ and (2) follows from Theorem 3.7. $\square$

A 2-component boundary link L may also be characterized as one for which there is a connected closed surface C in $X(L)$ which separates the components of L in S^3 and such that each component is nullhomologous in $X(L) \setminus C$. Such a surface represents a generator of $H_2(X(L); \mathbb{Z})$ and lifts to a generator of $H_2(X'; \mathbb{Z}) = H_2(X; \Lambda_2)$. If L is a 2-component homology boundary link there is a map $f : X(L) \to S^1 \vee S^1$ which induces an epimorphism from πL to $F(2)$.

Does the inverse image of the wedge point serve as a singular separating surface for L? Is there a geometrically significant generator for $H_2(X; \Lambda_2)$ which projects nicely? (See also Theorem 11.12.)

8.5. 2-Component $Z/2Z$-boundary links

If L is a 2-component $Z/2Z$-homology boundary link then L is a $Z/2Z$-boundary link. This follows from the algebraic characterization of such links, as given in the next theorem, but can be seen more directly. Let U_1 and U_2 be disjoint singular spanning hypersurfaces corresponding to the components L_1 and L_2 of such a link L. Then $U_1 \cap \partial X(L_1)$ and $U_2 \cap \partial X(L_1)$ are unions of odd and even numbers of copies of longitudes of L_1, respectively. Since the total number of boundary components on $\partial X(L_1)$ is odd, there must be a pair of adjacent components from the same hypersurface, U_i say. By pushing a copy of the annular region between these longitudes off $\partial X(L_1)$ we reduce the number of components of ∂U_i by two. After finitely many such steps (applied also near $\partial X(L_2)$) we obtain disjoint spanning hypersurfaces which each have just one boundary component.

Whether a 2-component link is a $Z/2Z$-boundary link is determined by its Murasugi nullity, since $(Z/2Z) * (Z/2Z)$ is metabelian.

THEOREM 8.7. *Let L be a 2-component link with group $\pi = \pi L$ and linking number ℓ. Then the following are equivalent*

(1) $\eta(L) = 2$;
(2) $\Delta_1(L)(-1, -1) = 0$;
(3) L is a $Z/2Z$-boundary link.

Moreover, if these conditions hold ℓ is divisible by 4.

PROOF. Let $B = \pi'/\pi''$. Since $E_1(L) = \Delta_1(L)I_2$, the equivalence (1) $\Leftrightarrow$ (2) is immediate. If L is a $Z/2Z$-boundary link then $\eta(L) = 2$, by Theorem 5.11. If $\eta(L) = 2$ there is a Λ_2-epimorphism from $p : \tilde{\mathbb{Z}} \otimes_\Lambda B \to \tilde{\mathbb{Z}}$ with finite kernel. Then $C = (\pi/\pi'')/\mathrm{Ker}(p)$ has a presentation $\langle x, y, u \mid [x, y] = u^r, xux^{-1} = u^{-1}, yuy^{-1} = u^{-1} \rangle$ for some r, which must be odd since $C/C' \cong \mathbb{Z}^2$. After replacing y by a conjugate, if necessary, we may assume that $r = \pm 1$, and the presentation is then $\langle x, y \mid x^2 y = yx^2, xy^2 = y^2 x \rangle$. The centre of C

is $\zeta C = \langle x^2, y^2 \rangle \cong Z^2$. Let K be the preimage of ζC in π. Then $\pi/K = C/\zeta C \cong (Z/2Z) * (Z/2Z)$, where each factor is generated by the image of a meridian, and so L is a $Z/2Z$-boundary link.

Let M_1 and M_2 be disjoint surfaces in S^3 such that $\partial M_i = L_i$, for $i = 1, 2$. As the image of $[L_i] = [\partial M_i]$ in $H_1(M_i; Z)$ is divisible by 2, the linking number $\mathrm{lk}(L_1, L_2)$ is divisible by 4. $\square$

COROLLARY 8.7.1. *If L is a 2-component link such that $\Delta_1(L) = 0$ then L is a $Z/2Z$-boundary link.* $\square$

The linking number condition follows also from the Torres conditions.

A similar argument shows that if $\mathcal{L}$ is a concordance between two 2-component links $L(0)$ and $L(1)$ such that one end is a $Z/2Z$-boundary link then $\mathcal{L}$ is a $Z/2Z$-boundary concordance, i.e., it extends to an embedding of disjoint $(n+1)$-manifolds which meet $S^{n+2} \times \partial[0,1]$ transversely in Seifert hypersurfaces for the links. Is there a geometric proof that 2-component slice links are $Z/2Z$-boundary links?

The group C defined in Theorem 8.7 is a universal quotient of the groups of 2-component $Z/2Z$-boundary links, since it may be obtained from π by passing to $(\pi/\pi'')/(x+1, y+1)(\pi'/\pi'')$ and then factoring out the torsion subgroup of this quotient. It may also be described as the fibre product in the following pullback diagram:

$$
\begin{array}{ccc}
C & \longrightarrow & Z^2 \\
\downarrow & & \downarrow {\scriptstyle mod\ (2)} \\
(Z/2Z) * (Z/2Z) & \xrightarrow{ab} & (Z/2Z)^2
\end{array}
$$

Similarly, the pullback of C and $F(2)/F(2)_3$ over Z^2 is a universal quotient for 2-component $Z/2Z$-boundary links with linking number 0. The ring $Z[C]$ is a twisted quadratic extension of Λ_3, and is a noetherian domain of global dimension 4.

The above argument can be extended to show that if $L_1 = \partial M_1$ and $L_2 = \partial M_2$ where M_1 is a Z/pZ-manifold in S^3 and M_2 is a Z/qZ-manifold in $S^3 - M_1$ then $\mathrm{lk}(L_1, L_2)$ is divisible by pq. For then L_1 is homologous to $p\beta(M_1)$ in $S^3 - M_2$ and L_2 is homologous to $q\beta(M_2)$ in $S^3 - M_1$, where $\beta(M_i)$ is the closed curve representing the Bockstein of the characteristic class of the singular manifold M_i.

8.6. Topological concordance and F-isotopy

One of the triumphs of Freedman's work on 4-dimensional TOP surgery was his demonstration that a classical knot with Alexander polynomial 1 bounds a disc $D \subset D^4$ such that $\pi_1(D^4 \setminus D) \cong \mathbb{Z}$. All classical link groups other than 1, $\mathbb{Z}$ or $\mathbb{Z}^2$ have non-abelian free subgroups, and thus are outside the range of current surgery techniques. J.F.Davis has extended Freedman's construction of slice discs for such knots to obtain concordances from 2-component links with Alexander polynomial 1 to the Hopf link Ho [Da06]. (This was prompted by an incomplete earlier attempt, in [Hi85].)

Let L be a 2-component link with linking number 1 and let $T = S^1 \times S^1$. Let $V = X(L) \cup X(Ho)$, where the boundary components of $X(L)$ and $X(Ho)$ are identified by meridian and longitude preserving homeomorphisms. Let $a_H : X(Ho) \to T \times [0,1]$ be a homeomorphism. Since $T \times [0,1]$ is aspherical there is a map $a_L : (X(L), \partial X(L)) \to T \times ([0,1], \{0,1\})$ which induces an isomorphism on homology and restricts to a homeomorphism on the boundary. We may assume that these maps carry meridians to the standard generators of $\pi_1(T)$. Then we also have $V \cong X(L) \cup T \times [0,1]$, and the maps a_H and a_L together determine a map $\alpha : V \to T \times S^1$ which induces an isomorphism on first homology and hence has degree 1. Let β be the composite of α with projection onto T.

Suppose now that $\Delta_1(L) \doteq 1$. Then the components of L also have Alexander polynomial 1.

LEMMA 8.8. *If $\Delta_1(L) \doteq 1$ then α is a Λ_2-homology equivalence.*

PROOF. If $\Delta_1(L) \doteq 1$ then $X(L)'$ is acyclic, and so a_L is a homology equivalence over $\Lambda_2 = \mathbb{Z}[\pi_1(T)]$. Since a_H is a homeomorphism the lemma follows by a Mayer-Vietoris argument. $\square$

The strategy is to show first that V may be framed so that the image of β in $\Omega_3^{fr}(T)$ is trivial. (An Atiyah-Hirzebruch spectral sequence computation shows that $\Omega_3^{fr}(T) \cong \Omega_3^{fr} \oplus (\Omega_2^{fr})^2 \oplus \Omega_1^{fr}$. The middle terms are determined by the Arf invariants of the components of L, which are trivial since $\Delta_1(L) \doteq 1$, and the framing on V is easily modified to make the extreme terms 0.)

A null-cobordism determines a degree-1 normal map (F, B), where $F : (N, \partial N) \to T \times (D^2, S^1)$, N is a 4-manifold with $\partial N = V$ and $F|_{\partial N} = \alpha$, and B is a bundle map covering F. Since α is a Λ_2-homology equivalence, (F, B) has a surgery obstruction (rel ∂) in $L_4(\mathbb{Z}^2)$. The group $L_4(\mathbb{Z}^2)$ is isomorphic to the sum $L_4(1) \oplus L_2(1) = \mathbb{Z} \oplus (Z/2Z)$, by Shaneson's Splitting Theorem. The first summand is detected by the signature difference $\sigma(N) - \sigma(T \times D^2)$. The second summand is detected by a codimension 2 Kervaire invariant.

After replacing N by the connected sum with copies of the E_8-manifold, if necessary, we may assume that $\sigma(N) = 0$. If the codimension 2 Kervaire invariant is nontrivial, let Σ be a closed surface in N such that the image of $[\Sigma]$ generates $H_2(T;\mathbb{Z})$ and replace a product neighbourhood $\Sigma \times D^2$ by $\Sigma \times (T \setminus int\, D^2)$, where T is given the Arf invariant 1 framing. Thus there is an unobstructed surgery problem, and so there is a homotopy equivalence $h : P \to T \times D^2$, where P is a TOP 4-manifold with $\partial P = V$. The union $P \cup_{X(Ho)} D^4$ is a 1-connected homology ball with boundary S^3, and so is homeomorphic to D^4. The components of L bound locally flat discs which intersect in one point. On deleting a small ball centred on this point of intersection we obtain a TOP concordance from L to Ho.

A strict elementary F-isotopy on the ith component of a link L is the result of composition of L_i with a knot K in $S^1 \times D^2 = X(U)$ such that $\Delta_1(K \amalg U) \doteq 1$, where U is a trivial knot. If the Kervaire obstruction is always 0 for such links $K \amalg U$ then strictly F-isotopic links are TOP concordant.

8.7. Some examples

There is an algorithm due to J.H.C.Whitehead which makes it possible to decide whether a given set of elements of $F(\mu)$ generates the group. (See page 166 of [MKS].) If $\mu = 2$ a theorem of Nielsen allows us to replace elements by their conjugates, and leads to a criterion for showing that a homology boundary link is not a boundary link which is apparently independent of conditions in terms of ideals.

An element w_1 in $F(\mu)$ is *primitive* if there are elements $w_2, \ldots w_\mu$ such that $\{w_1, \ldots, w_\mu\}$ generates $F(\mu)$; equivalently, if there is an automorphism ψ of $F(\mu)$ such that $\psi(w_1) = x_1$, where $\{x_1, \ldots, x_\mu\}$ is a basis for $F(\mu)$.

THEOREM (Nielsen). *There is at most one conjugacy class of primitive elements of $F(2)$ with given image in $F(2)/F(2)' = \mathbb{Z}^2$.* □

See Theorem 3.9 of [**MKS**] for a proof.

It follows that if w_1 and w_2 in $F(2)$ represent a basis for $\mathbb{Z}^2$ then some conjugates of w_1 and w_2 generate $F(2)$ if and only if they are each primitive. This is clearly necessary. If w_1 and w_2 are each primitive then (after applying an automorphism of $F(2)$ if necessary) we may assume that $w_1 = x_1$ and $w_2 \equiv x_1^a x_2^b \bmod F(2)'$, for some a, b. Since the images of w_1 and w_2 generate $\mathbb{Z}^2$ we must have $b = \pm 1$. But the element $x_1^a x_2^{\pm 1}$ is clearly primitive, and so $w_2 = z x_1^a x_2^{\pm 1} z^{-1}$ for some z. Hence w_1 and $z^{-1} w_2 z$ generate $F(2)$. There is a simple procedure for finding a primitive word in the coset of $x^m y^n \bmod F(2)'$, whenever $(m, n) = 1$ [**OZ81**]. (See also [**GAR99**].)

The link L of Figure 1.5 is a ribbon link, and its ribbon group has a presentation $\langle a, w, x, y, z \mid axa^{-1} = y, wyw^{-1} = z, zwz^{-1} = x \rangle$. This presentation is Tietze-equivalent to $\langle a, w, z \mid wazw = zwaz \rangle$, and hence to $\langle b, w \mid \emptyset \rangle$, where $b = azw^{-1}$, and so L is a homology boundary link. This is the simplest non-splittable link with Alexander polynomial 0, since the non-splittable links with diagrams having at most 9 crossings all have nonzero Alexander polynomial [**Rol**]. The meridian a has image $bw^2 bw^{-1} b^{-1} w^{-1}$ in $\langle b, w \mid \emptyset \rangle$, and so a is not conjugate to b. Since a and b have the same image in $\mathbb{Z}^2$ it follows that the image of a is not primitive, and so L is not a boundary link. If $f : X(L) \to S^1 \vee S^1$ is a map realizing the projection of πL onto $\langle b, w \mid \emptyset \rangle$ the preimages of generic points of the circles corresponding to b and w are disjoint singular Seifert surfaces for L. We may arrange that one has three boundary components, all parallels of L_1, while the other has five boundary components, four of which are parallels of L_1 and the fifth is a parallel of L_2. (Note also that $E_2(L) = (x - 1, y^2 - y + 1)$, so $E_\mu(H(R))$ principal need not imply that $E_\mu(L)$ is principal.)

The 2-component 2-link with group $H(R) \cong F(2)$ constructed as in §7 of Chapter 1 from the ribbon of Figure 1.5 cannot be trivial, for a slice of a trivial link must be a boundary link. This gives a simple illustration of the result of [**Po71**]. Smythe's original homology boundary link may be obtained by giving the knotted ribbon of this link three half twists. It has a 13 crossing diagram, but the singular Seifert surfaces are easier to visualize than those for Figure 1.5 as they each have three boundary components. (The words corresponding to the meridians are a and $ax^2a^{-1}x^{-1}$.) As one of its longitudes is not in π'' it is not a boundary link [**Sm66**]. Moreover, for this link $E_2(L) \neq \overline{E_2(L)}$.

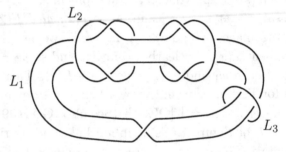

Figure 1.

The 2-component link $L = L_1 \cup L_2$ of Figure 1 is a $Z/2Z$-boundary link, with one spanning surface orientable. It extends to a ribbon map R with four throughcuts, and $H(R)$ has a presentation

$$\langle x_1, x_2, x_3, y_1, y_2, y_3 \mid y_1^{-1}x_1y_1 = x_2,\ y_3x_2y_3^{-1} = x_3,\ x_1^{-1}y_1x_1 = y_2,$$

$$x_3^{-1}y_2x_3 = y_3 \rangle.$$

This is Tietze-equivalent to $\langle x, y, a \mid xy^{-1}xyx^{-1} = (yxa)^{-1}xayxa \rangle$. (Here $x = x_1$, $y = y_1$ and $a = x_3x_1^{-1}$.) From this we see that $E_2(H(R)) = (x + 1, 2y - 1)$. As $E_2(H(R))$ is not principal, $H(R)$ cannot map onto $F(2)/F(2)''$, by Theorem 4.3. Therefore L is not a homology boundary link. (The link obtained by untwisting the lower ribbon disc is a boundary link.)

Further calculation shows that the longitudes of L are in π'', but $E_2(L) = (x+1, 2y-1)(x+1, 2-y)$, $\tau(E_2(L)) = (3, t+1)^2$ and $TA(L) = \Lambda_2/(x+1, 2-y)$ is pseudozero. Thus L is not a homology boundary link, by Theorem 4.15 or by Corollary 8.5.1. The 3-component link $\tilde{L}$ of Figure 1 is a homology boundary link such that $H_2(\tilde{L}; \Lambda_3)$ is not free.

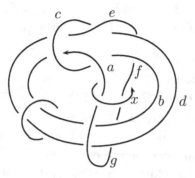

Figure 2.

The 2-component link L represented by the labeled arcs in Figure 2 is a ribbon link with trivial components. Its group π has a presentation

$$\langle a, b, c, d, e, f, g, x \mid g^{-1}cac^{-1}g = b,\ aba^{-1} = c,\ g^{-1}cec^{-1}g = d,$$

$$db^{-1}fbd^{-1} = e,\ ede^{-1} = c,\ db^{-1}x^{-1}fxbd^{-1} = g,\ xgx^{-1} = a,$$

$$fa^{-1}xaf^{-1} = x\rangle$$

where a, b, c, d, e, f, g, x are Wirtinger generators associated with the arcs so labeled in Figure 2. The words $c^{-1}ga^{-1}eg^{-1}cdb^{-1}xbd^{-1}x^{-1}$ and fa^{-1} represent longitudes commuting with a and x, respectively.

On replacing the generators b, c, d, e, f and g by $\beta = ba^{-1}, \gamma = ca^{-1}, \delta = da^{-1}, \varepsilon = ea^{-1}, \phi = fa^{-1}$ and $\theta = ga^{-1}$ we find that ϕ and $\delta\beta^{-1}x\beta\delta^{-1}x^{-1}$ represent the images of the longitudes, and π/π'' has a presentation

$$\langle a, x, \beta, \gamma, \delta, \varepsilon, \phi, \theta \mid x\phi = \phi x,\ a\beta = \gamma a,\ \phi a = a\varepsilon,\ \phi ax = x\varepsilon^{-1}\theta a\varepsilon,$$

$$\gamma a = \theta a\theta^{-1}\beta\gamma,\ \delta = \beta\varepsilon,\ \varepsilon a\delta = \gamma a\varepsilon,\ x\theta a = ax,\ F(a, \ldots, \theta)''\rangle.$$

These relations imply that ε, ϕ and $\delta\beta^{-1}$ represent $1 \in \pi/\pi''$, and so the longitudes of L are in π''. We may eliminate in turn the generators β, γ and θ to obtain the equivalent presentation $\langle a, x, | \ F(a,x)''\rangle$. Hence $\pi/\pi'' \cong F(2)/F(2)''$. In particular, L has the same Alexander module as a trivial 2-component link, and so cannot be proven nontrivial by the usual metabelian invariants.

This link extends to a ribbon map R with four throughcuts. The corresponding ribbon group $H(R)$ has the presentation

$$\langle a, b, c, g, x \mid aba^{-1} = c, cac^{-1}gbg^{-1}, \ xgx^{-1} = a\rangle.$$

This is equivalent to $\langle a, x, t \mid tx^{-1}a^{-1}xat^{-1}ata^{-1}\rangle$, which is a presentation for the group $G(-1,1)$ of Baumslag. Theorem 1.15 implies that $\pi/\pi_\omega \cong H(R) \cong G(-1,1)$, since $G(-1,1)$ is residually nilpotent [**Ba69**]. Hence L is not a homology boundary link, since $G(-1,1)$ is not free. On comparing the presentations of π and $H(R)$ we see that π is a semidirect product $\pi \cong \pi_\omega \rtimes H(R)$ (where π_ω is the normal closure in π of the subgroup generated by the image of ε).

The 3-component link of Figure 2 is a homology boundary link whose longitudes are all in π'', but which is not a boundary link.

CHAPTER 9

Symmetries

The word "symmetry" may be interpreted in several ways: we shall consider both questions dealing with symmetries of link *types* such as whether a link is invertible, amphicheiral or interchangeable, and also those dealing with group actions on a particular link. In §2 and §4 we sketch some results on amphicheirality and invertibility of simple $(2k-1)$-knots. Most of the rest of this chapter is devoted to group actions on classical links. The basic definitions and consequences of the Smith Conjecture are given in §3. In §5, §6 and §7 we assume that the action is defined by rotation about a disjoint axis, and in §8 we consider free actions. In the final section we return to knots invariant under rotations.

Tables of symmetries of knots and knot types for knots with diagrams having at most 10 crossings may be found in [**Ka2**].

9.1. Basic notions

If K is a 1-knot changing the ambient orientation inverts the meridians and changes the signs of intersection numbers in S^3, and so $H_1(X(rK); \Lambda) = \overline{H_1(X(K); \Lambda)}$ and $b_{rK}(\alpha, \beta) = -\overline{b_K(\alpha, \beta)} = -b_K(\beta, \alpha)$. Changing the string orientation also inverts meridians, but does not change intersection numbers, and so $H_1(X(K\rho); \Lambda) = \overline{H_1(X(K); \Lambda)}$ and $b_{K\rho}(\alpha, \beta) = \overline{b_K(\alpha, \beta)} = b_K(\beta, \alpha)$. Thus if K is invertible there is a bijection f from $H_1(X; \Lambda)$ to itself which is anti-linear ($f(r\alpha + s\beta) = \bar{r}f(\alpha) + \bar{s}f(\beta)$) and such that $b_K(f(\alpha), f(\beta)) = b_K(\beta, \alpha)$; if K is +amphicheiral there is an antilinear bijection g from $H_1(X; \Lambda)$ to itself such that $b_K(g(\alpha), g(\beta)) = -b_K(\beta, \alpha)$ while if K is −amphicheiral there is an automorphism h of $H_1(X; \Lambda)$ such that $b_K(h(\alpha), h(\beta)) = -b_K(\alpha, \beta)$. There are very similar results in

189

all odd dimensions. (Symmetries of simple even-dimensional knots may be studied through the Farber classification of such knots. We shall not consider them here.)

For links there are the additional features that a permutation of the components induces a permutation of the meridians, and for 1-links the longitudes become important. If $\gamma = (\epsilon_0, \ldots, \epsilon_\mu, \sigma)$ is in the link-symmetric group $LS(\mu)$ let $\gamma(t_i) = t_{\sigma(i)}^{\epsilon_0 \epsilon_i}$, for $1 \leq i \leq \mu$. If M is a Λ_μ-module let γM be the module with the same underlying abelian group and Λ_μ-action given by $\lambda.m = \gamma(\lambda)m$ for all $\lambda \in \Lambda_\mu$ and $m \in M$. If $b : M \times M \to \mathbb{Q}(t_1, \ldots, t_\mu)/\Lambda_\mu S$ is a bilinear pairing let $\gamma b(m, n) = \epsilon_0 \gamma(b(m, n))$ for all $m, n \in M$. We may then extend and summarize the conditions given earlier for knots as follows.

THEOREM 9.1. *If a 1-link L admits γ then there is an isomorphism $f : A(L) \cong \gamma A(L)$ such that $b_L(f(\alpha), f(\beta)) = \gamma b_L(\alpha, \beta)$ for all $\alpha, \beta \in \hat{t}A(L)_S$. Moreover $f(m_i) = u_i \epsilon_0 \epsilon_i m_{\sigma(i)}$ and $f(\ell_j) = v_j \epsilon_0 \epsilon_j \ell_{\sigma(j)}$, for some units $u_i, v_j \in \Lambda_\mu^\times$ and for all $1 \leq i, j \leq \mu$.* $\square$

The sign-determined ω-torsion satisfies $\overline{\omega_+(L)} = (-1)^\mu u \omega_+(L)$ for some "positive unit" $u \in \mathbb{Z}^\mu$, by Lemma 5.3.2 of [**Tu86**]. Since reversing all the components inverts all the meridians and so changes the homological orientation of $X(L)$ by a factor $(-1)^\mu$ we have $\omega_+(L) = (-1)^\mu \overline{\omega_+(L)} = \omega_+(L)$, up to positive units. Similarly, reflection changes the orientation of $X(L)$ and of its boundary components, and inverts the meridians. Therefore

$$\omega_+(rL) = (-1)^{2\mu-1}(-1)^{\alpha(L)}\overline{\omega_+(L)} = (-1)^{\alpha(L)+\mu-1}\omega_+(L),$$

up to positive units.

9.2. Symmetries of knot types

If K is a $(2k-1)$-knot let $B(K) = H_k(X(K); \Lambda)/T$, where T is the $\mathbb{Z}$-torsion submodule. Let $\Delta_i(K) = \Delta_{i-1}(B(K))$ (so the indexing is consistent with our earlier usage for 1-links). Then $\lambda(K) = \Delta_1(K)/\Delta_2(K)$ generates $Ann(B(K))$. The Blanchfield pairing $b_K : B(K) \times B(K) \to \mathbb{Q}(t)/\Lambda$ is $(-1)^{k+1}$-hermitean. If $k > 1$ then the isometry class of b_K is a complete invariant for the subclass of *simple* $(2k-1)$-knots (those for which $X(K)'$ is $(k-1)$-connected). Moreover the Witt class of b_K is a complete invariant for the concordance class of K.

The confirmation that certain classical knots are not invertible is notoriously delicate. As every $\delta \in \Lambda$ such that $\overline{\delta} \doteq \delta$ and $\varepsilon(\delta) = 1$ is the first Alexander polynomial of some invertible 1-knot K [**Sa83**], the Alexander polynomial conveys no useful information about invertibility of classical knots. The SFS invariants may be used to show that the pretzel knot $K(25, -3, 13)$ is not invertible [**FS64**].

THEOREM 9.2. *Let K be an invertible $(2k - 1)$-knot and δ a simple factor of $\lambda(K)$ such that $\overline{\delta} \doteq \delta$. Then the SFS row class of $B(K)/\delta B(K)$ is invariant under the involution. If moreover the power of δ that divides $\Delta_1(K)$ is odd there is an element u in the field $\mathbb{Q}\Lambda/(\delta)$ such that $u\overline{u} = (-1)^{k+1}$.*

Conversely, if K is simple, $k \geq 2$, $\Delta_1(K)$ is irreducible, the SFS row class of $B(K)$ contains an ideal J such that $\overline{J} = J$ and $u \in \Lambda/(\Delta_1(K))$ is such that $u\overline{u} = (-1)^{k+1}$ then K is invertible.

PROOF. If K is invertible then $B(K) \cong \overline{B(K)}$ and the first assertion follows easily.

Suppose that δ is a simple factor of $\lambda(K)$ such that $\overline{\delta} \doteq \delta$. We may assume in fact that $\overline{\lambda(K)} = \lambda(K)$ and that $\overline{\delta} = \delta$. Let $F = \mathbb{Q}\Lambda/(\delta)$. The Blanchfield pairing b_K takes values in $\lambda(K)^{-1}\Lambda/\Lambda < \mathbb{Q}(t)/\Lambda$ and so determines a $(-1)^{k+1}$-hermitean pairing c with values in $\Lambda/(\delta)$ by $c(u, v) = \lambda(K)b_K(u, v) \ mod \ (\delta)$. This induces a nonsingular $(-1)^{k+1}$-hermitean pairing (which we also denote by c) on the F-vector space $V = F \otimes_\Lambda B(K)$. The dimension of V is the power of δ that divides $\Delta_1(K)$. If f is an anti-automorphism of $B(K)$ such that

$$b_K(f(x), f(y)) = \overline{b(x, y)} = (-1)^{k+1}b(y, x),$$

for all x, y in $B(K)$, we must have (on choosing a basis for V) $\det(f)\det(c)\overline{\det(f)} = (-1)^{(k+1)s}\det(c)$. Thus $u = \det(f)$ satisfies $u\overline{u} = (-1)^{(k+1)s}$; in particular, $u\overline{u} = -1$ if s is odd and k is even.

Suppose that $\Delta_1(K)$ is irreducible. Then $B(K)$ is a torsion free $\Lambda/(\delta)$-module of rank 1. Hence there is an embedding j of $B(K)$ in

$\Lambda/(\delta)$ with image an ideal J. If $J = \overline{J}$ and $u \in \Lambda/(\delta)$ is such that $u\overline{u} = (-1)^{k+1}$ then $f(x) = j^{-1}(u\overline{j(x)})$ is an antiautomorphism of $B(K)$. The Blanchfield pairing is necessarily of the form $b_K(x, y) = mj(x)\overline{j(y)}$ for all $x, y \in B(K)$, and for some $m \in \mathbb{Q}\Lambda/(\delta)$ such that $\overline{m} = (-1)^{k+1}m$. We then see that $b_K(f(x), f(y)) = \overline{b(x, y)}$ and so K is invertible. $\qquad\square$

The strongest results on the polynomials of amphicheiral knots known at present are due to Hartley [**Ha80, Ha80'**]. We say that $r(t) \in \Lambda$ is *of type* X if $r(t) \doteq f_0(t)^2 f_1(t) \cdots f_n(t)$ for some $n \geq 0$, where $f_i \doteq \overline{f_i}$ for all i and $f_i \doteq_2 f_0^{2^i} \nu_\ell^{2^{i-1}}$ for some fixed odd ℓ and all $i > 0$, where $\doteq_2$ means the images in $\mathbb{F}_2\Lambda$ are equal up to multiplication by a unit. Then he proved

THEOREM. [**Ha80'**] *Let K be a 1-knot.*

(1) *If K is $-$amphicheiral then $\Delta_1(K)(t^2) \doteq f(t)f(t^{-1})$ for some f such that $f(t) \doteq f(-t^{-1})$.*

(2) *If K is $+$amphicheiral then there are polynomials $r_i(t)$ of type X and odd integers $\alpha(i)$, for $1 \leq i \leq m$, such that $\Delta_1(K)(t^2) \doteq r_1(t^{\alpha(1)}) \ldots r_m(t^{\alpha(m)})$.* $\qquad\square$

An amphicheiral invertible knot satisfies both conditions. The proof uses the Jaco-Shalen-Johansson decomposition of an irreducible 3-manifold along essential tori and Mostow rigidity to reduce to cases already considered by Kawauchi. (The congruence conditions derive ultimately from the Murasugi congruences, given below.)

Any knot polynomial $\Delta_1(K)$ satisfying (1) is the Alexander polynomial of some strongly $-$amphicheiral knot. (See the Figure in the announcement [**Va79**]!) The following theorem of Coray and Michel may be used to show that (1) does not follow from consideration of the Blanchfield pairing alone.

THEOREM 9.3. [**CM83**] *Let K be a $-$amphicheiral $(2k-1)$-knot. Suppose that $\Delta_1(K) \doteq \delta\overline{\delta}\Pi\gamma_i$, where the γ_i are distinct irreducible polynomials such that $\overline{\gamma_i} \doteq \gamma_i$. Then for each i there is an element u_i in the integral closure of $\Lambda/(\gamma_i)$ such that $u_i\overline{u_i} = -1$.*

Conversely, if K is simple, $k \geq 2$ and $u \in \Lambda/(\Delta_1(K))$ is such that $u\overline{u} = -1$ then K is $-$amphicheiral.

PROOF. (Sketch.) We shall assume that $\lambda(K)$ is irreducible. Then $\Delta_1(K) = \lambda(K)^s$, for some $s \geq 1$. We may also assume that s is odd, for otherwise there is nothing to prove. Let $F = \mathbb{Q}\Lambda/(\lambda(K))$ and $V = \mathbb{Q} \otimes B(K) \cong F^s$. The Blanchfield pairing determines a $(-1)^{k+1}$-hermitean pairing from $b : V \times V \to F$. Since K is $-$amphicheiral there is an automorphism h of $B(K)$ such that $b_K(h(x), h(y)) = -b_K(x, y)$ for all $x, y \in B(K)$. Let $u = \det(\mathbb{Q} \otimes h)$. Then u is in the integral closure of $R = \Lambda/(\lambda(K))$ since it is an automorphism of the finitely generated R-module $\wedge_s B(K)$. Moreover, $u.\det(b).\bar{u} = (-1)^s \det(b) = -\det(b)$, and so $u\bar{u} = -1$. Conversely, multiplication by such a u gives an automorphism of $B(K)$ which changes the sign of b_K.   □

The proof in the general case proceeds by first restricting the Blanchfield pairing to the δ_i-primary submodule $B(K)(\delta_i)$ to obtain a non-singular pairing. Then, there are non-singular pairings induced on the subquotients $\mathrm{Ker}(\delta_i^j)/(\mathrm{Ker}(\delta_i^{j-1}) + \delta_i\mathrm{Ker}(\delta_i^{j+1}))$, for each $j \geq 0$ (see [Mi69]) and at least one of these modules must have odd rank.

The polynomial $\Delta_1(8_1) = 3t^2 - 7t + 3$ does not satisfy condition (1) of Hartley's Theorem so 8_1 is not $-$amphicheiral, but $-1 = (\frac{3+\sqrt{13}}{2})(\frac{3-\sqrt{13}}{2})$ in $\Lambda/(\Delta_1(8_1)) = \mathbb{Z}[\frac{1+\sqrt{13}}{6}]$.

There is a corresponding result for $+$amphicheiral knots.

THEOREM 9.4. *Let K be a $+$amphicheiral $(2k-1)$-knot and δ a simple factor of $\lambda(K)$ such that $\bar{\delta} \doteq \delta$. Then the SFS row class of $B(K)/\delta B(K)$ is invariant under the involution. If moreover the power of δ that divides $\Delta_1(K)$ is odd there is an element v in the field $\mathbb{Q}\Lambda/(\delta)$ such that $v\bar{v} = (-1)^k$.*

Conversely, if K is simple, $k \geq 2$, $\Delta_1(K)$ is irreducible, the SFS row class of $B(K)$ contains an ideal J such that $\bar{J} = J$ and $v \in \Lambda/(\Delta_1(K))$ is such that $v\bar{v} = (-1)^k$ then K is $+$amphicheiral.

PROOF. The argument is similar to that of Theorem 9.2.   □

We may modify the notion of link symmetry by relaxing the equivalence relation of ambient isotopy to concordance or, if $\mu > 1$, to isotopy or even homotopy. If K is concordant to $-K$ then either K is slice or its concordance class has order 2. It is unknown whether every class of order 2 in the classical knot concordance group contains a $-$amphicheiral 1-knot, but Coray and Michel gave counterexamples in higher dimensions, based on the following theorem.

THEOREM. [**CM83**] *Let K be a $(2k-1)$-knot such that $\Delta_1(K)$ is irreducible. Let O_F be the ring of algebraic integers in $F = \mathbb{Q}\Lambda/(\Delta_1(K))$. Then*

(1) *the Witt class b_K has order 2 in $W_\epsilon(\mathbb{Q}(t), \Lambda, -)$ (with $\epsilon = (-1)^{k+1}$) if and only if there is an element $\alpha \in F$ such that $\alpha\overline{\alpha} = -1$.*

(2) *If K is concordant to some $-$amphicheiral knot there is a $\beta \in O_F[1/\delta(0)]$ such that $\beta\overline{\beta} = -1$. If $\delta(0)$ is prime there is such a β in O_F.* □

The simplest irreducible polynomials satisfying (1) but not (2) are perhaps $367t^2 - 735t + 367$ and $t^4 - 72t^3 + 143t^2 - 72t + 1$.

A classical knot represents an element of infinite order in the concordance group if and only if some Milnor signature is nonzero. No such knot can be $-$amphicheiral.

THEOREM 9.5. *Let K be a $(2k-1)$-knot with $k \geq 2$ and let δ be an irreducible factor of $\lambda(K)$ such that $\overline{\delta} \doteq \delta$ and such that the power of δ dividing $\Delta_1(K)$ is odd. If K is concordant to an invertible, $+$amphicheiral or $-$amphicheiral knot then there is an element $w \in \mathbb{Q}\Lambda/(\delta)$ such that $w\overline{w} = (-1)^{k+1}$, $(-1)^k$ or -1, respectively.*

PROOF. Let K' be a knot concordant to K with the asserted symmetry, and let K_b be the simple knot with Blanchfield pairing $b_{K'}$. Then K_b is concordant to K and has the same symmetry as K'.

A Blanchfield pairing b on a module M is Witt equivalent to one on a subquotient of M with squarefree annihilator, in such a way that any (anti)automorphism of M which transposes or changes the sign of b induces a similar (anti)automorphism of the new pairing. If δ is as stated then the power of δ dividing $\Delta_1(K')$ must be odd, for any K' concordant to K. The result now follows from Theorems 9.2, 9.3 and 9.4.   □

The following result on $-$amphicheiral 1-links is due to Traldi [**Tr84**], who showed that the Milnor invariants of such a link must vanish. (See Chapter 12.) This implies that all the nilpotent quotients of πL are free, and hence $\alpha(L) = \mu$.

THEOREM 9.6. *Let L be a $-$amphicheiral μ-component 1-link. Then $E_{\mu-1}(L) = 0$.*

PROOF. Let $h : S^3 \to S^3$ be an orientation reversing homeomorphism such that $h \circ L_i = L_i\rho = L_i \circ r_1$ for $1 \leq i \leq \mu$. We may assume that h fixes a point $* \in X(L)$ and points $P_i \in \partial X(L_i)$, and choose paths γ_i from $*$ to P_i, for $1 \leq i \leq \mu$. Then h induces an automorphism θ of $A(L)$ such that $\theta(m_i) = U_i m_i$ and $\theta(\ell_i) = -U_i\ell_i$, where $U_i = \Pi t_j^{e(ij)}$ is the image in π/π' of the loop $\gamma_i \cup h(\gamma_i)$, for $1 \leq i \leq \mu$.

We shall show that $\ell_i \in (I_\mu)^n A(L)$ for all $1 \leq i \leq \mu$ and $n \geq 0$, by induction on n. This is clear when $n = 0$. Suppose that ℓ_i is in $(I_\mu)^n A(L)$, for $1 \leq i \leq \mu$. Then $A(L)/(I_\mu)^{n+1} A(L) \cong (\Lambda_\mu/(I_\mu)^{n+1})^\mu$. (See Theorem 4.13.) Moreover $2\ell_i = 0$ in this quotient, since $U_i \equiv 1 \ mod \ I_\mu$, and so ℓ_i is in $(I_\mu)^{n+1} A(L)$. It now follows as in Theorem 4.14 that the longitudes are in $\pi(\infty)$ and so $E_{\mu-1}(L) = 0$. $\square$

Traldi shows also that if L is a $+$amphicheiral 2-component link then $E_1(L) = 0$. (This was first observed in [**Co70**].) Turaev has used the properties of $\omega_+(L)$ to show that if L is concordant to rL or $-L$ then $\alpha(L) \equiv \mu + 1 \ mod \ (2)$ [**Tu86**]. (Note that the Borromean rings Bo is a strongly $+$amphicheiral 3-component link with nonzero first Alexander ideal.)

Is there an unsplittable μ-component link L with $\mu > 1$ and symmetry group $\Sigma(L) = LS(\mu)$? This is not known even for $\mu = 2$. (See [**CCMP12**] for examples illustrating 21 of the 27 conjugacy classes of subgroups of $LS(2)$.)

9.3. Group actions on links

If p is a prime the fixed point set of a Z/pZ-action on a homology n-sphere is a Z/pZ-homology sphere of even or odd codimension (the empty set $S^{-1} = \emptyset$ having codimension $n + 1$) according as whether the map preserves or reverses the orientation, by Smith theory. (Consideration of orthogonal matrix actions on spheres shows that there are no further constraints on the codimension.) If the fixed point set has dimension > 2 it need not be simply connected. The Smith theory makes no tameness assumptions, and the fixed point set can be wild, even for an involution of S^3. However, we shall always assume that the fixed point sets of the actions are locally flat submanifolds.

In particular, if the fixed point set of an orientation preserving homeomorphism h of finite order q on a homology 3-sphere M is nonempty it is homeomorphic to S^1, and is unknotted if $\Sigma = S^3$, by the resolution of the celebrated Smith Conjecture [**BM**]. (Hence a finite group G which acts effectively on S^3 and leaves invariant a nontrivial knot K must be cyclic or dihedral.)

Suppose that L is a μ-component link in M which is invariant under such a homeomorphism h. There are three cases of particular interest. If the cyclic group $\langle h \rangle \cong Z/qZ$ acts freely on M then L has *free period* q. If it fixes pointwise an axis $A \cong S^1$ disjoint from L then L has *semifree period* q. (For q an odd prime these are the only possibilities.) If each component of L meets the fixed point set A in two points (equivalently, if $h|_{L_i}$ is an orientation reversing self map of L_i, for all $1 \le i \le \mu$) then $q = 2$ and L is *strongly invertible*.

If h is orientation reversing then it has even order, and there are further possibilities. In particular, if $h^2 = id_M$ and $h(L_i) = L_i$ for all $1 \le i \le \mu$ then L is *strongly +amphicheiral* if h preserves the orientations of each component of L and is *strongly −amphicheiral* if h reverses the orientations of each component of L. If moreover h fixes just two points of S^3 then in the latter case L must be a knot.

A link L is *periodically +amphicheiral* if it is componentwise invariant under an orientation reversing homeomorphism h of finite order q such that $h|_{L_i}$ is orientation preserving, for all $1 \le i \le \mu$. On replacing h by an odd power of h if necessary we may assume that h has order 2^r for some $r \geq 1$. Such a link is +amphicheiral, and is strongly +amphicheiral if we may take $r = 1$. The analogous definitions of "periodically −amphicheiral" and "periodically invertible" lead to nothing new. For if $h|_{L_i}$ is orientation reversing for all i then h^2 fixes L pointwise. Therefore either $\mu = 1$ and L is unknotted (by the Smith conjecture for involutions) or $h^2 = 1$. In either case L is strongly −amphicheiral or strongly invertible, respectively. Similarly, a periodically +amphicheiral link is either strongly +amphicheiral or has semifree period $q/2$. However, there are invertible knots which are not strongly invertible, and amphicheiral knots which are not strongly amphicheiral [**Ha80**].

The torus links provide useful examples of many of these types of symmetry. Let m, n be integers ≥ 1. The (m, n)-torus link $L(m, n)$ has image the subset $\{(u, v) \in \mathbb{C}^2 \mid u^m + v^n = 0\}$ of the unit sphere in $\mathbb{C}^2$. If $(m, n) = 1$ then $L(m, n)$ is a knot, which we denote by $K_{m,n}$ below. It is nontrivial if $m, n > 1$. The S^1-action on $S^3 \subset \mathbb{C}^2$ given by $z(u, v) = (z^n u, z^m v)$ for all $z \in S^1$ and $(u, v) \in S^3$ leaves $L(m, n)$ invariant. This induced action of the qth roots of unity is free if $(q, mn) = 1$, and is semifree if q divides m or n. We shall see below that these are the only free or semifree periods of $K_{m,n}$. Complex conjugation induces an orientation preserving involution of S^3 which leaves $L(m, n)$ invariant and reverses the string orientations, so torus links are strongly invertible. (However, nontrivial torus knots are never amphicheiral.)

If a knot K is not a torus knot then it has only finitely many finite group actions, up to conjugacy in $\mathrm{Diff}(S^3, K)$ [**Fl85**].

9.4. Strong symmetries

We shall give here some necessary conditions for odd-dimensional knots to admit strong symmetries of order 2.

THEOREM 9.7. *Let K be a strongly −amphicheiral $(2k − 1)$-knot. Then there is an automorphism θ of $H_k(X; \Lambda)$ such that $\theta^2 = id$ or $t.id$ and $b(\theta(x), \theta(y)) = -b(x, y)$ for all x, y. If $\theta^2 = id$ then $\mathbb{Z}[\frac{1}{2}] \otimes b$ is hyperbolic. Conversely, if K is simple, $q > 1$ and b is hyperbolic* *then K is amphicheiral. If $\Delta(K) = \Delta_0(H_k(X;\Lambda))$ is an irreducible quadratic, $\Delta(K) = at^2 + (1-2a)t + a$ then $-a$ is a perfect square.*

PROOF. Let h be an involution of S^{n+2} which leaves an n-knot K invariant. Let t be the meridianal generator of the covering group $Aut(X'/X)$. Then $h|_X$ lifts to a map $\tilde{h}: X' \to X'$ such that $\tilde{h} \circ t^\epsilon = t \circ \tilde{h}$, where $\epsilon = 1$ or -1 as $h|_X$ preserves or reverses the meridians. Moreover $(\tilde{h})^2 = t^m$ for some m, as it lifts id_X. If h has nonempty fixed point set we may assume that $(\tilde{h})^2 = id_{X'}$; if $\epsilon = +1$ we may assume that $m = 0$ or 1. The first assertion follows from $n = 2k-1$ and $\epsilon = 1$ on taking θ to be the automorphism induced by $\tilde{h}$.

If $\theta^2 = id$ let $A_\pm$ be the ± 1 eigenspaces of $\mathbb{Z}[\frac{1}{2}] \otimes \theta$. They are self-orthogonal with respect to $\mathbb{Z}[\frac{1}{2}] \otimes b$, and $H_k(X;\Lambda[\frac{1}{2}]) \cong A_- \oplus A_+$. Conversely, if b is hyperbolic then the map which multiplies one of a pair of complementary self-orthogonal summands of $H_k(X;\Lambda)$ by -1 induces an isometry of b with $-b$, and so if K is simple and $q > 1$ then K is $-$amphicheiral.

If $\Delta(K)$ is an irreducible quadratic then $\mathbb{Q} \otimes b$ and hence $\mathbb{Z}[\frac{1}{2}] \otimes b$ cannot be hyperbolic, so we must have $\theta^2 = t.id$. Therefore the image of t in $F = \mathbb{Q}\Lambda/(\Delta(K))$ is a perfect square in F. Since $a(t-1)^2 = -t$ in F it follows that $-a$ is a square in F. The discriminant of F divides $1 - 4a$ and cannot be 1, since $F \neq \mathbb{Q}$. Therefore $-a$ is a square in $\mathbb{Q}$ and hence in $\mathbb{Z}$. $\square$

The Blanchfield pairing of an odd dimensional knot is also hyperbolic if the knot is doubly null concordant. A simple $(2k-1)$-knot (with $k > 1$) with hyperbolic Blanchfield pairing is doubly null concordant. When is a strongly $-$amphicheiral knot doubly null concordant, or vice-versa? For any knot K the sum $K\sharp - K$ is both doubly null concordant and strongly $-$amphicheiral. On the other hand 4_1 is strongly $-$amphicheiral but not slice, while 9_{46} is doubly null concordant but not amphicheiral.

A closer connection between concordance and strong symmetries is given by the following result of Long [**Lo84**].

THEOREM 9.8. *Let b be an ϵ-linking pairing on a knot module M with an antilinear involution θ such that $b(\theta(m), \theta(n)) = -b(n, m)$ for all $m, n \in M$. Then the Witt class of b is trivial.*

PROOF. By assumption, $\theta^2(m) = m$ and $\theta(tm) = t^{-1}\theta(m)$ for all $m \in M$. Let $\eta = \pm 1$ and $M_\eta = \mathrm{Ker}(\theta - \eta id)$. If $M_\eta = 0$ then $\theta - \eta id$ is 1-1 so $\theta + \eta id = 0$. Hence $tm = t^{-1}m$ for all $m \in M$ and so $(t^2 - 1)M = 0$. Since M is a pure knot module this implies that $M = 0$. Thus we may assume that neither M_+ nor M_- is 0. If $m \in M_\eta$ then

$$b(m, m) = -b(\theta(m), \theta(m)) = -b(\eta m, \eta m) = -b(m, m)$$

and so is 0. Therefore $W = \Lambda m$ is a self orthogonal submodule of M as $\theta(W) = W$, $W^\perp$ and $W^{\perp\perp}$ are also preserved by θ. Then b induces a Witt equivalent pairing b_1 on $M_1 = W^\perp/W^{\perp\perp}$, and θ induces an antilinear involution θ_1 of M_1 which changes the sign of b_1. If $M \neq 0$ then $dim_\mathbb{Q}\mathbb{Q} \otimes M_1 < dim_\mathbb{Q}M$, and so by a finite induction we conclude that b is Witt equivalent to the trivial pairing. $\square$

COROLLARY 9.8.1. *Every strongly $+$amphicheiral 1-knot is algebraically slice. Every strongly invertible $(4k - 1)$-knot or strongly $+$amphicheiral $(4k + 1)$-knot (with $k > 0$) is slice.*

PROOF. The involution must fix a point in $X(K)$, and so induces an antilinear involution θ as in the theorem. $\square$

If a 1-knot K is not concordant to its reverse $K\rho$ then $K \sharp rK$ is strongly $+$amphicheiral but is not slice. Is a Blanchfield pairing admitting an antilinear involution as in Theorem 9.7 hyperbolic?

9.5. Semifree periods – the Murasugi conditions

Let L be a ν-component link in an (oriented) homology 3-sphere M which is invariant under a rotation h of order q about a disjoint axis A. The orbit space $\overline{M} = M/\langle h \rangle$ is again a homology sphere, and the orbit map $c : M \to \overline{M}$ is a q-fold cyclic branched cover, branched over $\overline{A} = c(A)$. Moreover, $c(L)$ is the image of a μ-component link $\overline{L}$ in $\overline{M}$, for some $\mu \leq \nu$. We shall assume that $\overline{A}$, $\overline{L}$ and $\overline{M}$ are compatibly oriented with A, L and M.

Let $X = M \backslash L$, $Y = M \backslash (L \cup A)$, $\overline{X} = \overline{M} \backslash \overline{L}$ and $\overline{Y} = \overline{M} \backslash (\overline{L} \cup \overline{A})$, and choose basepoints $* \in Y$ and $\overline{*} = c(*) \in \overline{Y}$. The map on homology induced by $c|_X$ carries meridians to meridians and so is onto. Hence it determines an epimorphism $\gamma : \Lambda_\nu \to \Lambda_\mu$. Let m and $\overline{m}$ be the meridians of A and $\overline{A}$, and let $u \in Aut(\overline{Y}'/\overline{Y})$ be the covering transformation determined by $\overline{m}$. Let $x : X^\gamma \to X$ and $y : Y^\gamma \to Y$ be the covers induced from the maximal abelian cover of $\overline{X}$ by $c|_X$ and $c|_Y$. Then x is the q-fold cyclic branched cover of $\overline{X}'$, branched over the lifts of A, while $Y^\gamma = \overline{Y}'/\langle u^q \rangle$ and cy is an (unbranched) abelian covering of $\overline{Y}$. In what follows we shall view Λ_μ as both a subring and a quotient of $\Lambda_{\mu+1} = \Lambda_\mu[u, u^{-1}]$.

THEOREM 9.9. *Let L be a ν-component link in a homology 3-sphere M with a semifree period q. Then*

$$w\gamma(\Delta_1(L)) \doteq \Delta_1(\overline{L}) Res(\Delta_1(\overline{L} \cup \overline{A}), \nu_q),$$

where $w = t - 1$ if $\nu > \mu = 1$ and $w = 1$ otherwise, and $\nu_q = (u^q - 1)/(u - 1)$.

PROOF. We shall show that each side of the equation equals $\Delta_0(H_1(X^\gamma; \mathbb{Z}))$. Let $\gamma A(L) = H_1(X, *; c^* \Lambda_\mu) = H_1(X^\gamma, x^{-1}(*); \mathbb{Z})$ be the corresponding relative module. Then

$$\Delta_0(H_1(X^\gamma; \mathbb{Z})) = \Delta_0(H_1(X; c^* \Lambda_\mu)) = \Delta_1(\gamma A(L)),$$

by Theorem 3.4. Since $E_1(\gamma A(L)) = \gamma(E_1(L))$, which is $\gamma(\Delta_1(L))$, if $\nu = 1$, and $\gamma(I_\nu)\gamma(\Delta_1(L))$, if $\nu > 1$, by part (1) of Theorem 4.12, it follows that $\Delta_0(H_1(X^\gamma; \mathbb{Z})) \doteq w\gamma(\Delta_1(L))$.

Let $H = H_1(\overline{Y}; \Lambda_{\mu+1})$. The Wang sequence for the projection of $\overline{Y}'$ onto $\overline{Y}'/\langle u \rangle$ gives a short exact sequence

$$0 \to H/(u - 1)H \to H_1(\overline{Y}'/\langle u \rangle; \mathbb{Z}) \to \mathbb{Z} \to 0,$$

in which the right-hand term is generated by the image of $\overline{m}$. There is a similar exact sequence

$$0 \to H/(u^q - 1)H \to H_1(Y^\gamma; \mathbb{Z}) \to \mathbb{Z} \to 0,$$

in which the right-hand term is generated by the image of m. We may complete Y and $\overline{Y}$ to X and $\overline{X}$ by adding a 2-cell along meridians of the axes, and then adding a 3-cell. Similarly, we may obtain $\overline{X}'$ from $\overline{Y}'/\langle u \rangle$ and X^γ from $Y^\gamma = \overline{Y}'/\langle u^q \rangle$ by adding 2-cells along the lifts of meridians and then 3-cells. The transfer tr from $H_1(\overline{Y}'/\langle u \rangle; \mathbb{Z})$ to $H_1(Y^\gamma; \mathbb{Z})$ sends the class of h to the class of $\nu_q h$, for $h \in H$, while $tr[\overline{m}] = [m]$. We obtain the exact sequence

$$H_1(\overline{X}; \Lambda_\mu) \xrightarrow{tr} H_1(X^\gamma; \mathbb{Z}) \to H/\nu_q H \to 0.$$

If $\mu = 1$ then H has a square presentation matrix, and so $E_0(H)$ is principal. If $\mu > 1$ then $I_{\mu+1\wp} = \Lambda_{\mu\wp}[u, u^{-1}]$ for any principal prime ideal $\wp$ of Λ_μ, as $t_1 - 1$ and $t_2 - 1$ are relatively prime. Thus $E_0(H)_\wp$ is generated by the image of $\Delta_0(H) \doteq \Delta_1(\overline{L} \cup \overline{A})$, by Theorem 4.12. In all cases $\Delta_0(H/\nu_q H) \doteq Res(\Delta_1(\overline{L} \cup \overline{A}), \nu_q)$, by Theorem 3.13.

If $\Delta_0(H/\nu_q H) = 0$ then $H/\nu_q H$ has positive rank as a Λ_μ-module. Hence so does $H_1(X^\gamma; \mathbb{Z})$, which maps onto $H/\nu_q H$, and both sides of the equation are 0.

If $\Delta_0(H/\nu_q H) \neq 0$ then $E_0(H_\wp) = (\Delta_0(H)) \neq 0$, for $\wp$ any principal prime ideal of Λ_μ. Hence the $\Lambda_{\mu+1\wp}$-module $H_\wp$ has a square presentation matrix with nonzero determinant, by Theorem 3.7, and has no nontrivial pseudonull submodule, by Theorem 3.8. As ν_q and $\Delta_0(H)$ have no common factors, multiplication by ν_q is injective on $H_\wp$. It follows easily that the localization of the transfer homomorphism is injective. Thus we may apply Theorem 3.12 to the above sequence to get $\Delta_0(H_1(X^\gamma; \mathbb{Z})) \doteq \Delta_1(\overline{L})\Delta_0(H/\nu_q H)$ in $\Lambda_{\mu\wp}$. As this holds for all principal prime ideals of Λ_μ, this completes the proof.   □

This result was first proven by Murasugi, for knots in S^3 ([**Mu71**] – see the Corollary below), and in general by Sakuma [**Sk81**]; an independent proof requiring the hypothesis that $lk(A, L_i) \neq 0$ for at least one component of L was given in [**Hi81'**]. The present argument combines ideas from the latter two proofs. (Compare also the discussion of cyclic branched covers of homology 3-spheres in §7 of Chapter 5.) The proof can be simplified when $\mu = 1$ and $|lk(A, K)| = 1$, for then $[m]$ generates an infinite cyclic direct summand in $H_1(Y^\gamma; \mathbb{Z})$, so that $H_1(X^\gamma; \mathbb{Z}) \cong H/(u^q - 1)H$, and we may apply Theorem 3.13 directly to this quotient.

Davis and Livingston conjecture that any knot polynomial satisfying the *Murasugi conditions* (i.e., with a factorization as in the theorem) is the polynomial of some knot K with semifree period q, and with $\ell = |\text{lk}(A, K)|$ as determined from this equation in Corollary 1 below. They have confirmed this when $\ell = 1$, by surgically modifying one of the components of the Hopf link [**DL91**]. Thus for knots the Murasugi conditions may well be best possible in terms of the Alexander polynomial.

COROLLARY 9.9.1. *If K is a 1-knot which has a semifree period of prime power order $q = p^r > 1$ then $\Delta_1(K) \equiv (\nu_\ell)^{q-1}(\Delta_1(\overline{K}))^q$ mod (p), where ℓ is prime to p. Therefore either $\Delta_1(K) \equiv 1$ mod (p) or the degree of $\Delta_1(K)$ is at least $q - 1$. Furthermore $\Delta_1(K)$ is divisible by $\Delta_1(\overline{K})$.*

PROOF. This follows immediately from the theorem and the Torres conditions, since every p^{th} root of unity is congruent to 1 mod (p). On reducing the equation *mod* $(t - 1, p)$ we get $1 \equiv \ell^{q-1}$ *mod* (p), and so $(\ell, p) = 1$. (This is also clear for topological reasons, since $\ell = \text{lk}(\overline{A}, \overline{K})$ and the link K covering $\overline{K}$ is connected.) □

We do not need the full strength of Theorem 3.13 to obtain this *Murasugi congruence*. For we may take R to be the PID $\mathbb{F}_p\Lambda$ and $q = p^r$, so that $u^q - 1 = (u - 1)^q$. If $N = H_1(\overline{Y}; \mathbb{F}_p\Lambda_2)$ then $E_0(N)$ is principal. The module $Q = N/(u^q - 1)N$ is an R-torsion module if and only if $\varepsilon(\Delta_0(N)) = \Delta_0(N)(t, 1) \neq 0$, in which case multiplication by $(u - 1)$ is injective on N and so $N/(u - 1)N \cong (u-1)^j N/(u-1)^{j+1} N$ for all $j > 0$. Hence $\Delta_0(Q)(t) \doteq \Delta_0(N)(t, 1)^q$ in $\mathbb{F}_p\Lambda$.

This congruence extends to higher dimensional knots invariant under a $Z/p^r Z$ action which is free off a 1-dimensional fixed point set S^1 [**DL91'**].

COROLLARY 9.9.2. *Let L be a 2-component link with linking number ℓ which is componentwise invariant under a rotation h of order q about a disjoint axis A. Then q divides ℓ.*

PROOF. We may assume that $q = p^r$ for some prime p and $r \geq 1$, and induct on r. Let $\overline{A}, \overline{L}, \overline{L_1}$ and $\overline{L_2}$ be the images of A, L and its component knots in $\overline{M} = M/\langle h^{q/p}\rangle$, and let $\ell_i = \mathrm{lk}(\overline{A}, \overline{L}_i)$. These linking numbers ℓ_1 and ℓ_2 are relatively prime to p, by Corollary 9.9.1. In particular, they are nonzero, so

$$|\ell| = |\Delta_1(L)(1,1)| = |\Delta_1(\overline{L})(1,1)|\Pi_{j=1}^{j=p-1}\Delta_1(\overline{L} \cup \overline{A})(1,1,\zeta_p^j),$$

by the theorem and the Torres conditions. Moreover,

$$\Delta_1(\overline{L} \cup \overline{A})(1,1,z) \doteq (z-1)\nu_{\ell_1}(z)\nu_{\ell_2}(z)\Delta_1(\overline{A})(z).$$

Therefore $|\ell| = p|\overline{\ell}|\Pi_{j=1}^{j=p-1}|\Delta_1(\overline{A})(\zeta_p^j)| = p|\overline{\ell}|$, since $\overline{M}$ is a homology 3-sphere. This proves the Lemma if $q = p$, i.e., if $r = 1$. Now h induces a rotation of order $q/p = p^{r-1}$ on $\overline{M}$ about the axis $\overline{A}$, leaving the components of $\overline{L}$ invariant, and so p^{r-1} divides $\overline{\ell}$, by the inductive hypothesis. Hence $q = p^r$ divides ℓ.   □

In general a rotation of a 2-component link may interchange the components, and so its order need only divide twice the linking number. (The Hopf link in S^3 has a semifree period 2.)

If $\nu > \mu = 1$ it is easy to see that $(q, \ell) > 1$, where $\ell = \mathrm{lk}(\overline{A}, \overline{L})$.

COROLLARY 9.9.3. *If $m, n > 1$ the (m, n)-torus knot $K_{m,n}$ has semifree period q if and only if q divides m or n.*

PROOF. We have already seen that $K_{m,n}$ has such symmetries.

Suppose $K_{m,n}$ has a semifree period q with axis of rotation A, and let $\lambda = \mathrm{lk}(A, K_{m,n})$. The knot has semifree periods with the same axis, for all divisors of q. Let p be a prime divisor of q which does not divide n. Let $m = p^r a$ where $r \geq 0$ and $(a, p) = 1$ and let $f = \nu_\lambda(t)\Delta_1(\overline{K}_{m,n})$. The congruence of Corollary 1 gives

$$((t^\lambda - 1)/((t^n - 1))(t^{an} - 1)/(t^a - 1))^{p^r} \equiv f^p \ mod \ (p)$$

and f has degree $\geq \lambda - 1$. Moreover $(\lambda, p) = 1$, by Corollary 9.9.1. If $(k, p) = 1$ then $t^k - 1$ has no repeated roots in $\mathbb{F}_p[t]$, since its derivative is kt^{k-1}. Therefore we must have $\lambda = n$ and $r > 0$. In particular, p must divide m. Similarly, if p does not divide m then it divides n and $\lambda = m$. It follows that if q is divisible by two distinct primes $p \neq p'$ then either both divide m or both divide n.

Suppose now that p^{r+1} divides q, where $m = p^r a$ with $(a, p) = 1$. Then we obtain the congruence $((t^{an} - 1)/t^a - 1)^{p^r} \equiv f^{p^{r+1}} \ mod \ (p)$, and so the quotient $(t^{an} - 1)/(t^a - 1)$ is a pth power in $\mathbb{F}_p(t)$. But the roots of $(t^{an} - 1)/(t^a - 1)$ are distinct $mod\ (p)$, and so p^{r+1} cannot divide q. Together these observations give the Corollary.   □

This was first proven by Conner [**Cn59**]. The above argument is a variation on that of Murasugi.

Edmonds used minimal surfaces to show that a knot K of genus g which is invariant under a rotation of order q admits a Seifert surface F of genus g which is invariant under the action. The quotient of F under the action is a Seifert surface $\overline{F}$ for the orbit knot $\overline{K}$, and the projection from F to $\overline{F}$ is a branched covering. It then follows easily from the Riemann-Hurwitz formula that if $g > 0$ then $q \leq 2g + 1$. Moreover, $q \leq g$ unless $\overline{K}$ is trivial and $\overline{F}$ is a disc [**Ed84**]. In particular, no nontrivial knot has infinitely many semifree periods. This result was first due to Flapan [**Fl85**]. We shall give her argument and extend it to the case of links in §6 below.

Naik combined the results of Edmonds and Murasugi with an elementary estimate for the genus to obtain the following result.

THEOREM. [**Na94**] *Let K be a knot with an n-crossing diagram and which is not a $(2, n)$-torus knot. Let g be the genus of K. If K has a semifree period of order $q = p^r \geq \lceil n/2 \rceil$ where p is prime then $\overline{K}$ is trivial and either*

(1) $q = 2g + 1$, $\Delta_1(K) \equiv (t + 1)^{q-1}\ mod\ (p)$ *and*
    $deg(\Delta_1(K)) = 2g$; *or*
(2) $q = g + 1 = \lceil n/2 \rceil$, $\Delta_1(K) \equiv (t^2 + t + 1)^{q-1}\ mod\ (p)$ *and*
    $deg(\Delta_1(K)) = 2g$; *or*
(3) $q = g + 1 = \lceil n/2 \rceil$ *and* $\Delta_1(K) \equiv 1\ mod\ (p)$.   □

Sakuma has used the Equivariant Sphere Theorem to show that the only semifree periods of a composite knot are the obvious ones (rotating one possibly trivial summand and permuting the others). He also showed that the semifree periods of a composite link are sharply constrained by the multiplicities of the prime factors [**Sk81**].

In another approach to Murasugi's results, Ko and Song show that the Seifert matrix for an equivariant spanning surface can be chosen to have a preferred form which exhibits the periodicity [**KS07**].

9.6. Semifree periods and splitting fields

If $f \in \mathbb{Q}[t]$ we shall let $Split(f/\mathbb{Q})$ denote the splitting field of f over $\mathbb{Q}$. Trotter showed that if K is a 1-knot such that $(\pi K)'$ is free and $\Delta_1(K)$ has no repeated roots then the qth roots of unity are in $Split(\Delta_1(K)/\mathbb{Q})$ [**Tr61**]. This is an immediate consequence of the following theorem.

THEOREM 9.10. *Let $\Delta \in \mathbb{Z}[t]$ satisfy the Murasugi conditions for some prime power $q = p^r$. Suppose that Δ has irreducible factorization $\Delta = \Pi \delta_i^{e(i)}$. Then either $\Delta \equiv 1$ mod (p) or the degree of $\mathbb{Q}(\zeta_q)$ over $\mathbb{Q}(\zeta_q) \cap Split(\Delta/\mathbb{Q})$ is at most $m = \max\{e(i)\}$.*

PROOF. Let $G = Gal(\mathbb{Q}(\zeta_q)/\mathbb{Q})$. The Murasugi conditions assert that there is an integer λ, a knot polynomial $\widehat{\Delta}$ and a polynomial $f(t) \in \mathbb{Z}[\zeta_q, t]$ such that $\Delta = \widehat{\Delta}\Pi_{g \in G}f^g$ and $f(t) \equiv \nu_\lambda \widehat{\Delta}$ mod (p). We may assume that $\Delta \not\equiv 1$ mod (p). Then f is nontrivial and so has a nontrivial irreducible factor h in $\mathbb{Q}(\zeta_q)[t]$. Let $S = \{g \in G \mid h^g = h\}$ and let T be a set of coset representatives for the subgroup S in G. Then $H = \Pi_{\tau \in T}h^\tau$ is irreducible in $\mathbb{Z}[t]$ and $H^{|S|} = \Pi_{g \in G}h^g$ divides Δ, so $|S| \leq m$. Let $M = \mathbb{Q}(\zeta_q)^S$ be the subfield fixed by S. Then M is generated over $\mathbb{Q}$ by the coefficients of h, and $[\mathbb{Q}(\zeta_q) : M] = |S|$. Since the coefficients of h are elementary symmetric functions of the roots of h, which are among the roots of Δ, they are in $Split(\Delta/\mathbb{Q})$. The theorem now follows easily. □

Burde weakened Trotter's hypotheses to requiring that $\Delta_1(K)$ does not represent a unit in $\mathbb{F}_p\Lambda$ and that $\Delta_2(K) = 1$ [**Bu78**]. This condition is relaxed further in Theorem 9.12 below.

LEMMA 9.11. *Let C be a finitely generated $\mathbb{Q}\Lambda$-torsion module with an automorphism τ of prime power order $q = p^r$. Suppose that n is the least integer such that $\Delta_n(C) \doteq 1$. Then the degree of the qth roots of unity over $Split(\Delta_0(C)/\mathbb{Q})$ is at most n.*

PROOF. The module C is the direct sum of its δ-primary submodules, over irreducible factors δ of $\delta_0(C)$, and this decomposition is preserved by τ. The order of τ is the least common multiple of the orders of its restrictions to such summands. Hence the order must be exactly q for at least one such summand, and so we may assume that $\Delta_0(C) = \delta^e$, where δ is irreducible. Let $L = \mathbb{Q}\Lambda/(\delta)$. The kernel of the natural homomorphism from $Aut(C/\delta^{m+1}C)$ to $Aut(C/\delta^m C)$ is isomorphic to $Hom_L(C/\delta^{m+1}C, \delta^m C/\delta^{m+1}C)$, and so is an L-vector space, if $m \geq 1$. In particular, it is torsion free. Therefore the image of τ in $Aut(C/\delta C) \cong GL(n, L)$ has order exactly q. Let $A = L[\tau]$ be the subalgebra of the algebra of $n \times n$ matrices over L generated by this element. Then $dim_L A \leq n$, by the Cayley-Hamilton Theorem.

On the other hand, the image of τ in the semisimple algebra $A/radA$ again has order exactly q, as the kernel of the map on units is an iterated extension of L-vector spaces. The algebra $A/rad\,A$ is a product of fields containing L (see page 452 of [**Lan**]), so some factor is an extension L' of L which contains the qth roots of unity and for which $[L' : L] \leq dim_L A \leq n$. This implies the lemma, since L is contained in $Split(\Delta_0(C)/\mathbb{Q})$.　　　　　　　　$\square$

THEOREM 9.12. *Let K be a knot in a homology 3-sphere M which has a semifree period $q = p^r$ for some prime p. Then either*

(1) $\Delta_1(K) \equiv 1 \mod (p)$; or
(2) *the degree of the qth roots of unity over $Split(\Delta_1(K)/\mathbb{Q})$ is at most n, where $\Delta_n(K) \neq 1$ but $\Delta_{n+1}(K) \doteq 1$.*

PROOF. Let h be a self homeomorphism of M of order q which leaves K invariant and fixes a circle A disjoint from K. Let $\pi = \pi K$ and let τ be the induced automorphism of the Λ-module π'/π''. Suppose that (2) does not hold. Since π'/π'' is $\mathbb{Z}$-torsion free $\tau^n = 1$ if and only if $(id_\mathbb{Q} \otimes \tau)^n = 1$. By Lemma 9.11 the order of $id_\mathbb{Q} \otimes \tau$ is a proper divisor of q. Thus on replacing h by $h^{q/p}$ we may assume that K is invariant under a self homeomorphism h of M of prime order $p > 1$ which induces the identity on π'/π''.

Let $\overline{M}$ be the orbit space of h and let $c : M \to \overline{M}$, $\overline{K} = c(K)$, X and $\overline{X}$ be as above. (Here we shall assume the base point $*$ is fixed by h.) Let $\bar{\pi} = \pi\overline{K}$. Then $\bar{\pi} = \pi/\langle\langle g^{-1}h_*(g) \mid g \in \pi \rangle\rangle$, where h_* is the automorphism of π induced by h, and so $c|_X : X \to \overline{X}$ induces an isomorphism from π/π'' to $\bar{\pi}/\bar{\pi}''$.

The map $c|_X$ extends to a p-fold cyclic branched cover $d : Z \to \overline{Z}$, where Z and $\overline{Z}$ are the closed 3-manifolds obtained by 0-framed surgery on K and $\overline{K}$. The map d is of degree p, and induces an isomorphism of $\pi/\pi'' = \pi_1(Z)/\pi_1(Z)''$ with $\overline{\pi}/\overline{\pi}'' = \pi_1(\overline{Z})/\pi_1(\overline{Z})''$, since the longitudes of a knot lie in the second commutator subgroup of the knot group. Let Z' and $\overline{Z}'$ be the maximal abelian covering spaces of Z and $\overline{Z}$, respectively, and let $d' : Z' \to \overline{Z}'$ be a lift of d.

The modules $H_q(Z; \mathbb{F}_p\Lambda)$ are finitely generated and $H_q(Z; \mathbb{F}_p) \cong \mathbb{F}_p$ or 0, for all q. It follows from the exact sequence of homology associated to the coefficient sequence

$$0 \to \mathbb{F}_p\Lambda \to \mathbb{F}_p\Lambda \to \mathbb{F}_p \to 0$$

and the structure theorem for finitely generated modules over the PID $\mathbb{F}_p\Lambda$ that $H_q(Z'; \mathbb{F}_p) = H_q(Z; \mathbb{F}_p\Lambda)$ is a torsion module, and so is finitely generated over $\mathbb{F}_p$. If F is a field the functors $Hom_F(-, F)$ and $Hom_{F\Lambda}(-, F(t)/F\Lambda) \cong Ext^1_{F\Lambda}(-, F\Lambda)$ are naturally isomorphic on the category of finitely generated torsion $F\Lambda$-modules. It follows that $H^q(Z'; \mathbb{F}_p) \cong H^{q+1}(Z; \mathbb{F}_p\Lambda)$, for all q.

Let $d_i = H_i(d; \mathbb{F}_p\Lambda)$ and $d^i = H^i(d; \mathbb{F}_p\Lambda)$, for $i \geq 0$. Since $d_1 = H_1(d'; \mathbb{F}_p)$ is an isomorphism so is $H^1(d'; \mathbb{F}_p)$. Hence d^2 is also an isomorphism, by naturality. Now $d_1(d^2(\alpha \cap [Z])) = \alpha \cap d_3[Z]$ for all $\alpha \in H^2(Z; \mathbb{F}_p\Lambda)$. But $d_3 = 0$, since d has degree p, and so $H^2(Z; \mathbb{F}_p\Lambda) = 0$. Therefore $H_1(Z; \mathbb{F}_p\Lambda) = 0$ also, and so $\Delta_1(K) \equiv 1$ mod (p). $\square$

The Alexander polynomial of $\#^3 4_1$ satisfies the Murasugi conditions with $\ell = 1$. This knot clearly has a semifree period 3. However, Theorem 9.10 implies that no knot with this polynomial and which is invariant under a rotation of order 3 about a disjoint axis can have a cyclic knot module. (In particular, no such knot can have unknotting number 1.) The polynomial $3t^2 - 5t + 3$ satisfies the Murasugi conditions with $q = 3$ and $\lambda = 1$, and hence is the Alexander polynomial of some knot with a semifree period 3, by Theorem 1.1 of [**DL91**]. As the splitting field for this quadratic is $\mathbb{Q}(\sqrt{-11})$, which does not contain ζ_3, we see that in general the alternative "$\Delta_1(K) \equiv 1$ mod (p)" is necessary.

Theorem 9.12 may be extended to the Hosokawa polynomial of a link, if $H_1(X^\tau; \mathbb{Z})$ is $\mathbb{Z}$-torsion free, as is the case for knots and fibred links.

9.7. Links with infinitely many semifree periods

Trivial knots are the only knots with infinitely many distinct semifree periods. Flapan's argument uses the Equivariant Characteristic Variety Theorem to reduce to the cases of torus knots and hyperbolic knots [Fl85]. We shall extend this to obtain the corresponding result for links by reduction to the cases where the complements are Seifert fibred or hyperbolic. (The Seifert case uses Corollary 9.9.3 and the Burde-Murasugi characterization of links with Seifert fibred complement.) The equivariant Dehn Lemma then implies that the axis of a rotation must link each component in the simplest nontrivial fashion, as in a key-ring.

LEMMA 9.13. *Let K be a knot invariant under a homeomorphism h of order q, and let N be a regular neighbourhood of K. Then h is conjugate to a homeomorphism of order q which leaves K and N invariant.*

PROOF. Let $N_h = \cap_{j \geq 0} h^j(N)$. Then N_h is a regular neighbourhood of K which is invariant under h. By the uniqueness of regular neighbourhoods there is an isotopy ϕ_t of S^3 fixing K and such that $\phi_1(N_h) = N$. Therefore $\phi_1 h \phi_1^{-1}$ is a homeomorphism of S^3 of order q which leaves K and N invariant. $\square$

LEMMA 9.14. *Let $U \subset S^3$ be a solid torus which is invariant under a rotation h of order $q > 2$ about an axis A. Then either $S^3 \setminus U \cong D^2 \times S^1$ or $A \cap U = \emptyset$.*

PROOF. We may assume that U is knotted. If $A \subset \text{int } U$ then A has order 0 with respect to U (since the unknot has no companions) and so there is a meridianal disc D in U which misses A. Let h' be a power of h of prime order p. By a disc swapping argument we may assume that $D \cap h'(D) = \emptyset$ and hence that all the translates of D by powers of h' are disjoint. The action of h' permutes the components of $U \setminus \cup (h')^j(D)$ and thus has no fixed points, which is a contradiction.

If $A \cap \partial U \neq \emptyset$ then the action of h on ∂U has fixed points, and so, by the Riemann-Hurwitz formula relating $\chi(\partial U)$ and $\chi(\partial U / \langle h \rangle)$, the order of $h|_{\partial U}$ is 1, 2, 3, 4 or 6. Since h is orientation preserving this is also the order of h. But an automorphism of the torus of order 3, 4 or 6 with nonempty fixed point set cannot have eigenvalue ± 1 on homology, and so cannot extend to the solid torus.

Thus A must be disjoint from U. $\qquad\square$

Notice that if h is an involution ($h^2 = id$) then the axis of h may meet U. In this case h must reverse both meridian and longitude of U. If U is an invariant neighbourhood of a knot K then the axis also meets K, and the knot is said to be "strongly irreducible".

THEOREM 9.15. *Let $T \subset S^3$ be a torus which is invariant under rotations of infinitely many orders. Then T is unknotted, i.e., the closure of each component of $S^3 \setminus T$ is a solid torus.*

PROOF. By Alexander's torus theorem at least one component N of $S^3 \setminus T$ has closure a solid torus. Let $X = S^3 \setminus N$. Suppose that X is not a solid torus. Then X is irreducible and has incompressible boundary, and so has a characteristic family of tori $\mathcal{T} = \{T_j \mid 1 \leq j \leq m\}$. This is a minimal family of disjoint incompressible non-parallel tori in X, one of which is ∂X, such that the closures of the components of $X \setminus \cup_{j=1}^{j=m} T_j$ are either Seifert fibred or simple, and $\mathcal{T}$ is unique up to ambient isotopy of X. If h is a homeomorphism of X of finite order then there is a characteristic family of tori for X whose union is invariant under h. Since there is an ambient isotopy carrying such an invariant family to $\mathcal{T}$, we may find (as in Lemma 9.13) a conjugate homeomorphism $\tilde{h}$ which leaves $\cup T$ invariant. Furthermore $(\tilde{h})^{m!}$ leaves each torus in $\mathcal{T}$ invariant. Thus after conjugating and raising to the $m!^{th}$ power if necessary, we may assume that there is a family $\{h_i \mid i \in \mathbb{N}\}$ of rotations of finite order $n_i > 2$ of S^3, each of which leaves each torus in $\mathcal{T}$ invariant, and such that $\{n_i\}_{i \in \mathbb{N}}$ is an unbounded sequence.

There is a component X_1 of $X \setminus \cup T_i$ whose closure has connected boundary. (This follows by induction on m and the fact that each closed surface in S^3 separates.) This component is either simple or a torus knot exterior. If X_1 is simple then the mapping class group of X_1 is finite, of order f say, and so each homeomorphism $(h_i|_{X_1})^f$ is isotopic to the identity. With only minor modifications the argument of [**Gi67**] then implies that $(h_i|_{X_1})^f$ is the identity, since $\pi_1(X_1)$ has trivial centre. But then $h_i^f = id$ for all $i \in \mathbb{N}$, contradicting the unboundedness of the orders of the rotations. Thus X_1 must be the exterior of a torus knot.

If X_1 is not a solid torus then it contains the axes of all the rotations h_i, which therefore act freely on $\overline{S^3 \setminus X_1} \cong S^1 \times D^2$. After conjugating each h_i by a map ϕ_i isotopic to the identity $rel\ X_1$, we may assume that the core $S^1 \times \{0\}$ is invariant under all the rotations. But the only semifree periods of a (p,q)-torus knot with $p, q > 1$ divide pq. Therefore X_1 must be a solid torus, and so ∂X is compressible in X, contradicting the definition of characteristic family. Thus our supposition was wrong and X must also be a solid torus. $\square$

COROLLARY 9.15.1. [**F185**] *A 1-knot K with infinitely many semifree periods is trivial.*

PROOF. Let N be a regular neighbourhood of K. By Lemma 9.13 we may assume that there is a family $\{h_i \mid i \in \mathbb{N}\}$ of rotations of finite order of S^3, each of which leaves K, N and hence $X = X(K)$ and $\partial N = \partial X$ invariant and has axis disjoint from K, and such that the sequence of orders is unbounded. By Theorem 9.15 the exterior X is a solid torus, and so K is trivial. $\square$

A similar reduction to the Seifert fibred case extends this result to links.

THEOREM 9.16. *Let L be a μ-component 1-link with infinitely many semifree periods. Then L is trivial.*

PROOF. As before we may assume that there are infinitely many rotations $\{h_i \mid i \in \mathbb{N}\}$ leaving L componentwise invariant. We may also assume that L is unsplittable, and hence that $X(L)$ is irreducible. By Corollary 9.15.1 the components of L are trivial. We shall show that $\mu = 1$.

Suppose that X has an incompressible boundary, i.e., that $\mu > 1$. Let $\mathcal{T} = \{T_j \mid 1 \leq j \leq m\}$ be a characteristic family of tori for X. Again, we may assume that each torus in $\mathcal{T}$ is invariant under each of the rotations h_i, and that each component of $X \setminus \cup_{1 \leq j \leq m} T_j$ is Seifert fibred. At least one such component, X_1 say, contains the axes of infinitely many of the rotations. The complement $S^3 \setminus X_1$ is a union of solid tori, since by Theorem 9.15 each torus in $\mathcal{T}$ is unknotted, and so $X_1 = X(L_1)$ for some link L_1. By the definition of characteristic family L_1 is not the unknot. It is clear from the work of Burde and Murasugi on links whose exteriors are Seifert fibred [**BM70**] that L_1 must have a sublink L_2 which is either a nontrivial torus knot or a $(2, 2\alpha)$-torus link, for some $\alpha \geq 1$. By filling in the solid tori corresponding to the other components of L_1 we obtain a submanifold X_2 which is homeomorphic to $X(L_2)$ and which is invariant under all of the rotations. Moreover, since each rotation acts freely on each component of $S^3 \setminus X_2$ we may assume that L_2 is componentwise invariant under all of the rotations. Hence L_2 has unknotted components and so must be a $(2, 2\alpha)$-torus link. But the semifree periods of such a link divide α, by Corollary 9.9.2. Therefore our supposition was wrong and L is trivial. $\qquad \square$

The axis of a rotation leaving a trivial link componentwise invariant must be in the "obvious" position with respect to the link, as the next result shows.

THEOREM 9.17. *Let L be a trivial μ-component 1-link which is componentwise invariant under a rotation of finite order q about a disjoint axis A. Then there are disjoint discs $D_i^2 \subset S^3$ which meet A transversely in one point and such that $\partial D_i = L_i$, for each $1 \leq i \leq \mu$.*

PROOF. Suppose that there are invariant discs $D_1, \ldots, D_{k-1}$, with $\partial D_i = L_i$ and which are disjoint from one another and from the remaining components of L. Let N be an open regular neighbourhood of $L \cup (\cup_{i<k} D_i)$ which is invariant under h. Let $Y = S^3 \setminus N$

and let Z be the component of ∂Y nearest L_k. Since L is trivial Z is compressible in Y, which is invariant under $h|_Y$. Therefore by the Equivariant Dehn's Lemma of [MY81] there is a disc D in Y with ∂D essential in Z and such that for each $j \geq 0$ either $h^j(D) = D$ or $h^j(D) \cap D = \emptyset$. But L_k must link A, by Corollary 9.9.1, and so $D \cap A \neq \emptyset$. Therefore D is invariant under h, and so may be extended to an invariant spanning disc D_k for L_k, which is disjoint from $\cup_{1<k} D_i$ and from the other components of L. An application of the Riemann-Hurwitz formula to the discs D_i and their images D_i^* in $S^3/\langle h \rangle$ shows that each map $: D_i \to D_i^*$ has just one branch point, and hence that A meets each D_i in one point. The intersections must be transverse for (as remarked above) A must link each component of L. $\qquad\square$

In higher dimensions, if an n-knot K has semifree periods of every finite order then $X(K)'$ is acyclic, and so if also $\pi K \cong Z$ then K is trivial [Su75]. A simple $(2k-1)$-knot (with $k > 1$) has infinitely many semifree periods if and only if it is fibred and the homological monodromy has finite order [CN96].

9.8. Knots with free periods

If a knot K in a homology 3-sphere M has free period q then a suitable $1/n$-surgery on K will give another knot $\hat{K}$ in another homology 3-sphere $\hat{M}$ which is the fixed point set of a Z/qZ-action and such that $X(K)$ and $X(\hat{K})$ are equivariantly homeomorphic.

THEOREM 9.18. *Let K be a knot in a homology 3-sphere M which has free period q (or which is fixed pointwise under a Z/qZ-action on M). Then $\Delta_1(K)(t^q) = \Pi_{j=1}^{j=q} \delta(\zeta_q^j t)$, where δ is a knot polynomial.*

PROOF. We shall assume that K is freely periodic. Let $X = M \setminus K$ and $\overline{X} = \overline{M} \setminus \overline{K}$ and suppose that X and $\overline{X}$ and K and $\overline{K}$ are compatibly oriented. Then $H_1(f)$ is injective, so X' is also the maximal abelian covering space of $\overline{X}$. The automorphism of X' which generates $Aut(X'/X) \cong H_1(X;\mathbb{Z}) = \mathbb{Z}$ is the qth power of the automorphism which generates $Aut(X'/\overline{X})$.

Let $\Lambda_u = \mathbb{Z}[u, u^{-1}]$ and $\Lambda_v = \mathbb{Z}[v, v^{-1}]$, where u and v correspond to meridians for K and for $\overline{K}$, respectively, and let $g : \Lambda_u \to \Lambda_v$ be the homomorphism sending u to v^q induced by f. Let $H_u = H_1(X; \Lambda_u)$ and $H_v = H_1(X; \Lambda_v)$. (Of course $H_u = H_1(X'; \mathbb{Z}) = H_v$ as abelian groups, while $H_u = g^* H_v$ as a Λ_u-module.) Let $\Lambda_t = \mathbb{Z}[t, t^{-1}]$ be yet another copy of the Laurent polynomial ring, let $\psi : \Lambda_u \to \Lambda_t$ be the homomorphism sending u to t^q and let $\Lambda_{t,v} = \mathbb{Z}[t, t^{-1}, v, v^{-1}]$. Then

$$\Lambda_t \otimes_{\Lambda_u} H_u = (\Lambda_t \otimes_{\Lambda_u} \Lambda_v) \otimes_{\Lambda_v} H_v = (\Lambda_{t,v}/(v^q - t^q)) \otimes_{\Lambda_v} H_v,$$

as Λ_t-modules. Clearly $E_0(\Lambda_t \otimes_{\Lambda_u} H_u)$ is generated by $\Delta_1(K)(t^q)$, since $H_u = H_1(X; \Lambda_u)$, while to make the corresponding computation for the module on the right we may extend coefficients from $\mathbb{Z}$ to $\mathbb{Z}[\zeta_q]$. But then $\mathbb{Z}[\zeta_q] \otimes ((\Lambda_{t,v}/(v^q - t^q)) \otimes_{\Lambda_v} H_v)$ has a filtration with subquotients isomorphic to $\mathbb{Z}[\zeta_q] \otimes ((\Lambda_{t,v}/(v - \zeta_q^j t)) \otimes_{\Lambda_v} H_v)$, and the theorem follows from Theorem 3.12. $\qquad\square$

This theorem was first proven by Kinoshita, for knots fixed point-wise [**Ki58**]. The connection with freely periodic knots is due to Hartley [**Ha81**]. A similar argument may be used to show that if a μ-component link L is componentwise invariant under a free $\mathbb{Z}/q\mathbb{Z}$-action then there are integers $0 < b_i < q$ such that $(b_i, q) = 1$, for $2 \leq i \leq \mu$ and an $f \in \Lambda_\mu$ such that

$$\Delta_1(L)(t_1 \Pi_{2 \leq i \leq \mu} t_i^{b_i}, t_2, \ldots, t_\mu) = \Pi_{j=1}^{j=\mu} f(\zeta_q^j t_1, t_2, \ldots, t_\mu).$$

(There are analogous results for the other components of L.) Such conditions obviously imply the corresponding ones for sublinks of L. There appears to be no such result for the Hosokawa polynomial, for $Aut(X^\tau / \overline{X})$ may not be infinite cyclic.

Every polynomial satisfying the above condition is the Alexander polynomial of some freely periodic knot in a homology 3-sphere. Let $\overline{K}$ be a knot with $\Delta_1(\overline{K}) = \delta$ and let K be the branch locus in $M = M_q(\overline{K})$. Then some $1/n$ surgery on K produces such a knot. Are there always such knots in S^3?

The generator of $Aut(X'/\overline{X})$ induces an isometry V of b_K such that $V^q = u.id_{H_u}$ and $b_{\overline{K}}(\alpha, \beta) = \Sigma_{j=1}^{j=q} v^j b_K(\alpha, V^j \beta)$ in $\mathbb{Q}(v)/\Lambda_v$. In particular, the pairing on the right-hand side is non-singular.

Theorem 9.18 has the following immediate consequences [**Fo62'**]:

(1) $\Delta_1(K)$ and δ have the same degree;

(2) the leading coefficient of $\Delta_1(K)$ is a qth power;

(3) the roots of $\Delta_1(K)$ are the qth powers of roots of δ.

Condition (2) implies that if K has infinitely many free periods then $\Delta_1(K)$ is monic. Using the following lemma from algebraic number theory, it is not hard to show that conditions (1) and (2) then imply that $\Delta_1(K)$ must be a product of cyclotomic polynomials.

LEMMA 9.19. *For each n there is an $\epsilon_n > 0$ such that any algebraic integer of degree n and such that all of its conjugates in $\mathbb{C}$ have absolute value less than $1 + \epsilon_n$ must be a root of unity.*

PROOF. (Sketch.) Let N be the number of algebraic integers of degree $\leq n$ and with all conjugates of absolute value ≤ 2. If α is an algebraic integer of degree n whose conjugates all have absolute value $\leq 1 + \epsilon_n = 2^{\frac{1}{N+1}}$ then two of $\alpha, \ldots, \alpha^{N+1}$ must be equal. $\square$

In particular, if the Mahler measure of a polynomial is 1 then the factors of the polynomial are cyclotomic.

Hartley refined this Lemma to find explicit bounds on the possible free periods of knots whose Alexander polynomials have non-cyclotomic factors. He also used Theorem 9.18 to show that if K has free period q and $\Delta_1(K) = \Pi_{1 \leq i \leq n} \phi_{a_i}^{e(i)}$ then there are integers b_i and p_i such that $(a_i, p_i) = 1$, b_i divides a_i and $b_i p_i = q$, for all $1 \leq i \leq n$ and strengthened another result of Fox by showing that if $\Delta_1(K)$ is quadratic but is not $t^2 - t + 1$ then K has no free period.

Since $\Delta_1(K_{m,n})(t)$ contains $\phi_{mn}(t)$ as a simple factor, $\phi_{mn}(t^q)$ is a simple factor of $\Delta_1(K_{m,n})(t^q)$. Multiplication by ζ_q permutes the primitive $(mnq)^{th}$ roots of unity, and so $\Delta_1(K_{m,n})(t^q)$ has no factorization as in Theorem 9.18. Thus $K_{m,n}$ has free period q if and only if $(q, mn) = 1$. Torus knots are the only knots in S^3 with infinitely many free periods [**Fl85**]. If a link L in a homology 3-sphere has infinitely many free periods must $X(L)$ be Seifert fibred?

The following result is analogous to Theorem 5.19.

THEOREM 9.20. [**Ha81**] *Let K be a knot in a homology 3-sphere M which has free period q (or which is fixed pointwise under a Z/qZ-action on M), and let δ be as in the preceding theorem. Then there is a homomorphism of $H_1(M_2; \mathbb{Z})$ onto some group E of order $\delta(-1)$ with kernel a direct double. Hence $H_1(M_2; \mathbb{Z})$ is not cyclic unless $|\delta(-1)| = |\Delta_1(K)(-1)|$. In particular, if $q = 2$ then $H_1(M_2; \mathbb{Z})$ is always a direct double.*

PROOF. Consider the case when K is fixed pointwise and apply Theorem 5.19 to the $2q$-fold cyclic branched cover of the orbit space $\bar{M}$, branched over the orbit knot $\bar{K}$. □

If $k > 1$ a $(2k - 1)$-knot K with $\pi K \cong \mathbb{Z}$ has semifree period m if and only if it has free period m, and the actions can be assumed to agree on $X(K)$. It follows easily by Smith theory that at most one of the components of a $(2k - 1)$-link is invariant under a free action of Z/mZ, the others being permuted [**Ni92**].

9.9. Equivariant concordance

Let $L(0)$ and $L(1)$ be μ-component n-links which are invariant under Z/qZ-actions determined by homeomorphisms h_0 and h_1, respectively. The links with these actions are *equivariantly concordant* if there is an action h of Z/qZ on $S^{n+2} \times [0,1]$ such that $h|_{S^{n+2} \times \{i\}} = h_i$ for $i = 0$ and 1, and which leaves invariant a concordance $\mathcal{L}$ from $L(0)$ to $L(1)$. Our interest is primarily in the classical case.

LEMMA 9.21. *Let h be a self-homeomorphism of $S^3 \times [0, 1]$ which is orientation preserving, leaves the boundary components invariant and has finite order $q > 1$. Then the fixed point set of h is empty or is an annulus meeting $S^3 \times \{i\}$ in a circle, for $i = 0$ and 1.*

PROOF. Let F be the fixed point set of h. If F is nonempty it is a submanifold of even codimension in $S^3 \times [0, 1]$, which meets the boundary components $S^3 \times \{i\}$ properly. The homeomorphism h may be extended by coning to a periodic self-homeomorphism $\hat{h}$

of $S^4 = D^4 \cup (S^3 \times [0,1]) \cup D^4$. The fixed point set of $\hat{h}$ is the union of F with the cones over $F \cap (S^3 \times \{i\})$ for $i = 0$ and 1, and so is nonempty. Moreover it is a Z/pZ-homology sphere of even codimension, for each prime p dividing q, by Smith theory. Therefore either F is empty or F is an annulus. $\square$

It follows that if two links with Z/qZ-actions are equivariantly concordant then one action is semifree if and only if the other is.

A link L with a semifree Z/qZ-action fixing L component-wise is *equivariantly slice* if the action bounds an action of Z/qZ on D^4 which leaves invariant a null-concordance of L. Smith theory (applied to the actions on D^4 and on the slice discs) implies that the fixed point set of the action on D^4 is a 2-disc which meets each invariant slice disc in 1 point. (The fixed point set may be knotted.) The orbit space is a 1-connected homology 4-ball with boundary S^3, and so is homeomorphic to D^4, by topological surgery. Clearly L is equivariantly slice if and only if it is equivariantly concordant to the trivial μ-component link with the standard Z/qZ-action.

THEOREM 9.22. *Let $L(0)$ and $L(1)$ be 1-links with semifree period q, determined by homeomorphisms h_0 and h_1 with axes $A(0)$ and $A(1)$, respectively. If these actions are equivariantly concordant the links $A_0 \amalg L(0)$ and $A(1) \amalg L(1)$ are concordant.*

PROOF. The fixed point set $\mathcal{A}$ of a periodic homeomorphism h extending h_0 and h_1 is an annulus, by Lemma 9.21. An analogous argument shows that $\mathcal{A} \cap \mathcal{L}$ is empty. Therefore $\mathcal{A} \cup \mathcal{L}$ is a concordance between $A_0 \amalg L(0)$ and $A(1) \amalg L(1)$. $\square$

COROLLARY 9.22.1. *If a 1-knot K has semifree period q, with axis A, then $\text{lk}(A, K)$ is an invariant of equivariant concordance.*

PROOF. This follows from the theorem and Corollary 4.17.1. $\square$

If K is a knot invariant under a rotation h of order q about a disjoint axis A let $\Delta_{Z/qZ}(K)$ be the image of $\Delta_1(\overline{L})$ in the ring $\mathbb{Z}[Z \times (Z/qZ)] = \Lambda_2/(u^q - 1)$, where $\overline{L} = \overline{A} \cup \overline{K}$ is the orbit link in $S^3/\langle h \rangle$ and u, t are the meridians corresponding to $\overline{A}$ and $\overline{K}$, respectively.

The next result was first proven for equivariant ribbon knots [**DN02**]. The full result is due to Cha, and we follow his proof.

THEOREM 9.23. [**Ch04**] *Let K be a 1-knot with semifree period q about an axis A and which is equivariantly slice. Then $|\mathrm{lk}(A, K)| = 1$ and we may normalize our choice of $\Delta_1(\overline{L})$ so that $\Delta_{Z/qZ}(K) = a\bar{a}$ for some $a = a(t, u) \in \Lambda_2/(u^q - 1)$ such that $a(1, u) = 1$.*

PROOF. The first assertion follows from the Corollary above (or on considering the fixed point set of the restriction of the action to an invariant slice disc). As a consequence we may use the Torres conditions (Corollary 5.3.1) to normalize $\Delta_{Z/qZ}(K)$ to be fixed under conjugation and augment to 1.

Since $\Delta_1(\overline{L}) \neq 0$ the homology module $H_1(X(\overline{L}); \Lambda_2)$ is a pure torsion module, by Lemma 4.5 and the subsequent remark, and since the linking number is nonzero the components of $\partial X'$ are contractible. The UCSS gives an exact sequence

$$0 \to \mathbb{Z} = e^2 H_0 \to H^2(X(\overline{L}); \Lambda_2) \to e^1 H_1(X(\overline{L}); \Lambda_2) \to 0.$$

Let S be the multiplicative system generated by some nonzero element s of the augmentation ideal. (For instance, s could be $t - 1$ or $u - 1$.) Then $\mathbb{Z}_S = 0$ and the Blanchfield pairing localized with respect to S is a $+1$-linking pairing over Λ_2. The argument of Theorem 2.4 applies without essential change, to show that this S-localized pairing is neutral. (Note that the orbit link $\overline{L}$ is concordant to the Hopf link Ho, and $X(Ho)'$ is contractible.) Hence $\Delta_1(\overline{L}) = \pm f \bar{f} s^k$, for some $f \in \Lambda_2$ and $k \in \mathbb{Z}$. Since $\varepsilon(s) = 0$ and $\varepsilon(\Delta_1(\overline{L})) = 1$ we must have $k = 0$. $\qquad\square$

In particular, $\Delta_1(K)(t) = \Pi_{\zeta^q=1} a(t, \zeta) a(t^{-1}, \zeta^{-1})$.

There is a converse: if $a(t, u) \in \Lambda_2/(u^q - 1)$ and $a(1, u) = 1$ then $a\bar{a} = \Delta_{Z/qZ}(K)$ for some equivariantly ribbon knot K with semifree period q [**DN02**]. The construction involves modifying the base link $\overline{L}$, starting from the Hopf link, considered as the boundary of two discs $D_{\overline{A}}$ and $D_{\overline{K}}$ given by the intersection of two orthogonal planes with the unit ball D^4 in $\mathbb{R}^4$. Attaching a cancelling pair of 1- and 2-handles to D^4, disjoint from $\overline{K}$, gives a 4-manifold $W \cong D^4$. If $\overline{A}$

bounds a disc in ∂W, disjoint from the handles, then the preimage of $\overline{K}$ in the q-fold branched cover of W, branched over $D_{\overline{A}}$ is an equivariantly slice knot. Careful choice of the attaching maps for the handles ensures that K is a ribbon knot with $\Delta_{Z/qZ}(K) = a\bar{a}$.

Equivariant concordance of $(2k-1)$-knots with free actions has been studied by Stolztfus. He defines the notion of *equivariant isometric structure*, and shows that equivalence classes of such objects under "algebraic concordance" form an abelian group $E^{\varepsilon}(q; \mathbb{Z})$, where $\varepsilon = (-1)^k$. This group is isomorphic to the subgroup of the algebraic knot concordance group C_{2k-1} consisting of classes with representatives whose Alexander polynomial Δ satisfies $|\prod_{i=1}^{q} \Delta(\lambda^i)| = 1$, where λ is a primitive qth root of unity. Its torsion subgroup is infinite, of exponent 4, and the torsion-free quotient is free of infinite rank. The corresponding knot invariant is defined for all $k \geq 1$, but is only a complete invariant of equivariant concordance in the higher dimensional cases [St79].

CHAPTER 10

Singularities of Plane Algebraic Curves

In this chapter we shall give an account of plane curve singularities from the point of view of elementary commutative algebra. The geometric and topological aspects of plane curves and their singularities are treated in much greater detail in the books [**BK**], [**EN**] and [**Mil**].

The only algebraic result that we use which is not in the text [**AM**] is the fact that the integral closure of a complete discrete valuation ring in a finite extension of its fields of fractions is again a complete discrete valuation ring, for which we refer to [**Ser**].

10.1. Algebraic curves

A non-constant polynomial $f \in \mathbb{C}[X,Y]$ defines a *plane curve* $V(f) = \{(x,y) \in \mathbb{C}^2 \mid f(x,y) = 0\}$ whose irreducible components of $V(f)$ correspond to the irreducible factors of f. The points of $V(f)$ correspond to the maximal ideals of the coordinate ring $O_{V(f)} = \mathbb{C}[X,Y]/(f)$, by the Nullstellensatz. (See Corollary 7.2.2 above.)

THEOREM 10.1. *Let* $f, g \in \mathbb{C}[X,Y]$ *have no common factors. Then*

$$|V(f) \cap V(g)| \le \dim_{\mathbb{C}} \mathbb{C}[X,Y]/(f,g) < \infty.$$

PROOF. The images of f and g in $\mathbb{C}(X)[Y]$ have no common factor there, since $\mathbb{C}[X,Y]$ is a UFD. Hence there are $m, n \in \mathbb{C}(X)[Y]$ such that $mf + ng = 1$, by the Euclidean algorithm. Clearing denominators, there are polynomials $M, N \in \mathbb{C}[X,Y]$ and a monic polynomial $h \in \mathbb{C}[X]$ such that $Mf + Ng = h$. Hence the first coordinate of any point of $V(f) \cap V(g)$ is a root of h, and for each such root α the number of points in $V(f) \cap V(g)$ with first coordinate α is bounded by $\dim_{\mathbb{C}}\mathbb{C}[X,Y]/(X-\alpha,f,g)$. Since $X-\alpha$ does not divide both f and g this dimension is finite, and so $V(f) \cap V(g)$ is finite.

More precisely, let $h = \Pi_{i=1}^{i=r}(X-\alpha_i)^{m(i)}$ be the factorization of h. Let $A_i = \mathbb{C}[X]/((X-\alpha_i)^{(m(i))})$ and let f_i, g_i be the images of f, g in $A_i[Y] = \mathbb{C}[X,Y]/((X-\alpha_i)^{(m(i))})$, for all $i \leq r$. Then $\dim_{\mathbb{C}}\mathbb{C}[X,Y]/(X-\alpha_i,f,g) \leq \dim_{\mathbb{C}}A_i[Y]/(f_i,g_i) < \infty$, for all $i \leq r$. As $\mathbb{C}[X,Y]/(f,g) \cong \oplus_{i=1}^{i=r}A_i[Y]/(f_i,g_i)$ we have $|V(f) \cap V(g)| \leq \dim_{\mathbb{C}}\mathbb{C}[X,Y]/(f,g) < \infty$. $\qquad \square$

Since $\mathbb{C}[X,Y]/(f,g)$ is finite dimensional and hence Artinian, it is the direct sum of its localizations at maximal ideals, by Theorem 8.7 of [**AM**]. Thus $\mathbb{C}[X,Y]/(f,g) \cong \oplus_{P\in\mathbb{C}^2}\mathbb{C}[X,Y]_{(P)}/(f,g)$, where $\mathbb{C}[X,Y]_{(P)}$ is the ring of rational functions $r/s \in \mathbb{C}(X,Y)$ with $s(P) \neq 0$, and $\mathbb{C}[X,Y]_{(P)}/(f,g) \neq 0$ if and only if $P \in V(f) \cap V(g)$.

If f and g have total degree m and n (respectively) then Bezout's Theorem gives mn as the intersection number of the projective completions of $V(f)$ and $V(g)$ in the projective plane $\mathbb{CP}^2$, with multiplicities for intersections at singularities and points where the curves are tangent. (The multiplicity of the intersection at P is $\dim_{\mathbb{C}}\mathbb{C}[X,Y]_{(P)}/(f,g)$.) The intersection points are all in the affine plane $\mathbb{C}^2$ if and only if the homogeneous parts of highest degree f_m and g_n have no common factors.

A plane curve is *smooth* or *non-singular* if it has a well-defined tangent at every point, i.e., if and only if the partial derivatives f_X and f_Y do not vanish simultaneously anywhere. Let $Sing(V(f)) = \{(x,y) \in V(f) \mid f_X(x,y) = f_Y(x,y) = 0\}$ be the set of singular points of $V(f)$. Then $V(f)$ is smooth if $Sing(V(f))$ is empty.

COROLLARY 10.1.1. *If f is square-free then $Sing(V(f))$ is finite.*

PROOF. If f is square-free then f has no common factors with either of its partial derivatives, and so

$$Sing(V(f)) = V(f) \cap V(f_X) \cap V(f_Y)$$

is finite. $\qquad \square$

The next theorem shows that smoothness depends only on the ring $O_{V(f)}$, and not on the embedding of $V(f)$ in $\mathbb{C}^2$.

THEOREM 10.2. *Let $f \in \mathbb{C}[X, Y]$ be irreducible. Then the following are equivalent:*

(1) $O_{V(f)} = \mathbb{C}[X, Y]/(f)$ *is integrally closed;*
(2) $f \notin M^2$, *for all maximal ideals $M < \mathbb{C}[X, Y]$ which contain (f, f_Y);*
(3) $(f, f_X, f_Y) = \mathbb{C}[X, Y]$;
(4) $V(f)$ *is non-singular.*

PROOF. The nonzero prime ideals of $O_{V(f)}$ correspond bijectively to the maximal ideals of $\mathbb{C}[X, Y]$ which contain f, via the canonical epimorphism from $\mathbb{C}[X, Y]$ to $O_{V(f)}$, and so $O_{V(f)}$ is a 1-dimensional noetherian domain. Therefore $O_{V(f)}$ is integrally closed if and only if each of its localizations is a discrete valuation ring. Let $A = \mathbb{C}[X]$, and let N be a maximal ideal of $O_{V(f)}$, with preimage M in $A[Y] = \mathbb{C}[X, Y]$. (Then $f \in M$.) Let $Q = M \cap A$, $B = A_Q$ and $C = A[Y]_M$. Then B is a discrete valuation ring with maximal ideal QB and C is a local ring with maximal ideal MC generated by Q and g, for some g representing an irreducible factor of the image of f in $(A/Q)[Y] = (B/QB)[Y]$. Since $0 < (f)C < MC$ is a chain of distinct prime ideals, MC cannot be principal. Therefore $MC/(MC)^2$ has dimension 2 as a vector space over $C/MC \cong \mathbb{C}$, by Nakayama's Lemma. The maximal ideal of the localization of $O_{V(f)}$ at N is $MC/(f)$ and so is principal if and only if there is some $t \in C$ such that $MC = (f, t)$. In this case the images of f and t in $MC/(MC)^2$ form a basis, so $f \notin M^2$. Thus (1) $\Rightarrow$ (2).

If $(f, f_X, f_Y) \leq M = (X - \alpha, Y - \beta)$ then on considering the Taylor expansion of f around $(\alpha, \beta) \in \mathbb{C}^2$, we see that $f \in M^2$. Hence (2) $\Rightarrow$ (3). Conversely, if $f \in M^2$ then f_X and f_Y are in M, by an easy application of the Leibniz formula for derivatives of products, so (3) $\Rightarrow$ (2). Moreover if (2) holds and M is any maximal ideal containing f then $f \notin M^2$. Hence $MC/((f) + (MC)^2)$ is 1-dimensional and so $MC/(f)$ is principal, by Nakayama's Lemma again. Therefore (2) $\Rightarrow$ (1).

The equivalence of (3) and (4) is an immediate consequence of the Nullstellensatz. $\square$

COROLLARY 10.2.1. $V(f)$ *is smooth if and only if* $O_{V(f)}$ *is integrally closed.*   □

The equivalence (1) ≡ (2) in Theorem 10.2 goes through with little change if $\mathbb{C}[X]$ is replaced by any Dedekind domain R. The equivalence (2) ≡ (3) holds if $\mathbb{C}$ is replaced by any perfect field. (The latter hypothesis is necessary. Let K be a field of characteristic $p > 0$ with an element b which is not a pth power. Then $f(X, Y) = X^p - b$ is irreducible in $K[X]$, so $K[X]/(f)$ is a field and $K[X, Y]/(f)$ is a PID. However (3) fails, as $f_X \equiv f_Y \equiv 0$ and $V(f)$ is everywhere singular.)

10.2. Power series

The localization of the polynomial ring $\mathbb{C}[X, Y]$ at O is too small to reflect the topology adequately. For instance, $Y^2 - X^2(X + 1)$ is irreducible in $\mathbb{C}[X, Y]_{(X,Y)}$ but factors as

$$Y^2 - X^2(X + 1) = (Y + X(1 + X)^{\frac{1}{2}})(Y - X(1 + X)^{\frac{1}{2}})$$

in $\mathbb{C}\{X, Y\}$, the ring of germs of holomorphic functions at the origin $O \in \mathbb{C}^2$. (This is the ring of power series which converge on some neighbourhood of O.)

From the algebraic point of view it is natural to pass to the completion of this ring with respect to powers of its maximal ideal, which is the formal power series ring $\mathbb{C}[[X, Y]]$. The ring $\mathbb{C}[[X_1, \dots, X_n]]$ of formal power series in n variables is a local domain, with maximal ideal $M = (X_1, \dots, X_n)$, residue field $\mathbb{C}$ and field of fractions $\mathbb{C}((X_1, \dots, X_n))$. It is complete and Hausdorff with respect to the M-adic topology. In particular, $\mathbb{C}[[X]]$ is a complete discrete valuation ring, with valuation

$$v(f) = \max\{n \mid f \in (X)^n\} = \dim_{\mathbb{C}} \mathbb{C}[[X]]/(f),$$

for $f \neq 0$. (See Chapter 10 of [**AM**].)

THEOREM 10.3. *Let I be an ideal in* $\mathbb{C}[[X, Y]]$. *Then the following are equivalent:*

    (1) $\dim_{\mathbb{C}} \mathbb{C}[[X, Y]]/I < \infty$;

(2) I contains a power of $M = (X, Y)$;

(3) I is not contained in any proper principal ideal.

PROOF. If $\dim_{\mathbb{C}} \mathbb{C}[[X, Y]]/I < \infty$ then $\mathbb{C}[[X, Y]]/I$ has a finite composition series with subquotients isomorphic to $\mathbb{C}[[X, Y]]/M \cong \mathbb{C}$. Hence $\mathbb{C}[[X, Y]]/I$ is annihilated by a power of M, and so (1) $\Rightarrow$ (2).

If $M^n \leq I$ holds then $\mathbb{C}[[X, Y]]/I$ is generated as a $\mathbb{C}$-vector space by the images of the monomials of total degree less than n. Moreover, if also $I \leq (f)$ then f divides X^n and f divides Y^n, so f is a unit in $\mathbb{C}[[X, Y]]$. Hence (2) $\Rightarrow$ (1) and (3).

If (3) holds then I is not contained in any proper ideal of the localization $\mathbb{C}((X))[[Y]]$, and so we may write 1 as a finite sum $1 = \Sigma X^{-n_j} r_j$ for some $r_j \in I$ and $n_j \geq 0$. Clearing denominators, we see that $X^n \in I$ for $n \geq n_X$. Similarly, $Y^n \in I$ for $n \geq n_Y$. Hence $M^n \subset I$ for $n \geq n_X + n_Y - 1$, and so (3) $\Rightarrow$ (2). $\square$

If $f \in \mathbb{C}[[X, Y]]$ the *multiplicity* of f is $\nu(f) = \min\{n \mid f \in M^n\}$. It is easily seen that multiplicity is additive $(\nu(fg) = \nu(f) + \nu(g))$, positive $(\nu(f) \geq 0)$, and is 0 if and only if f is a unit. If $f \neq 0$ its *initial term* is the sum of the terms of total degree $\nu(f)$, i.e., the lowest nonzero homogeneous part of f.

COROLLARY 10.3.1. *If* $\dim_{\mathbb{C}} \mathbb{C}[[X, Y]]/I = d$ *then* $M^d \leq I$. $\square$

THEOREM 10.4. *The ring* $\mathbb{C}[[X, Y]]$ *is a 2-dimensional noetherian local unique factorization domain.*

PROOF. The above properties of multiplicity imply easily that every element of $\mathbb{C}[[X, Y]]$ has a finite factorization into irreducibles. Let f be an irreducible element which divides a product gh in $\mathbb{C}[[X, Y]]$. If $f = X$ then it is clearly prime, since $\mathbb{C}[[X, Y]]/(X) = \mathbb{C}[[Y]]$ is a domain. Suppose that $f \neq X$. Then f remains irreducible in $\mathbb{C}((X))[[Y]]$, and divides gh there. Since $\mathbb{C}((X))[[Y]]$ is a PID, f must divide one of the factors, g say. On clearing denominators we see that f divides gX^k in $\mathbb{C}[[X, Y]]$, for some $k \geq 0$. It follows easily that f divides g in $\mathbb{C}[[X, Y]]$, and so irreducibles are prime.

Let I be a nonzero ideal in $\mathbb{C}[[X, Y]]$, and let h be the highest common factor of the elements of I. (This is well defined up to units, since $\mathbb{C}[[X, Y]]$ is factorial.) Then $I = hJ$, where J is contained in no proper principal ideal. Hence J contains M^n, for some $n \geq 0$, by Theorem 10.3, and so is generated by the monomials of degree n together with a basis for the finite dimensional vector space J/M^n. Therefore I is finitely generated and so $\mathbb{C}[[X, Y]]$ is noetherian.

In particular, if I is prime and $0 \neq I \neq M$ then I is principal. Thus $\mathbb{C}[[X, Y]]$ is 2-dimensional. $\qquad\square$

If R is a domain and $f, g \in R[X]$ there are unique polynomials q and r with r of degree less than the degree of g such that $f = gq + r$, by polynomial long division. This may be extended to formal power series over complete local rings, by the Weierstrass Division Theorem.

THEOREM 10.5. [WDT] *Let A be a complete local ring with maximal ideal M, such that $\cap_{n \geq 1} M^n = 0$. If $P, D \in A[[Y]]$ and $D \equiv Y^n$ modulo $MA[[Y]]$ there is a unique pair $q \subset A[[Y]]$ and $r \in A[Y]$ of degree $< n$ such that $P = Dq + r$.*

PROOF. Define a metric d on A by $d(a, b) = 2^{-n}$ if $a - b \in M^n$ but $a - b \notin M^{n+1}$ and extend this to a metric on $A[[Y]]$ by $d(\Sigma a_i Y^i, \Sigma b_j Y^j) = \sup_{k \geq 0} d(a_k, b_k)$. Then $A[[Y]]$ is complete with respect to this metric. Define A-linear functions E, F, T from $A[[Y]]$ to itself as follows. If $h \in A[[Y]]$ let $F(h)$ be the polynomial given by the terms of h of degree $< n$ in Y, and write $h = E(h)Y^n + F(h)$. Let $T(h) = h + E(P - Dh)$. Then it is easily seen that $d(T(h), T(k)) \leq \frac{1}{2} d(h, k)$ and so T is a contraction. Since $A[[Y]]$ is a complete metric space T has a unique fixed point q. Then $E(P - Dq) = 0$, so $r = P - Dq$ is a polynomial of degree $< n$. $\qquad\square$

The hypothesis $\cap_{n \geq 1} M^n = 0$ is clearly satisfied in $\mathbb{C}[[X, Y]]$, and holds more generally in any noetherian local ring A, by Corollary 10.20 of [**AM**]. See also the third edition of [**Lan**] for a short self-contained proof due to Manin.

The Weierstrass Preparation Theorem is a direct consequence. If A is a complete local ring a polynomial $g \in A[Y]$ of degree n is *distinguished* if $g = Y^n + g_1 Y^{n-1} + \cdots + g_n$, where the coefficients g_i are in the maximal ideal of A, for $1 \leq i \leq n$.

THEOREM 10.6. [WPT] *Let A be a complete local ring with maximal ideal M, such that $\cap_{n \geq 1} M^n = 0$. If $f \in A[[Y]]$ has nonzero image $\bar{f}$ in $A/M[[Y]]$ then $f = ug$ for some distinguished polynomial $g \in A[Y]$ and unit $u \in A[[Y]]^\times$, which are uniquely determined by f.*

PROOF. We may write $\bar{f} = Y^n \bar{v}$, where n is maximal and $\bar{v}$ is a unit in $A/M[[Y]]$. Let $v \in A[[Y]]$ have image $\bar{v}$ in $A/M[[Y]]$. Then v is a unit in $A[[Y]]$. Applying the WDT with $P = Y^n$ and $D = v^{-1}f$ gives q, r such that $Y^n = qv^{-1}f + r$. On reducing modulo M we see that q is a unit and $r \in MA[[Y]]$. Hence $u = q^{-1}v$ is a unit and $g = Y^n - r$ is a distinguished polynomial, such that $f = ug$. The uniqueness follows from the uniqueness in Theorem 10.5.   □

The above theorems hold also for $\mathbb{C}\{X, Y\}$. In particular, if P, D and f are holomorphic and $D(0, Y) = Y^n$ the power series q, r, g and u given by the WDT and WPT are holomorphic. These can be deduced from the formal versions by making suitable convergence estimates but it is more natural to use the Cauchy integral formula.

Square-free formal power series are equivalent to polynomials, up to a formal change of coordinates.

THEOREM 10.7. [Sa56] *Let $f \in M = (X, Y) < \mathbb{C}[[X, Y]]$ be square-free. Then there is an automorphism α of $\mathbb{C}[[X, Y]]$ such that $\alpha(f)$ is a polynomial.*

PROOF. If $f \notin M^2$ this is an immediate consequence of the Inverse Function Theorem for formal power series. For if $f_X(0) \neq 0$, say, then $\beta(X) = f$, $\beta(Y) = Y$ defines an automorphism of $\mathbb{C}[[X, Y]]$, and $\beta^{-1}(f) = X$.

Suppose that $f \in M^2$, and let $J = (f_X, f_Y)$. Then J contains some power of M, by Theorem 10.14 below, and so $M^{2k+1} \leq MJ^2$ for some $k \geq 1$. Let P be a polynomial in the coset $f + MJ^2$. (For instance, we may truncate f after terms of degree at most $2k$.) Then $f = P - A$ where $A = a_{11}f_X^2 + (a_{12} + a_{21})f_X f_Y + a_{22}f_Y^2$, for some $a_{ij} \in M$. We shall show that there is an automorphism α of the form

$$\alpha(X) = X + u_{11}f_X + u_{12}f_Y, \quad \alpha(Y) = Y + u_{21}f_X + u_{22}f_Y$$

with $u_{ij} \in M$ and such that $\alpha(f) = P$. We must solve the equation $f(x + u\nabla f) - f(x) = A$. The left hand side has the form

$$u_{11}(1+G_{11})f_X^2+u_{12}(1+G_{12})f_Xf_Y+u_{21}(1+G_{21})f_Yf_X+u_{22}(1+G_{22})f_Y^2,$$

where the G_{ij} are power series in the u_{ij} with coefficients in M and with constant term 0. Let $\theta_{ij}(u) = a_{ij}(1 + G_{ij})^{-1}$. Then θ defines a contraction mapping on the set of 2×2 matrices over $\mathbb{C}[[X,Y]]$, and its unique fixed point gives a solution.   □

10.3. Puiseux series

In studying algebraic links, there is no loss of generality in assuming that the defining function f is a polynomial, by Theorem 10.7. In particular, if f is irreducible as a polynomial in Y we may solve the equation $f(X, Y) = 0$ for Y in the algebraic closure of $\mathbb{C}((X))$. This algebraic closure may be constructed by adjoining n^{th} roots of the variable X, for all $n > 1$. Thus the roots of $f(X, Y) = 0$ are power series in fractional powers of X. The exponents of the nonzero terms determine the Puiseux invariants for f, and we outline how they may be interpreted in terms of Galois theory.

Let $f \in (X, Y) < \mathbb{C}[[X, Y]]$ be square-free. Then $\mathbb{C}[[X, Y]]/(f)$ is a 1-dimensional noetherian local ring without nilpotent elements. The uniform topology determined by powers of the maximal ideal is Hausdorff (hence metrizable), by Corollary 10.20 of [**AM**], and the ring is complete with respect to this uniform structure. After a linear change of coordinates, if necessary, we may assume that f is a distinguished polynomial of degree $n > 0$ in Y, by the WPT.

THEOREM 10.8. *Let $f \in \mathbb{C}[[X, Y]]$ be an irreducible distinguished polynomial of degree $n > 0$ in Y, and let $\widetilde{A}$ be the integral closure of $A = \mathbb{C}[[X, Y]]/(f)$ in its field of fractions $K = \mathbb{C}((X))[Y]/(f)$. Then*

(1) $\widetilde{A} = \mathbb{C}[[t]]$ *and* $K = \mathbb{C}((t))$ *for some* $t \in K$;
(2) $\widetilde{A}$ *is a free* $\mathbb{C}[[X]]$-*module of rank* n;
(3) $(X)\mathbb{C}[[t]] = (t^n)$;
(4) $\dim_{\mathbb{C}} \widetilde{A}/A < \infty$.

PROOF. Parts (1), (2) and (3) follow from Proposition II.3 of [**Ser**]. Since A is clearly free of rank n over $\mathbb{C}[[X]]$ the quotient $\widetilde{A}/A$ is a finitely generated torsion module and so $\dim_\mathbb{C} \widetilde{A}/A < \infty$. $\square$

We shall consider the invariant $\dim_\mathbb{C} \widetilde{A}/A$ further in §5 below.

Since the prime ideal (X) becomes an nth power in $\mathbb{C}[[t]]$ the extension is totally ramified. In fact we may assume that $t^n = X$, by the following theorem. Let $\widehat{\mathbb{Z}} = \varprojlim\{: Z/mnZ \to Z/nZ\}$ be the profinite completion of $\mathbb{Z}$, and let $\overline{\mathbb{C}((X))} = \cup_{n \geq 1}\mathbb{C}((X^{\frac{1}{n}}))$.

THEOREM 10.9. *The algebraic closure of* $\mathbb{C}((X))$ *is* $\overline{\mathbb{C}((X))}$, *with Galois group* $\widehat{\mathbb{Z}}$ *acting through multiplication of* $X^{\frac{1}{n}}$ *by* n^{th} *roots of unity.*

PROOF. Let K be a finite extension of $\mathbb{C}((X))$, of degree n. By Theorem 10.8 the integral closure of $\mathbb{C}[[X]]$ in K is $\mathbb{C}[[t]]$, and $(X) = (t)^n$ in R. Therefore $X = t^n u$, where u is a unit. Since u is a power series with nonzero constant term it has an nth root in $\mathbb{C}[[t]]$, which is also a unit. Then $t_1 = u^{1/n}t$ is another uniformizer for $\mathbb{C}[[t]]$, with $X = t_1^n$. Hence $K/\mathbb{C}((X))$ is a Galois extension, with cyclic Galois group, acting on t_1 via multiplication by roots of unity. The theorem follows easily. $\square$

By the theorem, if $f \in \mathbb{C}[[X]][Y]$ is an irreducible monic polynomial of degree n in Y and ζ is a primitive n^{th} root of unity then $f = \Pi_{i=1}^{i=n}(Y - h(\zeta^i X))$, for some $h \in \mathbb{C}[[X^{\frac{1}{n}}]]$. The fractional power series h is well defined up to the Galois action, and shall be called the *Puiseux series* for f. (In the holomorphic context, the algebraic closure of the field of germs of meromorphic functions $\mathbb{C}\{\{X\}\}$ is the union $\cup_{n \geq 1}\mathbb{C}\{\{X^{\frac{1}{n}}\}\}$. See Exercise 2.8 of Chapter IV of [**Ser**].)

LEMMA 10.10. *Let* $f \in \mathbb{C}[[X,Y]]$ *be nonzero, and let* f_ν *be its initial term. If* $f_\nu = g_m h_n$ *where* g_m *and* h_n *have no common factor then* $f = gh$ *for some* g *and* $h \in \mathbb{C}[[X,Y]]$ *with initial terms* g_m *and* h_n, *respectively.*

PROOF. We construct the homogeneous parts of $g = g_m + g_{m+1} + \cdots$ and $h = h_n + h_{n+1} + \cdots$ recursively. It suffices to show that given k homogeneous of degree $m + n + p$ for some $p > 0$ we can solve $g_m h_{n+p} + g_{m+p} h_n = k$ for g_{m+p} and h_{n+p}. On making the substitution $Y = XZ$ and dividing by X^{m+n+p} this reduces to an inhomogeneous equation in 1 variable, which may be solved by the Euclidean algorithm. (There are many possible solutions.)   □

We may relax the hypothesis in Theorem 10.8 that f be distinguished as follows. Let x and y denote the images of X and Y in $A = \mathbb{C}[[X, Y]]/(f)$.

THEOREM 10.11. *Let f be irreducible. Then $\widetilde{A} \cong \mathbb{C}[[t]]$ and $(x, y)\bar{A} = t^{\nu(f)}\bar{A}$.*

PROOF. The initial term of f is a product of linear terms, since it is a homogeneous polynomial in $\mathbb{C}[X, Y]$. Since f is irreducible the linear factors must all be equal, by Lemma 10.10, and so the initial term has the form $(dX + eY)^n$, where $n = \nu(f)$ and d and e are not both 0. After a linear change of coordinates we may assume that the initial term is Y^n. On applying the WPT we may then assume that $f = Y^n + \Sigma a_i X^i Y^{n-i}$, where $a_i \in (X)\mathbb{C}[[X]]$ for $1 \leq i \leq n$. Since $x = t^n$ in $\widetilde{A} \cong \mathbb{C}[[t]]$ and $f(y) = 0$ in $\bar{A}$ we see easily that y must be in $(t)^{n+1}$. Hence $(x, y)\widetilde{A} = t^n \widetilde{A} = t^{\nu(f)}\widetilde{A}$.   □

In particular, if f is a distinguished polynomial of degree n in Y and $\nu(f) = n$ then $(x, y)\widetilde{A} = x\widetilde{A}$.

Let $h(X^{1/n}) = \Sigma h_k X^{k/n}$ be a Puiseux series for f, and let $I = \{k \mid h_k \neq 0\}$ be the set of indices corresponding to nonzero coefficients in h. Then hcf$I$ is relatively prime to n, since $[K : \mathbb{C}((X))] = n$. Let $\beta(1) = \min\{k \in I \mid k \notin n\mathbb{Z}\}$ and write $\beta(1)/n = m_1/n_1$ where $(m_1, n_1) = 1$. If $n_1 = n$ then $(m_1, n_1) = (\beta(1), n)$ is the unique *characteristic pair* for f. If $n_1 < n$ then n_1 divides n, and there are indices $k \in I$ such that k/n cannot be expressed as a fraction with denominator n_1. Let $\beta(2)$ be the first such, and write $n_1 \beta(2)/n = m_2/n_2$, where $(m_2, n_2) = 1$. Continuing in this way, after finitely many steps we obtain characteristic pairs $(m_1, n_1), \ldots, (m_g, n_g)$ such that $m_{i-1}n_i < m_i$ for $1 < i \leq g$ and $n_1 \cdots n_g = n$. (Conversely, if a finite sequence $(m_1, n_1), \ldots, (m_g, n_g)$ satisfies $m_{i-1}n_i < m_i$ for $1 < i \leq g$ then it is the sequence of characteristic pairs associated to a series

$h \in \mathbb{C}((X^{\frac{1}{n}}))$, where $n = n_1 \cdots n_g$. The product of the Galois conjugates of $y - h$ gives a series $f \in \mathbb{C}[[X, Y]]$ corresponding to the Puiseux data. See also pages 56-58 of [**EN**].)

These pairs may be interpreted in terms of "higher ramification groups": let $G = Gal(K/\mathbb{C}((X))) \cong Z/nZ$ and $G_i = \{\sigma \in G \mid \sigma(y) \equiv y \mod (t^i)\}$. Then $\beta(1) = \max\{m \mid G_m = G\}$, and if $G_{\beta(i)+1} \neq 1$ then $\beta(i + 1) = \max\{m \mid G_m = G_{\beta(i)+1}\}$. Clearly $\beta(1) < \cdots < \beta(g)$ and $G_{\beta(g)+1} = 1$. The denominators n_i are given by $n_i = [G_{\beta(i)} : G_{\beta(i)+1}]$, and so $n_1 \cdots n_g = |G| = n$. (It may be helpful to contemplate a simple example, such as $Y = X^{\frac{1}{3}} + X^{\frac{1}{2}} + X^{\frac{3}{4}}$, with $n = 12$ and characteristic pairs (1,3), (3,2), (9,2).)

Suppose now that $f = \Pi_{i=1}^{i=r} f_i$ is square-free, with r irreducible factors. On applying Lemma 10.10 to each such factor we see that the homogenous part of f of lowest degree is a product of powers of linear terms. (These correspond to the tangent lines to the irreducible components of $V(f)$ at 0.) The extended Puiseux data for f consists of the Puiseux data for the factors together with the "linking numbers" $\ell_{ij} = \dim_{\mathbb{C}} \mathbb{C}[[X, Y]]/(f_i, f_j)$ (which are the topological linking numbers of the corresponding components of $L(f)$).

Let $A(i) = \mathbb{C}[[X, Y]]/(f_i)$ and let K_i be the field of fractions of $A(i)$, for $1 \leq i \leq r$. The total ring of quotients of A is the localization $K = A_S$, where $S = A - \cup_{i=1}^{i=r}(f_i)$ is the set of nonzerodivisors in A, and A embeds in K. Moreover $\mathbb{C}[[X]] - \{0\} \subseteq S$, since f is a distinguished polynomial in Y. It follows easily that $K = \mathbb{C}((X)) \otimes_{\mathbb{C}[[X]]} A \cong \Pi K_i$. The integral closure of A in K is $\widetilde{A} \cong \oplus \widetilde{A(i)}$. Let $v_i : A \to Z_{\geq 0}$ be the composite of projection from A onto $A(i)$ with the restriction of the canonical valuation on $\widetilde{A(i)} \cong \mathbb{C}[[t]]$. Then $v_i(g) < \infty$ for all i if and only if f and g have no common factors. The *singularity semigroup*

$$S(f) = \{(v_1(g), \ldots, v_r(g)) \mid g \in A, v_i(G) < \infty \; \forall i\}$$

is a subsemigroup of $Z_{\geq 0}^r$, which encodes the Puiseux data for f [**Wa72**]. (See also [**GDC99**].)

10.4. The Milnor number

Milnor showed that if $f \in \mathbb{C}\{\{X_1, \ldots, X_{n+1}\}\}$ has an isolated singularity at O then the exterior of the associated algebraic link fibres over S^1, and the fibre F is a wedge of n-spheres. The number of spheres is given by the codimension in $\mathbb{C}\{\{X_1, \ldots, X_{n+1}\}\}$ of the ideal generated by the partial derivatives of f [**Mil**]. In general this codimension $\mu(f)$ is finite if and only if f has an isolated singularity at O. We shall verify this for plane curve singularities ($n = 1$). In this case the fibre is a punctured surface F, say, and so $\mu(f)$ is the rank of $H_1(X; \Lambda) \cong H_1(F; \mathbb{Z})$ as an abelian group. We shall also compute $\mu(f)$ in an important special case (when f is a weighted homogeneous polynomial in two variables).

If R is a PID and N is a finitely generated torsion R-module let $\ell_R(N)$ denote its length. (Note that $\ell_{\mathbb{C}[[X]]}(N) = \dim_{\mathbb{C}} N$ when $R = \mathbb{C}[[X]]$.) If $f \in R[Y]$ is a monic polynomial the *discriminant* of f is $disc(f) = (-1)^{\binom{n}{2}} Res(f, f_Y)$.

LEMMA 10.12. *Let R be a PID and $\theta : R^n \to R^n$ a monomorphism. Then $\ell_R(R^n/\theta(R^n)) = \ell_R(R/det(\theta)R)$.*

PROOF. Since R is a PID we may apply elementary row operations, with composition E say, to reduce the matrix of θ to upper triangular form $E\theta$. The automorphism E of R^n induces an isomorphism $R^n/\theta(R^n) \cong R^n/E\theta(R^n)$, and $det(E)$ is a unit in R. The result now follows by induction on n.   □

If $f, g \in \mathbb{C}[[X, Y]]$ let $I(f, g) = \dim_{\mathbb{C}} \mathbb{C}[[X, Y]]/(f, g)$. Then $I(f, g) < \infty$ if and only if f and g have no common factor. Let $M(f) = I(f, f_Y)$.

LEMMA 10.13. *Let $f \in \mathbb{C}[[X]][Y]$ be monic of degree m (as a polynomial in Y). Then*

   (1) *If $g \in \mathbb{C}[[X]][Y]$ then $Res(f, g)$ is the determinant of the endomorphism of $\mathbb{C}[[X]][Y]/(f) \cong \mathbb{C}[[X]]^m$ given by multiplication by g;*

   (2) *If f and g have no common factor $I(f, g)$ is the highest power of X dividing $Res(f, g)$;*

(3) *if* $f = \Pi_{i=1}^{i=r} f_i$ *is square-free* $M(f) = \Sigma M(f_i) + 2\Sigma_{i<j} I(f_i, f_j)$.

PROOF. We may extend coefficients to a field containing $\mathbb{C}[[X]]$ in which f splits into linear factors, and (1) is then clear.

Part (2) follows from Lemma 10.12, since $\mathbb{C}[[X]][Y]/(f) \cong \mathbb{C}[[X]]^m$ and $I(f, g) = \dim_{\mathbb{C}} \mathbb{C}[[X]]/(Res(f, g))$, which is the highest power of X dividing $Res(f, g)$.

Part (3) follows by induction on r, since

$$Res(fg, (fg)_Y) = Res(fg, f_Y g + fg_Y) = Res(f, f_Y g) Res(g, fg_Y),$$

so $Res(fg, (fg)_Y) = Res(f, g)^2 Res(f, f_Y) Res(g, g_Y)$, and hence $M(fg) = M(f) + M(g) + 2I(f, g)$. $\qquad\square$

In particular, if $f \in \mathbb{C}[[X, Y]]$ is a distinguished polynomial in Y then $M(f)$ is the highest power of X dividing $disc(f)$.

The power series f is *non-singular* if f_X and f_Y do not both vanish at O. As in the global case, f is non-singular if and only if it is irreducible and $f \notin M^2$, i.e., $\nu(f) = 1$. This in turn is so if and only if $A = \mathbb{C}[[X]][Y]/(f)$ is a discrete valuation ring, and hence $A = \tilde{A}$. (See Theorem 10.2.) In general, let $J(f) = (f_X, f_Y)$ be the ideal in $\mathbb{C}[[X, Y]]$ generated by the partial derivatives of f. The *Milnor number* of f is

$$\mu(f) = \dim_{\mathbb{C}} \mathbb{C}[[X, Y]]/J(f) = I(f_X, f_Y).$$

Then f is non-singular if and only if $J(f) = (1)$, i.e., $\mu(f) = 0$.

THEOREM 10.14. *Let* $f \in M = (X, Y) < \mathbb{C}[[X, Y]]$. *Then the following are equivalent:*

(1) f *is square-free;*
(2) $J(f)$ *contains a power of* M*;*
(3) (f, f_X, f_Y) *contains a power of* M*;*
(4) $\mu(f) < \infty$.

PROOF. If (2) does not hold then $J(f) \leq (p)$ for some principal prime $(p) < \mathbb{C}[[X, Y]]$, by Theorem 10.3. If $(p) = (X)$ it is easily seen that X^2 divides f. Otherwise we may assume that p is a distinguished polynomial of degree $n > 0$ in Y, by Theorem 10.6. Suppose first that $p = Y - h(X)$ for some $h \in \mathbb{C}[[X]]$ with

$h(0) = 0$. Define a retraction $\rho : \mathbb{C}[[X, Y]] \to \mathbb{C}[[X]]$ with kernel (p) by $\rho(k) = k(X, h(X))$. Then $\rho(f)_X = \rho(f_X + f_Y h_X) = 0$, so $\rho(f) = f(0, 0) = 0$. Hence $f = gp$ for some $g \in \mathbb{C}[[X, Y]]$. Then $f_Y = g_Y p + g p_Y = g_Y p + g$, so $f_Y \in (p)$ implies that $g \in (p)$ and so $f \in (p^2)$. In general, p factors into Galois-conjugate linear terms $p_i = Y - h(\zeta^i X^{\frac{1}{n}})$, where $h \in \mathbb{C}[[X^{\frac{1}{n}}]]$. The factors p_i are distinct primes in $\mathbb{C}[[X^{\frac{1}{n}}]][Y]$, since p is irreducible in $\mathbb{C}[[X, Y]]$. Applying the earlier argument to each p_i we see that f is divisible by p^2, and so (1) $\Rightarrow$ (2).

The implications (2) $\Rightarrow$ (3) $\Rightarrow$ (1) and (2) $\equiv$ (4) are clear. $\square$

In particular, if f is square-free then $f^n \in J(f)$ for n sufficiently large. In fact $f^2 \in J(f)$ for any square-free f, by a difficult result of Briançon and Skoda [**LT81**].

A *weighted homogeneous polynomial of type* $(N; a, b)$ is a polynomial $f \in \mathbb{C}[X, Y]$ which is equivariant with respect to the actions of the multiplicative group $\mathbb{C}^\times$ on $\mathbb{C}^2$ and $\mathbb{C}$ given by $\lambda(x, y) = (\lambda^a x, \lambda^b y)$ and $\lambda(z) = \lambda^N z$, for all $x, y, z \in \mathbb{C}$ and $\lambda \in \mathbb{C}^\times$, where a, b and N are positive integers and a and b are relatively prime. In other words, $f(\lambda^a X, \lambda^b Y) = \lambda^N f(X, Y)$, for all $\lambda \in \mathbb{C}$. It is easy to see that any formal power series satisfying this equation is a polynomial, and is a sum of terms $c_j X^j Y^k$ where $aj + bk = N$. Differentiating each side of this equation with respect to λ at $\lambda = 1$ gives $Nf = aX f_X + bY f_Y$ and so $f \in J(f)$. Let $Z = X^{-\frac{b}{a}} Y$. Then we may write $f(X, Y) = X^{\frac{N}{a}} F(Z)$, where $F(Z) = f(1, Z) \in \mathbb{C}[Z]$. On considering the factorization of $F(Z)$ we see that the irreducible factors of f are of the form X, Y or $Y^a - cX^b$ for some $c \in \mathbb{C}$.

Suppose now that f is monic in Y and g is another weighted homogeneous polynomial, of type $(N'; a, b)$, and $Res(f, g) \neq 0$. Suppose also that neither f nor g is divisible by X or Y. Then $N = abd$ for some integer $d \geq 1$. The roots of f in $\overline{\mathbb{C}((X))}$ are of the form $\alpha_i = \gamma_i X^{\frac{b}{a}}$, where $\gamma_i \in \mathbb{C}$, and so $g(\alpha_i) = c_i X^{\frac{N'}{a}}$, for some $c_i \in \mathbb{C} - \{0\}$. Hence $I(f, g) = \frac{NN'}{ab}$. In particular, if f is a square free, weighted homogeneous distinguished polynomial in Y which is not divisible by Y then f_X and f_Y are weighted homogeneous of types $(N-a; a, b)$ and $(N-b; a, b)$, respectively, and so $\mu(f) = \frac{(N-a)(N-b)}{ab}$.

Reiffen showed that a power series f in two variables is weighted homogeneous up to change of coordinates if and only if $f \in J(f)$, and this was extended to the many-variable case by K.Saito [**Sa71**]. We shall outline a proof of Reiffen's theorem, based on Saito's argument.

THEOREM 10.15. *Let $f \in M = (X, Y) < \mathbb{C}[[X, Y]]$. Then there is a formal change of coordinates $(u, v) = \phi(X, Y)$ such that $f\phi^{-1}$ is weighted homogeneous in the variables u, v if and only if $f \in J(f)$.*

PROOF. Since the condition $f \in J(f)$ is invariant under formal change of coordinates it is clearly necessary. Suppose that it holds. Then there are $\alpha, \beta \in R$ such that $f = Df$, where D is the differential operator $\alpha \partial_X + \beta \partial_Y$. On writing $f = \Sigma_{n \geq 1} f_n$ as the sum of its homogeneous parts and working modulo powers of $M = (X, Y)$ we find that $\alpha = X(c_1 + m_1)$ and $\beta = Y(c_2 + m_2)$ for some $c_1, c_2 \in \mathbb{C}^\times$ and $m_1, m_2 \in M$.

Suppose first that $m_1 = m_2 = 0$, so $f = c_1 X \partial_X f + c_2 Y \partial_Y f$. If the coefficient of $X^k Y^{n-k}$ in f is nonzero we must have $(c_1 - c_2)k + c_2 n = 1$. Therefore either $c_1 = c_2 = \frac{1}{n}$ and $f = f_n$ is homogeneous of total degree n, or $c_1 \neq c_2$ and each homogeneous term f_n is monomial. Suppose that f is not homogeneous, and that $X^j Y^{m-j}$ and $X^k Y^{n-j}$ are monomials with nonzero coefficients in f and with $n \neq m$. Since $c_1 j + c_2(m - j) = 1$, $c_1 k + c_2(n - k) = 1$ and $n \neq m$ the vectors $(j, m - j)$ and $(k, n - k)$ cannot be proportional, and so the determinant $jn - km$ is nonzero. Hence this pair of equations determines c_1 and c_2 uniquely, as rational numbers $c_1 = \frac{a}{N}$ and $c_2 = \frac{b}{N}$, say. (We assume that these fractions are written in lowest terms. In particular, $(a, b) = 1$ if $c_1 c_2 \neq 0$.) If $c_1 > 0$ and $c_2 > 0$ then f is weighted homogeneous of type $(N; a, b)$. If $c_1 = 0$ then $c_2 = \frac{1}{N}$ and $f = Y^N g(X)$ for some 1-variable power series g. Let $g = X^p g_1$ where $g_1(0) \neq 1$ and let $(u, v) = \phi(X, Y) = (X, Y(g_1)^{\frac{1}{N}})$. Then $f\phi^{-1} = u^p v^N$. Similarly if $c_2 = 0$. If $c_1 c_2 < 0$ then $f = X^p Y^q h(X^{|b|} Y^{|a|})$ for some 1-variable power series h with $h(0) \neq 0$, and f becomes weighted homogeneous after a simple change of variables.

In general, we may reduce to the above case if there is a change of variables $\phi(X, Y) = (u, v)$ such that $u \equiv X$ and $v \equiv Y \mod M^2$ and $D = c_1 u \partial_u + c_2 v \partial_v$. The latter condition gives equations

$$\alpha \partial_X u + \beta \partial_Y u = c_1 u \quad \text{and} \quad \alpha \partial_X v + \beta \partial_Y v = c_2 v.$$

On writing $u = X + \Sigma_{m \geq 2} u_m$ and $v = Y + \Sigma_{n \geq 2} v_n$, where the u_m and v_n are homogeneous of total degree m and n, respectively, these lead to linear recursion equations for the coefficients of the terms u_m and v_n which are always solvable, and have unique solutions if $c_1 > 0$ and $c_2 > 0$. $\qquad\qquad\qquad\qquad\qquad\qquad\qquad\qquad\square$

The *Tjurina number* $\tau(f) = \dim_{\mathbb{C}} \mathbb{C}[[X, Y]]/(f, f_X, f_Y)$ is a closely related invariant. Clearly $\tau(f) \leq \mu(f)$, with equality if and only if $f \in J(f)$, and it follows easily from Theorem 10.14 that $\tau(f) < \infty$ if and only if $\mu(f) < \infty$. The polynomial $g = Y^5 + X^2 Y^2 + X^5$ is a simple example of a formal power series which cannot be so obtained, for $\tau(g) = 10 \neq \mu(g) = 11$.

10.5. The conductor

A distinguished polynomial $f \in \mathbb{C}[[X]][Y]$ is non-singular if and only if $A = \widetilde{A}$, where $A = \mathbb{C}[[X, Y]]/(f)$. Thus $\delta_A = \dim_{\mathbb{C}}(\widetilde{A}/A)$ is a numerical measure of the singularity of f at O. This invariant provides the local correction terms for the Plücker/Riemann-Roch formula $g = \binom{d-1}{2} + \Sigma_{z \in Z} \delta_z$ relating the genus g and degree d of a projective plane curve $V(f)$ with $Sing(V(f)) = Z$.

In this section we shall show first that δ_A is also the codimension in A of $\mathcal{C} = \{r \in A \mid r\widetilde{A} \leq A\}$, the *conductor* of $\widetilde{A}$ into A. (This is the largest ideal of A which is also an ideal of $\widetilde{A}$.) We shall then show that $\delta_A = (\mu(f) + 1 - r)/2$, a result found by Jung [**Jun**] long before the topological significance of $\mu(f)$ had been realised.

Let $S = \mathbb{C}[[X]]$ and let $S_0 = \mathbb{C}((X))$ be the field of fractions of S. Let $f = \Sigma_{j \leq n} f_j Y^j \in S[Y]$ be a square-free monic polynomial of degree n in Y. Let y be the image of Y in $A = S[Y]/(f)$. Then $K = S_0 \otimes_S A$ is the total ring of quotients of A and we may define an S_0-linear homomorphism $\tau : K \to S_0$ by $\tau(s y^j) = 0$ if $0 \leq j < n - 1$ and $\tau(s y^{n-1}) = s$, for all $s \in S_0$.

LEMMA 10.16. *Let $k \in K$. Then $k \in A$ if and only if $\tau(ky^j) \in S$, for all $0 \le j < n$.*

PROOF. Since $\{y^i \mid 0 \le i \le n - 1\}$ is a basis for A as an S-module, the condition is clearly necessary. Assume that it holds, and let $k = \Sigma_{j<n} s_j y^j$ where $s_j \in S_0$ for $0 \le j < n$. Then $s_{n-1} = \tau(k)$, $s_{n-2} = \tau(y(k - s_{n-1}y^{n-1}))$ and in general

$$s_{n-i} = \tau(y^{i-1}(k - \Sigma_{n-i<j<n} s_j y^j)),$$

for all $0 < i < n$. A finite induction on i shows that all the coefficients are in S, and so $k \in A$. $\qquad\square$

THEOREM 10.17. $\dim_\mathbb{C} A/\mathcal{C} = \dim_\mathbb{C} \widetilde{A}/A$.

PROOF. Let $D(N) - Hom_S(N, S_0/S)$, for N any S-module of finite length. Then $D(\mathbb{C}) = D(S/(X)) \cong \mathbb{C}$. Since every S-module of finite length has a finite composition series with subquotients $S/(X) \cong \mathbb{C}$, it follows easily that $\dim_\mathbb{C} D(N) = \dim_\mathbb{C} N$, for any such N. Define a pairing $\lambda : A \times \widetilde{A} \to S_0/S$ by $\lambda(a, \alpha) = [\tau(a\alpha)] \in S_0/S$, for all $a \in A$ and $\alpha \in \widetilde{A}$. As λ determines monomorphisms from $A/\mathcal{C}$ to $D(\widetilde{A}/A)$ and from $\widetilde{A}/A$ to $D(A/\mathcal{C})$, by Lemma 10.16, the result follows easily. $\qquad\square$

The function $f = Y^2 - X^3$ gives a simple but nontrivial and instructive example. We have $\nu(f) = 2$, $disc(f) = -4X^3$ and $f_Y = 2Y$, and $A = \mathbb{C}[[t^2, t^3]] < \widetilde{A} = \mathbb{C}[[t]]$. Hence $\delta_A = 1$, and $\mathcal{C} = t^2 \widetilde{A} = (x, y)A$. (Note that $\mathcal{C}$ is not principal as an ideal in A.)

Our exposition of Theorem 10.17 has its roots in the following observation, which goes back to Euler. (See Lemma III.2 of [**Ser**].)

LEMMA 10.18. *Let $f \in \mathbb{C}[[X]][Y]$ be irreducible. Then $\tau(a) = tr_{K/S_0}(a/f_Y(y))$.*

PROOF. If f is irreducible then $K \cong \mathbb{C}((t))$, where $t^n = x$, by Theorems 10.8 and 10.9, and we may identify tr_{K/S_0} with the $\mathbb{C}((x))$-linear homomorphism from $\mathbb{C}((t))$ to $\mathbb{C}((x))$ determined by $tr(1) = n$ and $tr(t^i) = 0$ for $1 \le i < n$.

Let $\{y_j \mid 1 \leq j \leq n\}$ be the roots of $f = f(Y)$ in $\mathbb{C}((t))$. The method of partial fractions gives $\frac{1}{f(Y)} = \Sigma \frac{1}{f_Y(y_j)(Y-y_j)}$. On expanding each side as a power series in $\frac{1}{Y}$ and comparing coefficients, we see that $tr(\frac{y^i}{f_Y}) = 0$ if $0 \leq i < n-1$ and $tr(\frac{y^{n-1}}{f_Y}) = 1$. $\qquad\square$

We shall show next that δ_A may be expressed in terms of $\mu(f)$ and r (the number of irreducible factors of f).

LEMMA 10.19. Let $A = \mathbb{C}[[X,Y]]/(f)$, where $f \in \mathbb{C}[[X,Y]]$ is irreducible, and let $\theta : A^n \to A^n$ be a monomorphism. Then $\dim_{\mathbb{C}} A^n/\theta(A^n) = \dim_{\mathbb{C}} \widetilde{A}^n/\theta(\widetilde{A}^n)$.

PROOF. The monomorphism θ extends to an endomorphism of $\widetilde{A}^n$, and gives rise to a commutative diagram

$$
\begin{array}{ccccccccc}
0 & \longrightarrow & A^n & \longrightarrow & \widetilde{A}^n & \longrightarrow & \widetilde{A}^n/A^n & \longrightarrow & 0 \\
& & \theta\downarrow & & \tilde{\theta}\downarrow & & [\theta]\downarrow & & \downarrow \\
0 & \longrightarrow & A^n & \longrightarrow & \widetilde{A}^n & \longrightarrow & \widetilde{A}^n/A^n & \longrightarrow & 0
\end{array}.
$$

The "Snake Lemma" gives an exact sequence

$$0 \to \mathrm{Ker}([\theta]) \to A^n/\theta(A^n) \to \widetilde{A}^n/\tilde{\theta}(\widetilde{A}^n) \to \mathrm{Cok}([\theta]) \to 0.$$

(See Proposition 2.10 of [AM].) The extreme terms have the same dimension, since $\widetilde{A}^n/A^n = (\widetilde{A}/A)^n$ is finite dimensional, by Theorem 10.8. The result follows easily. $\qquad\square$

LEMMA 10.20. Let $f = \Pi_{i=1}^{i=r} f_i$ be square-free, with r irreducible factors. Then

(1) $\delta_A = \Sigma \delta_{A(i)} + \Sigma_{i<j} I(f_i, f_j)$, where $A(i) = \mathbb{C}[[X,Y]]/(f_i)$ for $1 \leq i \leq r$;
(2) if $f \in \mathbb{C}[[X]][Y]$ is irreducible and monic in Y then $M(f) = 2\delta_A + n - 1$.

PROOF. The natural homomorphism from A to $\oplus A(i)$ is injective and its cokernel is $\oplus_{i<j}(\mathbb{C}[[X,Y]]/(f_i, f_j))$. Since $\widetilde{A} = \oplus\widetilde{A(i)}$ part (1) follows immediately,

Assume now that f is in $\mathbb{C}[[X]][Y]$ and is irreducible. Then $\widetilde{A}$ is a domain. Let $tr = tr_{K/S_0}$. If P is an A-submodule of $\mathbb{C}((t))$ let

$P^* = \{k \in \mathbb{C}((t)) \mid tr(kp) \in A \ \forall p \in P\}$. Now $tr(t^{-n}) = \frac{n}{x}$ and $tr(t^i) \in \mathbb{C}[[X]]$ if $i > -n$, so $\widetilde{A}^* = t^{1-n}\widetilde{A}$. The $n \times n$ matrix with (i,j) entry $tr(\frac{y^{i+j-2}}{f_Y})$ is 0 above the secondary diagonal, and is 1 along that diagonal. Since $\{y^i \mid 0 \le i \le n-1\}$ is a basis for A as a $\mathbb{C}[[X]]$-module it follows that $A^* = f_Y^{-1}A$. Hence

$$c \in \mathcal{C} \Leftrightarrow c\widetilde{A} \le A \Leftrightarrow cf_Y^{-1}\widetilde{A} \le A^* \Leftrightarrow tr(cf_Y^{-1}\alpha a) \in A \text{ for all } \alpha \in \bar{A}$$

and

$$a \in A \Leftrightarrow cf_Y^{-1} \in \widetilde{A}^* \Leftrightarrow c \in f_Y t^{1-n}\widetilde{A}.$$

Since $\mathcal{C} = t^{2\delta_A}\widetilde{A}$, by Theorem 10.17, and $M(f) = \dim_{\mathbb{C}} \widetilde{A}/f_Y \widetilde{A}$ (the t-adic valuation of f_Y), by Lemma 10.19, it follows that $M(f) = 2\delta_A + n - 1$. $\qquad\square$

The computation of $\mathcal{C}$ is based on Proposition III.11 of [**Ser**].

LEMMA 10.21. [**Ri71**] *If* $f \in \mathbb{C}[[X]][Y]$ *is monic (as a polynomial in* Y*) then* $I(f, f_Y) = I(mf + Xf_X, f_Y)$ *for all* $m \in \mathbb{Z}$.

PROOF. It shall suffice to prove that $I(f,g) = I(mf+Xf_X,g)$ for each irreducible factor g of f_Y. The integral closure of $\mathbb{C}[[X]][Y]/(g)$ in its field of fractions is isomorphic to $\mathbb{C}[[u]]$, by Theorem 10.8; let v be the associated valuation. Then $I(f,g) = v(f)$ and $I(mf + Xf_X, g) = v(mf + Xf_X)$, by Lemma 10.19. Now $mf + Xf_X = X^{1-m}((X^m f)_X)$, and

$$v(w_X) = v(dw/du) - v(dX/du) = v(w) - 1 - (v(X) - 1) = v(w) - v(X),$$

by the chain rule $dw/du = w_X.dX/du$ in $\mathbb{C}[[u]]$. Hence $v(mf + Xf_X) = (1-m)v(X) + v((X^m f)_X) = (m-1)v(X) + v(f)$, and so $I(mf + Xf_X, g) = v(f) = I(f,g)$. $\qquad\square$

Suppose that $f = Y^n + \Sigma a_i X^i Y^{n-i}$ is a distinguished polynomial of degree $n = \nu(f)$ in Y. Let $Y = XZ$ and define f^σ by $f(X, XZ) = X^n f^\sigma(X, Z)$. Then $f^\sigma = Z^n + \Sigma a_i Z^{n-i}$ is the *strict (quadratic) transform* of f. Moreover if f is irreducible then so is f^σ.

Quadratic transform is semi-local, in that it replaces the origin by a projective line, and separates the lines through the origin. If g is irreducible and has initial term $(Y - \lambda X)^m$ (i.e., has tangent line $Y - \lambda X = 0$) then $g^\sigma \equiv (Z - \lambda)^m \bmod (X)$ and g^σ determines the germ of a curve through $(X, Z) = (0, \lambda)$. It follows that if $f = \Pi_{i=1}^{i=r} f_i$ and $g = \Pi_{j=1}^{j=s} g_j$ then $I(f, g) = \nu(f)\nu(g) + \Sigma\Sigma I(f_i^\sigma, g_j^\sigma)$, (with only factors having common tangents contributing to the double sum). Let $\tilde{I}(f^\sigma, g^\sigma)$ denote the latter sum.

THEOREM 10.22. [**Jun, Ri71**] *Let* $f = \Pi_{i=1}^{i=r} f_i$ *be square-free, with* r *irreducible factors. Then* $\delta_A = (\mu(f) + 1 - r)/2$.

PROOF. We may assume that $f = Y^n + \Sigma a_i X^i Y^{n-i}$ is a distinguished polynomial of degree $n = \nu(f) > 1$ in Y, and (hence) that each factor f_i is also a distinguished polynomial of degree n_i in Y, for $1 \leq i \leq r$. Differentiating the equation $X^n f^\sigma(X, Z) = f(X, XZ)$ with respect to Z gives $X^n (f^\sigma)_Z = X f_Y$ (so $(f_Y)^\sigma = (f^\sigma)_Z = f_Z^\sigma$, say) and so $M(f) = I(f, f_Y) = n(n-1) + \tilde{I}(f^\sigma, f_Z^\sigma)$. Differentiation with respect to X gives $n X^{n-1} f^\sigma + X^n (f^\sigma)_X = f_X + Z f_Y$, and so

$$\mu(f) = I(f_X, f_Y) = I(f_X + Y f_Y, f_Y)$$
$$= (n-1)^2 + \tilde{I}(n f^\sigma + X(f^\sigma)_X + (X-1)Z f_Z^\sigma, f_Z^\sigma).$$

(Note that since f is square-free $f_X \neq 0$ and $\nu(f_X) = n - 1$.) This in turn equals $(n-1)^2 + \tilde{I}(f^\sigma, f_Z^\sigma)$, by Lemma 10.21. Hence $\mu(f) = M(f) + 1 - n$. Now

$$M(f) = \Sigma M(f_i) + 2\Sigma_{i<j} I(f_i, f_j)$$
$$= 2(\Sigma \delta_{A(i)} + \Sigma_{i<j} I(f_i, f_j)) + (\Sigma n_i) - r$$
$$= 2\delta_A + n - r,$$

by Lemmas 10.13 and 10.19. Therefore $\mu(f) = 2\delta_A - r + 1$ and so $\delta_A = (\mu(f) + 1 - r)/2$. $\square$

The invariant δ_A is the unknotting number of $L(f)$; this is a consequence of the Thom conjecture (see the survey article [**BW83**]), which has been proven by Kronheimer and Mrowka [**KM94**].

Milnor has given a topological argument for Theorem 10.22 (for f a polynomial), in which the projective completion of $V(f)$ is approximated by other related curves. (See Theorem 10.5 of [**Mil**].)

10.6. Resolution of singularities

To resolve the singularities of an algebraic variety V means to give a smooth variety V' and a morphism $p : V' \to V$ which restricts to an isomorphism over the non-singular points of V. This issue is usually considered in the context of projective varieties.

If $V(f)$ is an irreducible plane curve then $\widetilde{O}_{V(f)}$ is finitely generated as an $O_{V(f)}$-module, and hence as a $\mathbb{C}$-algebra. Hence it is the coordinate ring of a smooth algebraic curve $\widetilde{V}$ in a (possibly) higher-dimensional affine space $\mathbb{C}^m$. Maximal ideals of $\widetilde{O}_{V(f)}$ restrict to maximal ideals of $O_{V(f)}$, so there is a canonical map from $\widetilde{V}$ to $V(f)$. However it may not be possible to find such a "smooth model" which is also a plane curve.

For curves the singular set is finite, and the problem may also be considered locally. As observed above, resolution by quadratic transform is semi-local. Thus after a quadratic transform of f centred at $Y = 0$ we may have to apply further quadratic transforms centred at $Z = p_i$, where p_i is the slope of a tangent to $V(f)$ at 0.

The local approach actually gives more information than normalization (passage to the integral closure). Assume that f is a distinguished polynomial of degree $n > 0$ in Y and let $f^\sigma(X, Z) = X^{-n}f(X, XZ)$. Then $f_Z^\sigma = X^{1-n}f_Y$, by the chain rule, so $M(f^\sigma) = M(f) - n(n-1)$. Hence if $B = \mathbb{C}[[X]][Z]/(f^\sigma) \cong A[y/x] \leq \widetilde{A}$, then $\delta_B = \delta_A - (n(n-1))/2$. Thus after finitely many such quadratic transforms (possibly followed by linear changes of coordinate) we obtain a non-singular f^ω. This process of resolution by quadratic transformations is canonical, and if f is irreducible the sequence of multiplicities $\nu(f) \geq \nu(f^\sigma) \geq \cdots \geq \nu(f^\omega) = 1$ determines f up to topological equivalence. (It is conventional to truncate the sequence before the first "1".) The analogous invariant in the reducible case is somewhat more complex, as one must keep track of how the branches with common tangents interact. (See pages 502ff of [**BK**].)

Let $\psi_N : X_N \to X_{N-1} \to \cdots \to X_0 = Spec(\mathbb{C}[[X, Y]])$ be the sequence of quadratic transforms corresponding to the canonical resolution of f, let $E_N = \psi_N^{-1}(0)$ and let $V_N = \overline{\psi_N^{-1}(V)} \setminus E_N$ be the strict preimage of $V(f)$ in X_N. The *resolution graph* Γ_f is the weighted graph with one vertex for each irreducible component of $\psi_N^{-1}(V) = E_N \cup V_N$, an edge joining vertices corresponding to components which meet, and weights $i(e) = 1 + \max\{i \mid e \text{ has image } 1 \text{ } point \in X_i\}$ on each vertex e corresponding to a component of E_N. (These weights are the negatives of the self-intersection numbers of the components e in X_N.) An algorithm for determining the resolution graph for an irreducible f from its Puiseux data is given in [**CS00**]. Conversely, each of the multiplicity sequence, the resolution graph and the extended Puiseux data (i.e., including the linking numbers) determines the other sets of invariants. (See pages 512ff of [**BK**].)

10.7. The Gauß-Manin connection

Using the deep coherence theorem of Grauert, Brieskorn has shown that for an algebraic link the cohomology of the fibre $H^1(F; \mathbb{C})$ may be identified with the kernel of the topological Gauß-Manin connection over a deleted neighbourhood of 0 in $\mathbb{C}$, and that the Alexander polynomial may (in principle) be computed in terms of an algebraically defined Gauß-Manin connection on a relative de Rham cohomology module [**Br70**]. (His arguments apply also in the many variable cases.) We shall use elementary calculations to recover some of Brieskorn's results for the case when f is a weighted homogeneous polynomial in two variables.

Let $R = \mathbb{C}[[X, Y]]$ and let $f = \Sigma_{k \leq n} a_k(X) Y^k$ be a square-free distinguished polynomial of degree n in Y. Let $f^* : S = \mathbb{C}[[s]] \to R$ be the ring homomorphism sending s to f, defined by $f^*(g) = g \circ f$. Then $R \cong S[[X]][Y]/(f - s)$ and $A = \mathbb{C}[[X, Y]]/(f) \cong R/(s)$. Let $\Omega^1 = R dX \oplus R dY$ be the module of 1-differentials of R over $\mathbb{C}$ and let $\Omega^0 = R$ and $\Omega^2 = \Omega^1 \wedge \Omega^1 = R dX \wedge dY$ be the other nonzero exterior powers of Ω^1. (Thus Ω^p is the completion of the module of germs of holomorphic p-forms at O in $\mathbb{C}^2$.) The cochain complex Ω^* determined by the exterior derivatives $d : \Omega^p \to \Omega^{p+1}$ gives a resolution

$$0 \to \mathbb{C} \to R \to \Omega^1 \to \Omega^2 \to 0.$$

Let $\Omega^p_f = \Omega^p/df \wedge \Omega^{p-1}$. The exterior derivative on Ω^p induces S-linear differentials $d_f : \Omega^p_f \to \Omega^{p+1}_f$ (via f^*) and so we obtain a cochain complex Ω^*_f of S-modules. Let $H^1_f = H^1(\Omega^*_f)$, $H' = \Omega^1_f/dR = \Omega^1/dR + Rdf$ and $H'' = \Omega^2/df \wedge dR$. Then H^1_f, H' and H'' are S-modules, $H^1_f \leq H'$ and $H'' \cong \mathrm{Cok}(\delta)$, where $\delta : R \to R$ is the S-linear derivation given by $\delta(g) = g_X f_Y - g_Y f_X$, for all $g \in R$. Since δ is S-linear and $\mathrm{Im}(\delta) \leq J(f)$ it induces a $\mathbb{C}$-linear derivation $\delta_A : A \to A$, with $\mathrm{Im}(\delta_A) \leq J_A$, where J_A is the image of $J(f)$ in A.

Since f_X and f_Y are relatively prime wedge product with df induces a monomorphism $\kappa : H' \to H''$, with cokernel $\Omega^2_f \cong R/(f_X, f_Y)$. Moreover d induces a $\mathbb{C}$-linear bijection $d' : H' \to H''$, with $d'(H^1_f) = \kappa(H')$. (For if $d\eta = df \wedge dg$ then $d(\eta + gdf) = 0$ and so $\eta + gdf = dr$ for some $r \in R$, since Ω^* is exact above degree 0.) This induces an S-linear isomorphism $H'/H^1_f \cong \mathrm{Cok}(\kappa) \cong \Omega^2_f$.

A *meromorphic connection* on a $\mathbb{C}[[s]]$-module N is a $\mathbb{C}$-linear function $\Delta : N \to N[s^{-1}] = \mathbb{C}((s)) \otimes_{\mathbb{C}[[s]]} N$ such that $\Delta(gm) = g\Delta(m) + (\frac{d}{ds}g)m$ for all $g \in \mathbb{C}[[s]]$ and $m \in N$. Such a connection Δ is *regular* if $dim_{\mathbb{C}((s))} N[s^{-1}] < \infty$ and $s\Delta$ maps a lattice in $N[s^{-1}]$ into itself.

The $\mathbb{C}[[s]]$-module H^1_f supports a natural meromorphic connection, defined as follows. If $\eta \in \Omega^1$ represents a class in H^1_f then $d\eta = df \wedge \psi = \kappa([\psi])$, for some 1-form ψ. The class of ψ in H' is well-defined, and $d(f^k\psi) = f^kd\psi + kf^{k-1}df \wedge \psi$ is in $df \wedge \Omega^1 = (f_X, f_Y)dX \wedge dY$, for k large, since $f^k \in (f_X, f_Y)$ for k large, by Theorem 10.14. Hence $f^k\psi \in H^1_f$, and the function defined by $\nabla([\eta]) = s^{-k}[f^k\psi]$ gives a well-defined meromorphic connection on H^1_f. This is the *local Gauß-Manin connection* ∇ associated to f.

Brieskorn used the coherence theorem of Grauert to show that the S-modules H, H' and H'' are finitely generated and of rank $\mu(f)$, and that the local Gauß-Manin connection of an isolated singularity is always regular [**Br70**]. Moreover H'' is torsion free as an S-module [**Se70**]. Hence so are H^1_f and H'. If we identify H^1_f and H' with submodules of $H^1_f[s^{-1}]$ we have $\nabla([\eta]) = \kappa^{-1}d'([\eta])$, and so ∇ maps H^1_f bijectively onto H'.

The apparent contradiction of the injectivity of the local Gauß-Manin connection as just defined with the claim that the cohomology of the Milnor fibre may be identified with the kernel of the topological Gauß-Manin connection may be resolved by interpreting the equation $s\nabla(\eta) = 0$ as a linear system of 1st order ODEs and extending coefficients from S to a larger ring $\widehat{S}$, to include all possible solutions to such a linear system. Let $\widehat{S} = \overline{\mathbb{C}((s))}[\lambda]$, where $\overline{\mathbb{C}((s))} = S[s^{-1}, s^{\frac{1}{n}}; n > 1][\lambda]$ is the algebraic closure of the field of fractions of S, and extend the derivation $\frac{d}{ds}$ so that $\frac{d}{ds}(s^k) = ks^{k-1}$ for all rational exponents k and $\frac{d}{ds}(\lambda) = s^{-1}$. (Thus λ corresponds to $\log(s)$.) Fix a basis $\{v_1, \ldots, v_\mu\}$ for H_f^1 over S, and write $s\nabla(v_i) = \Sigma_{j \leq \mu} s_{ij} v_j$. Let $e_1, \ldots, e_\mu \in \widehat{S}$. Then

$$s\nabla(\Sigma_{i \leq \mu} e_i v_i) = \Sigma_{i \leq \mu}((s\frac{d}{ds}e_i)v_i + e_i\Sigma_{j \leq \mu} s_{ij} v_j),$$

and so the equation $s\nabla(\eta) = 0$ corresponds to the linear system

$$s\frac{d}{ds}e_i + \Sigma_{k \leq \mu} s_{ki} e_k = 0 \quad (1 \leq i \leq \mu)$$

over $\widehat{S}$. The monodromy of this linear system is essentially the cohomological monodromy (with coefficients $\mathbb{C}$) of the Milnor fibration [**Br70**].

10.8. The weighted homogeneous case

In this section we shall establish the results of Brieskorn and Sebastiani for the special case of weighted homogeneous polynomials in two variables.

THEOREM 10.23. *Let f be a square-free distinguished polynomial of degree n in Y which is weighted homogeneous of type $(N; a, b)$. Then $H'' \cong S^{\mu(f)}$.*

PROOF. Since $aXf_X + bYf_Y = Nf$ we have $\delta(Y^j) = -jY^{j-1}f_X$ if $0 < j < n$ and $\delta(X^iY^j) = \frac{ai+bj}{a}X^{i-1}Y^j f_Y - \frac{jN}{a}X^{i-1}Y^{j-1}f$ if $0 < i < \infty$. On passing to $A = R/(s)$ we get $x^iy^j f_Y = \delta_A(\frac{a}{ai+bj+a}x^{i+1}y^j)$ and $x^iy^j f_X = \delta_A(\frac{b}{ai+bj+b}x^iy^{j+1})$ for all $i, j \geq 0$. Hence $\text{Im}(\delta_A) = J_A$.

Since $f \in J(f)$ we have

$$R/J(f) \cong A/J_A = A/\text{Im}(\delta_A) \cong R/(sR + \text{Im}(\delta)).$$

Moreover $sR + \text{Im}(\delta) \leq J(f)$ and so $sR + \text{Im}(\delta) = J(f)$. Let $U = \{u_i \mid 1 \leq i \leq \mu(f)\} \subset R$ represent a basis for $R/J(f)$ as a $\mathbb{C}$-vector space. We may choose functions $h, k : R \to R$ so that if $g \in R$ then $g = \delta(h(g)) + sk(g) + \Sigma c_i(g)u_i$, for some coefficients $c_i(g) \in \mathbb{C}$. Applying a similar expansion to $k(g)$ and iterating, we conclude that $g = \sigma(g) + \delta(\tilde{h}(g))$, where $\sigma(g) = \Sigma_{m \geq 0} c_i(k^m(g))s^m$ and $\tilde{h}(g) = \Sigma_{0 \leq m} s^m h(k^m(g))$, and so $R = SU + \text{Im}(\delta)$. Hence $H'' \cong \text{Cok}(\delta)$ is generated by U as an S-module.

The ring R is a free $S[[X]]$-module with basis $\{1, \ldots, Y^{n-1}\}$, and so every element $g \in R$ is uniquely expressible as a sum $g = \Sigma_{0 \leq i} \Sigma_{j=0}^{j=n-1} g_{ij} X^i Y^j$, with coefficients in S. Since $\delta(1) = 0$ it follows that $\delta(g) = \Sigma' g_{ij} \delta(X^i Y^j)$, where Σ' denotes summation over indices (i, j) with $0 \leq i < \infty$, $0 \leq q < n$ and $0 < i + q$. Thus if H'' is not free of rank $\mu(f)$ as an S-module there is a nontrivial linear relation $\Sigma_{k=1}^{k=\mu(f)} e_k u_k = \Sigma' g_{i,j} \delta(X^i Y^j)$ in R, with coefficients $\{e_k\}$ and $\{g_{i,j}\}$ in S and not all divisible by s. Since $\Sigma_{k=1}^{k=\mu(f)} e_k(0) u_k = 0$ in $R/J(f)$ we see that $e_k(0) = 0$ for all $k \leq \mu(f)$. Hence $\Sigma' g_{i,j}(0) \delta_A(x^i y^j) = 0$ in A. Now $\delta_A(1) = 0$, $\delta_A(y^j) = -jy^{j-1} f_X$ if $0 < j < n$ and $\delta_A(x^i y^j) = \frac{ai+bj}{a} x^{i-1} y^j f_Y$ if $0 < i < \infty$, while $axf_X + byf_Y = 0$. Hence

$$0 = x(\Sigma' g_{i,j}(0) \delta_A(x^i y^j))$$

$$= \Sigma_{j=0}^{j=n-1} g_{0j}(0) x \delta(y^j) + \Sigma_{1 \leq i} \Sigma_{j=0}^{j=n-1} g_{ij}(0) x \delta(x^i y^j)$$

$$= (\Sigma' g_{i,j}(0) \frac{ai + bj}{a} x^i y^j) f_Y.$$

But $\{1, \ldots, y^{n-1}\}$ is a basis for A as a free $\mathbb{C}[[x]]$-module, and f_Y is a non-zerodivisor in A. Therefore the coefficients $g_{i,j}(0)$ are all 0 also. This contradicts our assumption and so $H'' \cong S^{\mu(f)}$. $\quad\square$

Since $f \in (f_X, f_Y)$ we have $f\Omega^2 \leq df \wedge \Omega^1$, and so H''/sH'' maps onto $\Omega_f^2 \cong R/J(f)$. Since these $\mathbb{C}$-vector spaces each have dimension $\mu(f)$ this epimorphism must be an isomorphism. Hence $sH'' = df \wedge \Omega^1 = \kappa(H')$. Since $H'/H_f^1 \cong \Omega_f^2$ also it follows that

$H^1_f = sH'$. Moreover $d(f\psi) \in df \wedge \Omega^1$ for all $\psi \in \Omega^1$ and so ∇ is clearly regular.

Suppose that $f = gf_X + hf_Y$ and let $\alpha = -hdX + gdY$ and $\omega = dX \wedge dY$. Then $\kappa(\alpha) = [df \wedge \alpha] = [f\omega]$. Let $U \subset R$ represent an S-basis for H''. Then $\{u\alpha \mid u \in U\}$ is a basis for H' and so $\{uf\alpha \mid u \in U\}$ is a basis for H^1_f. It is easily verified that $d(uf\alpha) = (u + (gu)_X + (hu)_Y)f\omega$ for all $u \in R$. (Note that we may always assume that U is a set of monomials $\{X^iY^j\}$.)

Suppose now that f is not divisible by X or Y. Then $\frac{N}{ab}$ is an integer, and $\mu(f) = \frac{(N-a)(N-b)}{ab}$. (See §4 above.) We may take $g = \frac{a}{N}X$ and $h = \frac{b}{N}Y$, to get $d\alpha = \frac{a+b}{N}\omega$. Then $d(X^iY^jf\alpha) = (1 + \frac{a(i+1)}{N} + \frac{b(j+1)}{N})X^iY^jf\omega$. The image of $\{X^iY^j|0 \le i \le (db-1), 0 \le j \le (ad-1)\}$ in $R/J(f)$ is linearly independent, and so is a basis.

In particular, if $f = Y^n+X^m$ we may take $\alpha = \frac{1}{n}(-YdX+XdY)$. Let $\eta_{i,j} = [fX^iY^j\alpha]$. Then $\{\eta_{i,j} \mid 0 \le i < m - 1, 0 \le j < n - 1\}$ is a basis for H^1_f, and we see easily that

$$\nabla(\eta_{i,j}) = [\lambda(i,j)X^iY^j\alpha] = s^{-1}\lambda(i,j)\eta_{i,j},$$

where $\lambda(i,j) = 1 + \frac{i+1}{m} + \frac{j+1}{n}$. The corresponding system of ODEs $s\frac{d}{ds}(e_{i,j}) + \lambda(i,j)e_{i,j} = 0$ has basic solutions $e_{i,j} = s^{-\lambda(i,j)}$. As s moves once around $0 \in \mathbb{C}$ these solutions change by $\zeta^{-m(j+1)-n(i+1)}$, where $\zeta = exp(2\pi i/mn)$. In particular, if $(m,n) = 1$ then as i,j vary in the ranges $0 \le i < m - 1$ and $0 \le j < n - 1$ these factors $\zeta^{-m(j+1)-n(i+1)}$ run through the $(mn)^{th}$ roots of unity which are neither m^{th} nor n^{th} roots of unity. Hence the characteristic polynomial of the monodromy is $(t^{mn} - 1)(t - 1)/(t^m - 1)(t^n - 1)$, which is the Alexander polynomial of the (m,n)-torus knot.

If f is a square-free product of such terms (with m and n fixed) the monodromy is again diagonizable and its eigenvalues are roots of unity, and so it has finite order. The corresponding links have $\frac{N}{mn}$ components, each of which is an (m,n)-torus knot, and the pairwise linking numbers are all mn.

Since finite generation of $Cok(\delta)$ is invariant under change of coordinates it follows from Theorem 10.15 that Theorem 10.23 and its consequences apply whenever f is a square-free distinguished polynomial in R such that $f \in J(f)$. Can the argument for this theorem be extended to cover all square-free distinguished polynomials?

If $f \notin J(f)$ then establishing regularity of ∇ may be harder. For if $A \in R$ is such that $A_X = X f_X$ then $d(A dY) = df \wedge (-X dY)$, so $\omega = A dY$ represents a class in H^1_f with $\nabla(\omega) = -X dY$, but $d(fX dY) = f dX \wedge dY + X df \wedge dY$ is not in $df \wedge \Omega^1$.

In general, the eigenvalues of the monodromy are always roots of unity, and the largest Jordan block size is 2. (See [Sc80] for a derivation of this from the Briançon-Skoda result that $f^2 \in J(f)$.) The monodromy has finite order if f is irreducible [Le72]. On the other hand the monodromy of $(Y^3 + X^2)(Y^2 + X^3)$ has infinite order and so is not diagonalizable [A'C73'].

10.9. An hermitean pairing

The Seifert form for an algebraic link may also be related to the local Gauß-Manin connection, via integration of C^∞ forms [Ba85]. Barlet defines a finitely generated $\mathbb{C}[[s, \bar{s}]]$-module $\mathcal{M}$ with an involution extending the involution $s \leftrightarrow \bar{s}$ on the coefficient ring and with a meromorphic connection ∂_s such that $\partial_s(\bar{s}.m) = \bar{s}.\partial_s m$, for all $m \in M$. He then uses integration of representative compactly supported C^∞ forms to define an hermitean pairing $\hat{H}$ on H' with values in $\mathcal{M}$ which is horizontal, i.e., $s^k \partial_s \hat{H}(\alpha, \beta) = \hat{H}(s^k \nabla \alpha, \beta)$, for all $\alpha, \beta \in H'$ and $k >> 0$.

He shows that $(H', \nabla, \hat{H})$ is essentially equivalent to the intersection form on $H_1(F; \mathbb{C})$ together with the monodromy automorphism $h_{\mathbb{C}}$, after localization away from the 1-eigenspace of the monodromy [Ba85]. The 1-eigenspace is trivial for any knot, and has dimension $r - 1$ if f has r irreducible factors [Du75]. The argument applies in all dimensions and does not require a resolution of the singularity in its application. He has since given an exposition of this pairing in terms of local commutative algebra, using the notion of "(a, b)-modules" [Ba05].

Part 3

Free Covers, Nilpotent Quotients and Completion

CHAPTER 11

Free Covers

From the homological perspective the significant feature of the group Z is that it is free, rather than that it is abelian. The Laurent polynomial ring $\Lambda_\mu = \mathbb{Z}[\mathbb{Z}^\mu]$ is a commutative factorial noetherian domain, but it has global dimension $\mu + 1$, and so homological arguments can become unwieldy for μ large. In contrast, while the group ring $\mathbb{Z}[F(\mu)]$ of a finitely generated non-abelian free group is no longer commutative or noetherian, it is coherent and has global dimension 2, for any value of μ.

In this chapter we shall concentrate on the algebra of modules over free group rings, and we shall refer to the papers of Sato, Duval and (particularly) Farber for the major applications to the study of boundary links.

11.1. Free group rings

Let $\Gamma_\mu = \mathbb{Z}[F(\mu)]$ and $K\Gamma_\mu = K[F(\mu)]$, for any field K. These group rings have canonical involutions, defined by $\bar{g} = g^{-1}$ for all $g \in F(\mu)$, and augmentations $\varepsilon : \Gamma_\mu \to \mathbb{Z}$ and $\varepsilon_K : K\Gamma_\mu \to K$, induced by the projection of $F(\mu)$ onto the trivial group. In particular, every free module over Γ_μ or $K\Gamma_\mu$ has a well-defined rank. Moreover $\Gamma_\mu = *_{\mathbb{Z}}^\mu \Lambda_1$ is the coproduct (in the category of augmented $\mathbb{Z}$-algebras) of μ copies of Λ_1, and similarly $K\Gamma_\mu = *_K^\mu K\Lambda_1$. Therefore every (left) ideal of $K\Gamma_\mu$ is free [**Coh**]. (This is of course false for Γ_μ, even if $\mu = 1$.) Hence every submodule of a free $K\Gamma_\mu$-module is free and so every (left) $K\Gamma_\mu$-module M has a short free resolution

$$0 \to (K\Gamma_\mu)^{(I)} \to (K\Gamma_\mu)^{(J)} \to M \to 0.$$

However, the index set I need not be finite even if J is finite. It follows that $K\Gamma_\mu$ has no zerodivisors and that finitely generated free modules are hopfian: every epimorphism from $(K\Gamma_\mu)^m$ to itself is an isomorphism. The latter results hold also for Γ_μ, since they hold for $\mathbb{Q}\Gamma_\mu$. The units of Γ_μ and $K\Gamma_\mu$ are all of the form $u\gamma$, where $u = \pm 1$ (respectively, $u \in K^\times$) and $\gamma \in F(\mu)$.

It follows easily from the characterizations of free groups and free modules in terms of universal properties, together with the correspondance between module homomorphisms from $I(G)$ to M and splitting homomorphisms for semidirect products (in §1 of Chapter 4) that $I(F(\mu)) = \mathrm{Ker}(\varepsilon)$ is freely generated by the elements $\{x_1 - 1, \ldots, x_\mu - 1\}$ as a left Γ_μ-module. Thus there is an exact sequence

$$0 \to (\Gamma_\mu)^\mu \xrightarrow{\ \lambda\ } \Gamma_\mu \xrightarrow{\ \varepsilon\ } \mathbb{Z} \to 0,$$

where $\lambda(\gamma_1, \ldots, \gamma_\mu) = \Sigma\gamma_i(x_i - 1)$, for all $(\gamma_1, \ldots, \gamma_\mu) \in (\Gamma_\mu)^\mu$. The ideal $I(F(\mu))$ is two-sided (and so is a (Γ_μ, Γ_μ)-bimodule) and the elements $\{x_1 - 1, \ldots, x_\mu - 1\}$ also form a basis for $\mathrm{Ker}(\varepsilon)$ as a free right Γ_μ-module. However, λ is *not* a bimodule homomorphism, and in fact $I(F(\mu))$ is not free as a (Γ_μ, Γ_μ)-bimodule. There are analogous sequences of left $K\Gamma_\mu$-modules.

Let $\widehat{\Gamma_\mu}$ be the $I(F(\mu))$-adic completion of Γ_μ. The canonical homomorphism from Γ_μ to $\widehat{\Gamma_\mu}$ is an embedding, since $\cap_{n\geq 0}I(F(\mu))^n = 0$ [**Fo53**]. The elements $X_i = x_i - 1$ generate $\widehat{\Gamma_\mu}$, and so we may identify $\widehat{\Gamma_\mu}$ with $\mathbb{Z}\langle\langle X_1, \ldots, X_\mu\rangle\rangle$, the ring of formal power series in the non-commuting variables. (The group homomorphism $M : F(\mu) \to \mathbb{Z}\langle\langle X_1, \ldots, X_\mu\rangle\rangle^\times$ determined by $M(x_i) = 1 + X_i$ for $1 \leq i \leq \mu$ is the *Magnus embedding*. See also Chapter 5 of [**MKS**].)

Let Ψ be the set of all square matrices over Γ_μ whose images under ε have determinant ± 1. A ring homomorphism $f : \Gamma_\mu \to R$ is Ψ-inverting if the image of every matrix in Ψ is invertible as an R-matrix. There is a universal Ψ-inverting homomorphism $\Gamma_\mu \to \Gamma_{\mu\Psi}$ (the *Cohn localization* of Γ_μ) through which every Ψ-inverting homomorphism factors. Moreover this homomorphism is injective, since $\mathbb{Q}\Gamma_\mu$ is a free ideal ring, and so has a universal field of fractions, and the involution of Γ_μ extends to an involution on $\Gamma_{\mu\Psi}$, since Ψ is stable under transposition and involution of the entries [**Coh**].

The Cohn localization of $\Gamma_1 = \Lambda_1$ is localization with respect to $S = \{f \in \Lambda_1 \mid \varepsilon(f) = \pm 1\}$. In general, $\Gamma_{\mu\Psi}$ may be identified with the subring of "rational functions" in $\widehat{\Gamma_\mu}$ [**DS78**]. (See [**Fa92'**, **FV92**] for a discussion of this ring in terms of finite automata.)

The rings Γ_μ and $K\Gamma_\mu$ are (left) coherent: every finitely generated submodule of a finitely presentable module is finitely presentable. Projective modules over these rings are free [**Ba64'**]. Hence every (left) Γ_μ-module N has a free resolution of length at most two:

$$0 \to P_2 \to P_1 \to P_0 \to N \to 0,$$

and the P_i may be assumed of finite rank if N is finitely presentable.

11.2. $\mathbb{Z}[F(\mu)]$-modules

We shall now look more closely at modules over these rings. If M is a left Γ_μ-module then the dual $M^* = Hom_{\Gamma_\mu}(M, \Gamma_\mu)$ and the extension modules $E^i M = Ext^i_{\Gamma_\mu}(M, \Gamma_\mu)$ for $i = 1, 2$ are naturally right Γ_μ-modules, since Γ_μ is a bimodule over itself. (We shall use similar notation for modules derived from right Γ_μ-modules and from left and right $K\Gamma_\mu$-modules.) There is a natural evaluation homomorphism $ev_M : M \to M^{**}$, given by $ev_M(m)(f) = f(m)$ for all $m \in M$ and $f \in M^*$.

We shall consider first modules over $K\Gamma_\mu$ (with K a field), where the situation is relatively simple. If H is a left $K\Gamma_\mu$-module and $f : H \to K\Gamma_\mu$ a nonzero homomorphism, then $\mathrm{Im}(f)$ is a left ideal, and so is free. Therefore $H \cong L \oplus \mathrm{Ker}(f)$, where L is free. If H is finitely generated then L is of finite rank not greater than the minimal number of generators of H. Thus after iterating the process of splitting off free factors finitely many times we find that $H \cong (K\Gamma_\mu)^r \oplus T$, where $T^* = 0$ and r is an integer. Clearly $H^* \cong (K\Gamma_\mu)^r \oplus T^* = (K\Gamma_\mu)^r$ and $H^{**} \cong (K\Gamma_\mu)^r$. Then the rank r of H is well defined and ev_H is an epimorphism, with kernel T. Moreover $E^1 H \cong E^1 T$. Since T has a short free resolution and $T^* = 0$, we find that $T \cong E^1 E^1 T$. Hence $\mathrm{Ker}(ev_H) \cong E^1 E^1 H$. Clearly T contains the $K\Gamma_\mu$-torsion elements of H (those elements $t \in H$ for which there is a nonzero $w \in K\Gamma_\mu$ such that $wt = 0$), but it may also contain a free submodule. (If $\mu = 1$ there is a much stronger structure result, for $K\Gamma_1 = K\Lambda_1$ is then a commutative PID and so T is a direct sum of cyclic modules.)

A consequence of this result is that if M is a finitely generated left $K\Gamma_\mu$-module such that $K \otimes_{\varepsilon_K} M = 0$ then $M^* = 0$.

We now turn to the more difficult case of Γ_μ-modules. A left Γ_μ-module is a *torsion* module if $M^* = 0$; it is a *pure* torsion module if also $E^2M = 0$, and it is *pseudonull* if $M^* = E^1M = 0$. (Note that these definitions agree with the earlier definitions for the commutative case $\mu = 1$, by Theorem 3.9.)

If $\mu > 1$ a torsion module may contain free submodules (in contrast to the commutative case). Let $\{x, y\}$ generate $F(2)$. Then $\{x - 1, y - 1\}$ is a basis for the augmentation ideal, and so $T = \Gamma_2/\Gamma_2(y - 1)$ contains the free submodule generated by the image of $x - 1$, while $T^* = 0$ (since a nontrivial homomorphism from Γ_μ to Γ_μ must be injective).

LEMMA 11.1. *Let M be a finitely presentable left Γ_μ-module. Then M^* is a finitely generated free right Γ_μ-module, and the modules E^1M and E^2M are finitely presentable.*

PROOF. Since M is finitely presentable it has a finite free resolution of length at most 2:

$$0 \to (\Gamma_\mu)^a \xrightarrow{\alpha} (\Gamma_\mu)^b \xrightarrow{\beta} (\Gamma_\mu)^c \xrightarrow{\gamma} M \to 0.$$

The modules M^*, E^1M and E^2M are the cohomology modules of the dual complex of right modules:

$$0 \to (\Gamma_\mu)^c \xrightarrow{\beta^*} (\Gamma_\mu)^b \xrightarrow{\alpha^*} (\Gamma_\mu)^a \to 0.$$

Let $A = \mathrm{Ker}(\alpha^*)$. Then there are exact sequences

$$0 \to M^* \xrightarrow{\gamma^*} (\Gamma_\mu)^c \xrightarrow{q} A \to E^1M \to 0$$

and

$$0 \to A \xrightarrow{p} (\Gamma_\mu)^b \xrightarrow{\alpha^*} (\Gamma_\mu)^a \to E^2M \to 0$$

with $pq = \beta^*$. Since Γ_μ is coherent of global dimension 2, A is a finitely generated free (right) Γ_μ-module, and therefore so is M^*. Moreover the latter two sequences are finite free resolutions for E^1M and E^2M. $\square$

Note that since $(E^2M) \otimes_{\mathbb{Z}} \mathbb{Q} = E^2(\mathbb{Q} \otimes_{\mathbb{Z}} M) = 0$ the sequence obtained on tensoring the above resolution of E^2M with $\mathbb{Q}$ splits, and so A has rank $a - b$.

The above resolutions of the right Γ_μ-modules E^1M and E^2M give rise to dual complexes of left Γ_μ-modules:

$$0 \to A^* \xrightarrow{q^*} (\Gamma_\mu)^c \xrightarrow{\gamma^{**}} M^{**} \to 0 \to \dots$$

and

$$0 \to (\Gamma_\mu)^a \xrightarrow{\alpha^{**}} (\Gamma_\mu)^b \xrightarrow{p^*} A^* \to 0 \to \dots$$

From these may be obtained four new exact sequences

$$0 \to (E^1M)^* \to A^* \to \mathrm{Ker}(\gamma^{**}) \to E^1E^1M \to 0,$$

$$0 \to \mathrm{Ker}(\gamma^{**}) \to (\Gamma_\mu)^c \xrightarrow{\gamma^{**}} M^{**} \to E^2E^1M \to 0,$$

$$0 \to (E^2M)^* \to (\Gamma_\mu)^a \to \mathrm{Ker}(p^*) \to E^1E^2M \to 0$$

and

$$0 \to \mathrm{Ker}(p^*) \to (\Gamma_\mu)^b \to A^* \to E^2E^2M \to 0.$$

Since the composition $(\Gamma_\mu)^a \to \mathrm{Ker}(p^*) \to (\Gamma_\mu)^b$ is $\alpha^{**} = \alpha$, it is injective, and so $(E^2M)^* = 0$. Since $q^*p^* = \beta^{**} = \beta$, $\mathrm{Ker}(p^*) \le \mathrm{Ker}(\beta) = \mathrm{Im}(\alpha)$, and so the homomorphism from $(\Gamma_\mu)^a$ to $\mathrm{Ker}(p^*)$ is an isomorphism. Therefore also $E^1E^2M = 0$.

LEMMA 11.2. *Let M be a finitely presentable left Γ_μ-module. Then E^2M is a finitely presentable pseudonull module. If M is pseudonull then $E^2E^2M \cong M$.*

PROOF. The first assertion follows from Lemma 11.1 and the above paragraph. If M is pseudonull the four new exact sequences above simplify to give an exact sequence

$$0 \to (\Gamma_\mu)^a \xrightarrow{\alpha} (\Gamma_\mu)^b \xrightarrow{\beta} (\Gamma_\mu)^c \to E^2E^2M \to 0,$$

whence it is clear that $E^2E^2M \cong \mathrm{Coker}(\beta) = M$. $\square$

We shall show that there is a natural isomorphism $E^2 E^2 M \cong M$.

LEMMA 11.3. *Let M be a finitely presentable left Γ_μ-module. Then M has a finite free resolution of length 1 if and only if $E^2 M = 0$. In this case $E^2 N = 0$ for any submodule $N \leq M$. The module M is free if and only if $E^1 M = E^2 M = 0$.*

PROOF. If $E^2 M = 0$ then α^* is an epimorphism and so splits, and therefore α is a split monomorphism.

Hence $\mathrm{Im}(\beta)$ is free and M has a finite free resolution of length (at most) 1. The argument in the other direction is clear. The second assertion follows from the long exact sequence of Ext^*, since $E^3(M/N) = 0$. The final assertion has a similar proof.   □

If M has a short free resolution we may assume that $a = 0$ and so the exact sequences following Lemma 11.1 reduce to

$$0 \to (\Gamma_\mu)^b \to \mathrm{Ker}(\gamma^{**}) \to E^1 E^1 M \to 0$$

and

$$0 \to \mathrm{Ker}(\gamma^{**}) \to (\Gamma_\mu)^c \to M^{**} \to E^2 E^1 M \to 0.$$

It follows that there is a natural embedding of $E^1 E^1 M$ into M, which is an isomorphism if and only if $M^{**} = 0$, and hence if and only if M is a pure torsion module.

LEMMA 11.4. *Let M be a finitely presentable left Γ_μ-module. Then*

(1) $(E^1 M)^* = 0$ *and $E^2 M$ has finite exponent as an abelian group;*

(2) *if $\mathbb{Z} \otimes_\varepsilon M = 0$ then $M^* = 0$ and $Hom_{\Gamma_\mu}(M, \mathbb{Z}/r\mathbb{Z}[F(\mu)]) = 0$ for all $r > 1$;*

(3) *if $\mathbb{Z} \otimes_\varepsilon M = 0$ then $E^1 M$ is $\mathbb{Z}$-torsion free.*

PROOF. The functor $\mathbb{Q} \otimes_\mathbb{Z} - = \mathbb{Q}\Gamma_\mu \otimes_{\Gamma_\mu} -$ from the category of left Γ_μ-modules to the category of left $\mathbb{Q}\Gamma_\mu$-modules is exact, and so $(\mathbb{Q} \otimes_\mathbb{Z} M)^* = M^* \otimes_\mathbb{Z} \mathbb{Q}$, $E^1(\mathbb{Q} \otimes_\mathbb{Z} M) = (E^1 M) \otimes_\mathbb{Z} \mathbb{Q}$ and $(E^2 M) \otimes_\mathbb{Z} \mathbb{Q} = E^2(\mathbb{Q} \otimes_\mathbb{Z} M) = 0$. If L is a finitely presentable left Γ_μ-module then L^* is free so $L^* = 0$ if and only if $(\mathbb{Q} \otimes_\mathbb{Z} L)^* = L^* \otimes_\mathbb{Z} \mathbb{Q} = 0$. An analogous argument for right modules shows that $(E^1 M)^* = 0$, since $E^1(\mathbb{Q} \otimes_{\mathbb{Z}} M)^* = 0$. (For this it suffices that M be finitely generated.)

If $K = \mathbb{Q}$ or $\mathbb{F}_p$ (for some prime $p \in \mathbb{Z}$) then $K \otimes_{\varepsilon_K} (K \otimes_{\mathbb{Z}} M) = K \otimes_{\mathbb{Z}} (\mathbb{Z} \otimes_{\varepsilon} M)$. Therefore if $\mathbb{Z} \otimes_{\varepsilon} M = 0$ then $M^* \otimes_{\mathbb{Z}} \mathbb{Q} = (\mathbb{Q} \otimes_{\mathbb{Z}} M)^* = 0$, so $M^* = 0$. Similarly, $\mathbb{F}_p \otimes_{\varepsilon} (\mathbb{F}_p \otimes_{\mathbb{Z}} M) = 0$, so $Hom_{\Gamma_\mu}(M, \mathbb{F}_p \Gamma_\mu) = (\mathbb{F}_p \otimes_{\mathbb{Z}} M)^* = 0$. A finite induction on the number of factors of r now gives $Hom_{\Gamma_\mu}(M, \mathbb{Z}/r\mathbb{Z}[F(\mu)]) = 0$ for all $r > 1$.

Since M is finitely presentable $E^2 M$ is finitely presentable. Let $\{w_i, \ldots, w_n\}$ generate $E^2 M$. Then there are nonzero integers $e_1, \ldots, e_n$ such that $e_i w_i = 0$, for $1 \leq i \leq n$, since $\mathbb{Q} \otimes_{\mathbb{Z}} E^2 M = 0$. Let $e = \Pi_{i=1}^{i=n} e_i$. Then clearly $ew = 0$ for all $w \in E^2 M$.

If $Hom_{\Gamma_\mu}(M, \mathbb{F}_p \Gamma_\mu) = 0$ multiplication by p induces an injective endomorphism of $E^1 M$. Therefore $E^1 M$ is $\mathbb{Z}$-torsion free if $Hom_{\Gamma_\mu}(M, \mathbb{F}_p \Gamma_\mu) = 0$ for all primes p. $\qquad\square$

We may now return to the four sequences following Lemma 11.1, which derive from a finite free resolution for an arbitrary finitely presentable left Γ_μ-module M. Since $(E^1 M)^* = 0$ they reduce to

$$0 \to A^* \to \mathrm{Ker}(\gamma^{**}) \to E^1 E^1 M \to 0,$$

$$0 \to \mathrm{Ker}(\gamma^{**}) \to (\Gamma_\mu)^c \xrightarrow{\gamma^{**}} M^{**} \to E^2 E^1 M \to 0$$

and

$$0 \to (\Gamma_\mu)^a \to (\Gamma_\mu)^b \to A^* \to E^2 E^2 M \to 0.$$

Bearing in mind the original resolution of M in Lemma 11.1, we obtain a natural isomorphism $\mathrm{Cok}(ev_M) \cong E^2 E^1 M$, and a monomorphism from $E^2 E^2 M$ to M with image in $\mathrm{Ker}(ev_M)$ and such that $\mathrm{Ker}(ev_M)/E^2 E^2 M \cong E^1 E^1 M$. Hence if M is a torsion module then $E^1 M$ is a pure torsion module. Let $fM = M/E^2 E^2 M$. The long exact sequence:

$$0 \to (fM)^* \to M^* \to (E^2 E^2 M)^* \to E^1 fM \to E^1 M \to$$

$$E^1 E^2 E^2 M \to E^2 fM \to E^2 M \to E^2 E^2 E^2 M \to 0$$

and the fact that the homomorphism $E^2 M \to E^2 E^2 E^2 M$ is the natural isomorphism for the pseudonull module $E^2 M$ give $(fM)^* \cong M^*$,

$E^1 fM \cong E^1 M$ and $E^2 fM = 0$. In particular, if M is a torsion module fM is a pure torsion module, and so is isomorphic to $E^1 E^1 fM \cong E^1 E^1 M$.

LEMMA 11.5. *Let $0 \to S \to P \to Q \to 0$ be an exact sequence of left Γ_μ-modules. If S and Q are both torsion (respectively, pseudonull or pure torsion) then so is P. Conversely if P is torsion then so is Q, and if P is finitely presentable and pseudonull then Q is pseudonull and S is torsion.*

PROOF. We shall only prove the last assertion, as the other proofs are similar and easier. From the long exact sequence of Ext^* it follows that if $P^* = E^1 P = 0$ then $Q^* = 0$ and $S^* = E^1 Q$. Since P is of finite exponent as an abelian group so is S and therefore $E^1 Q = S^* = 0$. $\square$

THEOREM 11.6. *Let M be a finitely presentable left Γ_μ-module.*

(1) $\mathrm{Ker}(ev_M)$ *is the maximal torsion submodule of M;*
(2) $E^2 E^2 M$ *is the maximal pseudonull submodule of M.*

PROOF. Let $N = \mathbb{Q} \otimes_{\mathbb{Z}} M$. Then $\mathrm{Ker}(ev_M)^* \otimes_{\mathbb{Z}} \mathbb{Q} = \mathrm{Ker}(ev_N)^* = 0$, and so $\mathrm{Ker}(ev_M)$ is a torsion module. If $T \leq M$ is any torsion submodule then $ev_M(T)$ is a torsion submodule of the free module M^{**} and so must be 0. Hence $\mathrm{Ker}(ev_M)$ is the maximal torsion submodule of M.

Let $j : N \to M$ be a monomorphism. Then $E^2 j : E^2 M \to E^2 N$ is an epimorphism, since $E^3(M/j(N)) = 0$. Since $E^1 E^2 N = 0$ it follows from the long exact sequence of Ext^* that $E^2 E^2 j$ is a monomorphism. Therefore if N is pseudonull $j(N) \leq E^2 E^2 M$, since the inclusion $E^2 E^2 L \to L$ is natural in L. Hence $E^2 E^2 M$ is the maximal pseudonull submodule of M. $\square$

If ev_M is an epimorphism then it splits, since M^{**} is free, and so $M \cong M^{**} \oplus \mathrm{Ker}(ev_M)$.

We have seen that a torsion module may contain a free module. Similarly a pseudonull module may contain a pure torsion module. (Consider the module $P = \Gamma_2/\Gamma_2(p, x-1)$ for $p \in \mathbb{Z}$ a prime, which is pseudonull but contains a submodule isomorphic to $\mathbb{F}_p\Gamma_2 = \Gamma_2/p\Gamma_2$, generated by the image of $y - 1$.)

11.3. The Sato property

A Γ_μ-module M is *of type* L if it is finitely presentable and $Tor_i^{\Gamma_\mu}(\mathbb{Z}, M) = 0$ for all $i \geq 0$. This is a key property of the homology modules of the maximal free covers of boundary link exteriors. We shall omit the proof of the following alternative characterizations as it is straightforward.

THEOREM 11.7. [Sa81'] *Let M be a finitely presentable Γ_μ-module. Then the following conditions are equivalent:*

(1) $Tor_i^{\Gamma_\mu}(\mathbb{Z}, M) = 0$ *for all $i \geq 0$;*

(2) *the homomorphism from M^μ to M sending $(m_1, \ldots, m_\mu)$ to $\Sigma(x_i - 1)m_i$ is an isomorphism;*

(3) *there is an exact sequence*

$$0 \to (\Gamma_\mu)^a \xrightarrow{\ \alpha\ } (\Gamma_\mu)^{a+b} \xrightarrow{\ \beta\ } (\Gamma_\mu)^b \xrightarrow{\ \gamma\ } M \to 0$$

such that the sequence

$$0 \to Z^a \xrightarrow{\ \varepsilon(\alpha)\ } Z^{a+b} \xrightarrow{\ \varepsilon(\beta)\ } Z^b \to 0$$

is also exact. □

The equivalence (1) $\Leftrightarrow$ (2) does not require that M be finitely presentable. If $\mu = 1$ and M is a finitely generated Λ_1-module such that $\mathbb{Z} \otimes_\varepsilon M = 0$ then $(t-1)id_M$ is an isomorphism, and M is of type L, by Lemma 4.8 or Theorem 11.7. In general these Tor-conditions are independent if $\mu > 1$. For instance, if $N = \Gamma_2/\Gamma_2(x^2 - x + 1, y - 1)$ then $\mathbb{Z} \otimes_\varepsilon N = 0$ while $Tor_1^{\Gamma_2}(\mathbb{Z}, N) \cong \mathbb{Z}$.

LEMMA 11.8. *Let M be a Γ_μ-module of type L. Then $M^* = 0$ and E^1M is of type L and is $\mathbb{Z}$-torsion free. Moreover E^2E^2M is the maximal $\mathbb{Z}$-torsion submodule, and M is $\mathbb{Z}$-torsion free if and only if $E^2M = 0$ if and only if $M \cong E^1E^1M$.*

PROOF. The first assertion follows from Lemma 11.4. Therefore E^1M has a short free resolution

$$0 \to (\Gamma_\mu)^b \xrightarrow{\ q\ } (\Gamma_\mu)^b \to E^1M \to 0,$$

by Lemma 11.1 and the subsequent remark. Moreover $\varepsilon(q)$ is an isomorphism, and so $E^1 M$ is a (right) module of type L. Hence $E^1 E^1 M$ is $\mathbb{Z}$-torsion free. Since M is a torsion module it is an extension of $E^1 E^1 M$ by $E^2 E^2 M$. The latter submodule has finite exponent, and so is the maximal $\mathbb{Z}$-torsion submodule. Hence M is $\mathbb{Z}$-torsion free if and only if $E^2 M = 0$ if and only if $M \cong E^1 E^1 M$. $\square$

A Γ_μ-module M is *separated* if $\cap_{q \geq 0} I(F(\mu))^q M = 0$.

LEMMA 11.9. [**Fa92**] *Let M be a Γ_μ-module of type L and let N be a separated $I(F(\mu))$-adically complete Γ_μ-module. Then*

$$Hom_{\Gamma_\mu}(M, N) = Ext^1_{\Gamma_\mu}(M, N) = 0.$$

PROOF. If $f : M \to N$ is a homomorphism then a simple induction shows that $f(M) \leq I(F(\mu))^q N$ for all $q \geq 0$, since $\mathbb{Z} \otimes_\varepsilon M = 0$ if and only if $M = I(F(\mu))M$. Therefore $Hom_{\Gamma_\mu}(M, N) = 0$.

Suppose

$$0 \to N \xrightarrow{\ f\ } P \xrightarrow{\ g\ } M \to 0$$

is an extension. Then we shall construct inductively a sequence of homomorphisms $h_q : P \to N/I(F(\mu))^q N$, for $q \geq 0$, such that $h_q \circ f$ is the canonical projection and h_q factors through h_{q+1}.

Let $h_0 = 0$. Suppose that h_j has been constructed for $j < q$ with the above properties. If $p \in P$ then $p = \Sigma_{i=1}^{i=\mu}(X_i - 1)p_i + f(n)$, for some $p_i \in P$ and $n \in N$, since M is of type L. It is easily verified that the function $h_q(p) = [\Sigma_{i=1}^{i=\mu}(X_i - 1)h_{q-1}(p_i) + n] \in N/I(F(\mu))^q N$ is a well-defined homomorphism, with the required properties.

Since N is complete the sequence converges to a homomorphism $h : P \to N$ which splits the extension. Hence $Ext^1_{\Gamma_\mu}(M, N) = 0$. $\square$

THEOREM 11.10. *Let M be a Γ_μ-module of type L. Then $E^1 M \cong Hom_{\Gamma_\mu}(M, \Gamma_{\mu\Psi}/\Gamma_\mu) \cong Hom_{\Gamma_\mu}(M, \widehat{\Gamma_\mu}/\Gamma_\mu)$.*

PROOF. If M is of type L and $E^2 M = 0$ it has a short free resolution with a square presentation matrix:

$$0 \to (\Gamma_\mu)^q \xrightarrow{\ P\ } (\Gamma_\mu)^q \to M \to 0,$$

where $det(\varepsilon(P)) = \pm 1$. Hence $Hom_{\Gamma_\mu}(M, \Gamma_{\mu\Psi}) = Ext^1_{\Gamma_\mu}(M, \Gamma_{\mu\Psi}) = 0$, since $\Gamma_{\mu\Psi} \otimes P$ is invertible. On applying $Ext^*_{\Gamma_\mu}(M, -)$ to the short exact sequence

$$0 \to \Gamma_\mu \to \Gamma_{\mu\Psi} \to \Gamma_{\mu\Psi}/\Gamma_\mu \to 0$$

we obtain $Hom_{\Gamma_\mu}(M, \Gamma_{\mu\Psi}/\Gamma_\mu) \cong E^1 M$.

Since $N = \widehat{\Gamma_\mu}$ is complete and separated a similar argument (using Lemma 11.9) shows that $Hom_{\Gamma_\mu}(M, \widehat{\Gamma_\mu}/\Gamma_\mu) \cong E^1 M$. $\quad\square$

The first isomorphism is from [**Du86**], and the second from [**Fa92**]. Farber showed also that

$$E^2 M \cong Hom_{\Gamma_\mu}(E^2 E^2 M, \widehat{\mathbb{Q}\Gamma_\mu}/(\widehat{\Gamma_\mu} + \mathbb{Q}\Gamma_\mu)).$$

Every $K\Gamma_\mu$-module of type L has a finite composition series whose subquotients are of type L and have no proper submodules of type L [**Fa92'**].

11.4. The Farber derivations

If M is a Γ_μ-module such that $Tor_i^{\Gamma_\mu}(\mathbb{Z}, M) = 0$ for all $i \geq 0$ then the inclusion of the augmentation ideal into Γ_μ induces an isomorphism $M \cong I(F(\mu)) \otimes M \cong M^\mu$. The projections of M to the summands determine *derivations* $\partial_i^F : M \to M$, for $1 \leq i \leq \mu$, such that $m = \Sigma_{i=1}^{i=\mu}(x_i - 1)\partial_i^F(m)$ and $\partial_i^F(\gamma m) = \frac{\partial \gamma}{\partial x_i}m + \varepsilon(\gamma)\partial_i^F(m)$ for all $\gamma \in \Gamma_\mu$ and $m \in M$. (Here $\frac{\partial \gamma}{\partial x_i}$ is the free derivative of γ with respect to x_i. See §6 of Chapter 12 below.) These derivations were introduced by Farber [**Fa91**]. The module M may be viewed as a module over the ring $D_\mu = \mathbb{Z}\langle\partial_1^F, \ldots, \partial_\mu^F\rangle$ of polynomials in non-commuting variables ∂_i^F. Every Γ_μ-homomorphism between modules M and M' of type L is a D_μ-homomorphism, and conversely: $Hom_{\Gamma_\mu}(M, M') = Hom_{D_\mu}(M, M')$.

Farber defined further operations $\overline{\partial}_i^F = -x_i\partial_i^F$, $\pi_i = (x_i-1)\partial_i^F = -\partial_i^F - \overline{\partial}_i^F$ and $z = -\Sigma_{1 \leq i \leq \mu}\partial_i^F$. Then $\{\pi_i \mid 1 \leq i \leq \mu\}$ is a family of orthogonal idempotents: $\Sigma\pi_i = Id$, $\pi_i^2 = \pi_i$ and $\pi_i\pi_j = 0$ if $j \neq i$. (Note that $\partial_i^F = \partial_i^F\pi_i = -z\pi_i$ and $\overline{\partial}_i^F = -\overline{z}\pi_i$, where $\overline{z} = 1 - z$.) There is a canonical homomorphism from D_μ to the ring

$P_\mu = \mathbb{Z}\langle \pi_1, \ldots, \pi_\mu, z \rangle / (\pi_i \pi_j - \delta_{ij} \pi_i \, (\forall i, j), \Sigma \pi_i - 1)$, sending ∂_i^F to $-z\pi_i$. A P_μ-module structure (and hence a D_μ-module structure) on an abelian group M may be given by a direct sum splitting $M = \oplus_{i=1}^{i=\mu} M_i$ and an endomorphism $z : M \to M$. (When $\mu = 1$ we have $\Gamma_1 = \Lambda_1$, $z = -\partial^F$ and $D_1 = P_1 = \mathbb{Z}[z]$ is a commutative ring, acting via the isomorphism $z.m = (1 - t)^{-1} m$.)

An important example is the quotient $\widehat{\Gamma_\mu}/\Gamma_\mu$. (This module is not of type L since it is not finitely generated.) We may define an additive map $\partial_i^F : \widehat{\Gamma_\mu} \to \widehat{\Gamma_\mu}$ which deletes the factor X_i from monomials beginning with X_i and is 0 on all other monomials. We then have $\gamma = \varepsilon(\gamma) + \Sigma_{i=1}^{i=\mu} X_i \partial_i^F(\gamma)$ for all $\gamma \in \Gamma_\mu$. These maps induce derivations ∂_i^F on $\widehat{\Gamma_\mu}/\Gamma_\mu$.

A *lattice* in a $\mathbb{Q}\Gamma_\mu$-module M is a $\mathbb{Q}D_\mu$-submodule which is finite dimensional as a $\mathbb{Q}$-vector space and which generates M as a Γ_μ-module. Every module of type L contains a unique minimal lattice. (This is not necessarily so for the integral analogue.) Hermitean forms on M correspond bijectively to pairings on the minimal lattice. (When $\mu = 1$ this reduces to the equivalence between rational Blanchfield pairings and rational Seifert forms.) See [Fa91].

Every $\mathbb{Q}\Gamma_\mu$-module M of type L determines a rational function $\chi_M \in \Gamma_{\mu\Psi}$ which encapsulates the "semisimple" part of M. (When $\mu = 1$ this invariant is equivalent to $\Delta_0(M)$, and so is a pseudoisomorphism invariant.) See [Fa92', GL02].

11.5. The maximal free cover and duality

Let L be a μ-component homology boundary n-link with group $\pi = \pi L$. Thus there is an epimorphism $f : \pi \to F(\mu)$ with kernel π_ω. After composing f with an automorphism of $F(\mu)$, if necessary, we may assume that the composition of f with the abelianization : $F(\mu) \to Z^\mu$ sends meridians to standard generators. (We may refine the notion of homology boundary link by defining an $hF(\mu)$-(n)-link to be a μ-component homology boundary n-link with a conjugacy class of such epimorphisms f.) If $\mu = 1$ there is a unique such isomorphism. If $\mu = 2$ it is determined up to composition with an inner automorphism of $F(2)$, by a theorem of Nielsen (Theorem 3.9

of [**MKS**]), and so the ambiguity is no worse than that involved in choosing a basepoint for X.

The *maximal free cover* of $X = X(L)$ is the cover $p_\omega : X^\omega \to X$ determined by the normal subgroup $\pi_\omega < \pi$. An epimorphism $f : \pi \to F(\mu)$ determines a basis for $Aut(X^\omega/X)$ which we may use to identify $\tilde{\Gamma} = \mathbb{Z}[Aut(X^\omega/X)]$ with $\Gamma_\mu = \mathbb{Z}[F(\mu)]$. We shall use the abbreviations $E^i M = Ext^i_{\tilde{\Gamma}}(M, \tilde{\Gamma})$ and $M^* = E^0 M$.

THEOREM 11.11. *Let L be a μ-component homology boundary n-link with exterior $X = X(L)$. Then*

(1) $H_j(X; \tilde{\Gamma})$ *and* $H^j(X; \tilde{\Gamma})$ *are finitely presentable, and are 0 if $j > n + 1$;*

(2) $H_j(X; \tilde{\Gamma})$ *is a module of type L, for $1 \le j < n$;*

(3) $H_{n+1}(X; \tilde{\Gamma})$ *is a free $\tilde{\Gamma}$-module;*

(4) $H_1(X; \tilde{\Gamma}) \cong \pi_\omega / (\pi_\omega)'$;

(5) $H_2(X; \tilde{\Gamma})$ *maps onto* $H_2(\pi; \tilde{\Gamma}) = H_2(\pi_\omega; \mathbb{Z})$;

(6) $\mathbb{Z} \otimes_\varepsilon H_n(X; \tilde{\Gamma}) = 0$, *and there is an exact sequence*

$$0 \to \mathbb{Z} \otimes_\varepsilon H_{n+1}(X; \tilde{\Gamma}) \to \mathbb{Z}^{\mu-1} \to Tor_1^{\tilde{\Gamma}}(\mathbb{Z}, H_n(X; \tilde{\Gamma})) \to 0.$$

Hence $H_{n+1}(X; \tilde{\Gamma})$ has rank $\le \mu - 1$, with equality if $H_n(X; \tilde{\Gamma})$ has type L;

(7) *there are short exact sequences*

$$0 \to E^2 H_{m-2}(X; \tilde{\Gamma}) \to H^m(X; \tilde{\Gamma}) \to E^1 H_{m-1}(X; \tilde{\Gamma}) \to 0,$$

for each $m \le n$, and a four-term exact sequence

$$0 \to J_{n,1} \to H^{n+1}(X; \tilde{\Gamma}) \to H_{n+1}(X; \tilde{\Gamma})^* \xrightarrow{d_2^{n+1,0}} E^2 H_n(X; \tilde{\Gamma}) \to 0,$$

where $J_{n,1}$ is an extension of $E^1 H_n(X; \tilde{\Gamma})$ by $E^2 H_{n-1}(X; \tilde{\Gamma})$.

PROOF. The first assertion holds since X is a compact $(n + 2)$-manifold with nonempty boundary and $\tilde{\Gamma}$ is a coherent ring, and so the homology and cohomology modules of a finite free $\tilde{\Gamma}$-complex are finitely presentable. The module $H_{n+1}(X; \tilde{\Gamma})$ is free since it is the kernel of a homomorphism between free modules. Parts (4) and (5) follow from the Hurewicz and Hopf Theorems.

The Cartan-Leray spectral sequence for p_ω collapses at the E^2 level, since it has the form $E^2_{pq} = Tor_p^{\tilde{\Gamma}}(H_q(X; \tilde{\Gamma}), \mathbb{Z}) \Rightarrow H_{p+q}(X; \mathbb{Z})$

and the augmentation module $\mathbb{Z}$ has a short free resolution. Hence $H_j(X;\tilde{\Gamma})$ is of type L if $j \neq 0$, 1, n or $n+1$, while $H_0(X;\tilde{\Gamma}) = \mathbb{Z}$, $Tor_1^{\tilde{\Gamma}}(\mathbb{Z}, H_1(X;\tilde{\Gamma})) = 0$ if $n > 1$, and $\mathbb{Z} \otimes_\varepsilon H_n(X;\tilde{\Gamma}) = 0$. Moreover there are exact sequences

$$0 \to \mathbb{Z} \otimes_\varepsilon H_1(X;\tilde{\Gamma}) \to H_1(X;\mathbb{Z}) \to Tor_1^{\tilde{\Gamma}}(\mathbb{Z}, \mathbb{Z}) \cong \mathbb{Z}^\mu \to 0$$

and

$$0 \to \mathbb{Z} \otimes_\varepsilon H_{n+1}(X;\tilde{\Gamma}) \to H_{n+1}(X;\mathbb{Z}) \to Tor_1^{\tilde{\Gamma}}(\mathbb{Z}, H_n(X;\tilde{\Gamma})) \to 0.$$

Since $H_1(X;\mathbb{Z}) \cong \mathbb{Z}^\mu$ we also have $\mathbb{Z} \otimes_\varepsilon H_1(X;\tilde{\Gamma}) = 0$, and so $H_1(X;\tilde{\Gamma})$ is also of type L, if $n > 1$.

Lemma 11.4 implies that $H_j(X;\tilde{\Gamma})^* = 0$ for all $j \neq n+1$. Hence the only differentials in the UCSS $E_2^{pq} = E^q H_p(X;\tilde{\Gamma}) \Rightarrow H^{p+q}(X;\tilde{\Gamma})$ which may be nonzero are the $d_2^{n+1,0} : E_2^{n+1,0} \to E_2^{n,2}$. Thus the spectral sequence breaks up into the exact sequences of part (7). $\square$

The module $H_1(X;\tilde{\Gamma})$ need not be of type L when $n = 1$.

The situation is simpler with coefficients in a field K. Let $K\tilde{\Gamma}$ denote the K-algebra $K[Aut(X^\omega/X)]$. Then $Ext_{K\tilde{\Gamma}}^2(M, N) = 0$ for all $K\tilde{\Gamma}$-modules M, N, and we obtain isomorphisms $H^j(X;K\tilde{\Gamma}) \cong E^1 H_{j-1}(X;K\tilde{\Gamma})$ for all $j \neq n+1$, and an exact sequence

$$0 \to E^1 H_n(X;K\tilde{\Gamma}) \to H^{n+1}(X;K\tilde{\Gamma}) \to H_{n+1}(X;K\tilde{\Gamma})^* \to 0.$$

Since ∂X^ω is a union of copies of $S^n \times R$ the natural homomorphism $H_j(X;\tilde{\Gamma}) \to H_j(X, \partial X;\tilde{\Gamma})$ is an isomorphism if $j \neq 0$, 1, n or $n+1$, a monomorphism if $j = 1$ or $n+1$ and an epimorphism if $j = 0$ or n. It is an isomorphism for $j = n$ if and only if the homomorphism induced by the inclusion $\partial X^\omega \subset X^\omega$ is 0. If L is a boundary link we may construct X^ω by splitting X along a family of disjoint Seifert hypersurfaces $\{U_i\}$. Since the n-spheres generating $H_n(\partial X;\tilde{\Gamma}) = H_n(\partial X^\omega;\mathbb{Z})$ bound lifts of these hypersurfaces in X^ω, this homomorphism is always 0 for boundary links.

The splitting construction also works for homology boundary links (although the spheres need no longer bound in X^ω). There is a corresponding Mayer-Vietoris sequence:

$$\cdots \to \tilde{\Gamma} \otimes H_j(U) \xrightarrow{d_j} \tilde{\Gamma} \otimes H_j(Y) \to H_j(X;\tilde{\Gamma}) \to \cdots$$

where $d_j(\gamma \otimes v_k) = \gamma x_k \otimes i_{k+*}(v_k) - \gamma \otimes i_{k-*}(v_k)$ for $\gamma \in \tilde{\Gamma}$ and $v_k \in H_j(U_k; \mathbb{Z})$. (See §2 of Chapter 2.) If $2 \leq k \leq n-1$ or if L is a boundary link $Y \simeq S^{n+2} \setminus U$ and so we may describe the maps $(i_{k\pm})_*$ in terms of Alexander duality in S^{n+2}.

Suppose that L is a boundary link. Since each component of U is a punctured $(n+1)$-manifold $H_{n+1}(U) = 0$ and $H_n(U)$ is free abelian, while $H_{n+1}(Y) \cong \tilde{H}^0(U) \cong \mathbb{Z}^{\mu-1}$ and $H_n(Y) \cong H^1(U)$ is free abelian, by Alexander duality. Therefore $\text{Ker}(d_n)$ is a free $\tilde{\Gamma}$-module, since d_n is a homomorphism between free modules. The Mayer-Vietoris sequence gives $H_{n+1}(X; \tilde{\Gamma}) \cong (\tilde{\Gamma})^{\mu-1} \oplus \text{Ker}(d_n)$. It now follows from Theorem 11.11 that $H_{n+1}(X; \tilde{\Gamma})$ has rank $\mu - 1$ and d_n is injective. (In general this need not be so.)

Given any sequence $H_2, \ldots, H_q$ of Γ_μ-modules of type L there is a boundary $(2q + 1)$-link L such that $H_j(X(L); \Gamma_\mu) \cong H_j$ for $1 < j \leq q$. If moreover $p.d._\Gamma H_q \leq 1$ there is a boundary $2q$-link with this homology [**Sa81'**].

As in the case of abelian coverings, Poincaré duality and the UCSS together give non-singular pairings between natural subquotients of the homology modules. An ε-*linking form over* Γ_μ is a pair (M, ϕ), where M is a Γ_μ-module of type L and $\phi : M \to \overline{E^1 M}$ is an isomorphism such that $\overline{E^1\phi} = \varepsilon\phi$. (It is *neutral* if there is a submodule J of type L such that $\phi(J) = \text{Ker}(\overline{E^1(j)})$, where $j : J \to M$ is the inclusion.) Let $B_j = H_j(X; \tilde{\Gamma})/E^2 E^2 H_j(X; \tilde{\Gamma})$ be the maximal $\mathbb{Z}$-torsion free quotient of the homology. Lemmas 11.4 and 11.8 together with Theorem 11.11 imply that duality determines an isomorphism $B_q \cong \overline{E^1 B_q}$ if $n = 2q - 1$ and $q \geq 2$; this can be shown to be a $(-1)^{q+1}$-linking form over Γ_μ on interpreting duality in terms of intersections of cells. It is the analogue for free covers of the Blanchfield pairing (and is the Blanchfield pairing for knots). If $q \geq 3$ every $(-1)^{q+1}$-linking form over Γ_μ is realized by some simple μ-component boundary $(2q - 1)$-link L [**Du86**], and the isomorphism class of the pairing is a complete invariant for such simple boundary links [**Fa91**]. If $q = 2$ there are constraints related to Rochlin's Theorem.

If L is a boundary link the image of $\oplus_{i=1}^{i=\mu} H_k(U_i; \mathbb{Z})$ in the module $H_k(X(L); \Gamma_\mu)$ is a lattice. The Farber derivations ∂_i^F have analogues for the stable homotopy pairings of "stable boundary links", and such links may be classified in terms of such pairings [**Fa92**].

Farber constructed invariants of $k[F(\mu)]$-modules of type L with values "non-commutative rational functions". When $\mu = 1$ these are equivalent to the usual Alexander polynomials Δ_0^q (although closer in form to the logarithmic derivative) [**Fa92'**]. His work has been reformulated in terms of Gelfand-Retakh quasideterminants [**RRV99**]. Is there a realization result analogous to Theorem 7.19 for the invariants of $H_1(X^\omega; k) = k \otimes_{\mathbb{Z}} (\pi_\omega/\pi_\omega')$?

11.6. The classical case

When $n = 1$ the top dimensional homology is $H_2(X; \tilde{\Gamma})$, which is free of rank $r \leq \mu - 1$, by parts (3) and (6) of Theorem 11.11. The UCSS gives $H^0(X; \tilde{\Gamma}) = 0$, $H^1(X; \tilde{\Gamma}) \cong E^1 \mathbb{Z}$ and an exact sequence

$$0 \to E^1(\pi_\omega/(\pi_\omega)') \to H^2(X; \tilde{\Gamma}) \to (\tilde{\Gamma}^r)^* \to E^2(\pi_\omega/(\pi_\omega)') \to 0.$$

If L is a boundary link the Mayer-Vietoris sequence becomes

$$0 \to (\tilde{\Gamma})^{\mu-1} \to H_2(X; \tilde{\Gamma}) \to (\tilde{\Gamma})^\nu \xrightarrow{d_1} (\tilde{\Gamma})^\nu \to \pi_\omega/(\pi_\omega)' \xrightarrow{\delta} .$$

The right hand end of this sequence continues to the standard resolution of $\mathbb{Z}$, and so $\delta = 0$. As observed in §5 above, the monomorphism from $(\tilde{\Gamma})^{\mu-1}$ to $H_2(X; \tilde{\Gamma})$ is an isomorphism. Therefore d_1 is a monomorphism and so $\pi_\omega/(\pi_\omega)'$ has a short free resolution. Furthermore, the longitudes of L are in $(\pi_\omega)'$ and so the long exact sequence of the pair $(X, \partial X)$ breaks up into two exact sequences:

$$0 \to H_2(X; \tilde{\Gamma}) \to H_2(X, \partial X; \tilde{\Gamma}) \to H_1(\partial X; \tilde{\Gamma}) \to 0$$

and

$$0 \to \pi_\omega/(\pi_\omega)' \to H_1(X, \partial X; \tilde{\Gamma}) \to H_0(\partial X; \tilde{\Gamma}) \to H_0(X; \tilde{\Gamma}) \to 0.$$

The right hand end of the latter sequence is just the obvious homomorphism from $\oplus(\tilde{\Gamma}/\tilde{\Gamma}(x_i - 1))$ to $\mathbb{Z}$, and so its kernel is free of rank $\mu - 1$. Therefore $\overline{H^2(X; \tilde{\Gamma})} \cong (\pi_\omega/(\pi_\omega)') \oplus (\tilde{\Gamma})^{\mu-1}$, by Poincaré duality. On substituting this into the sequence obtained from the UCSS we conclude that for L a boundary 1-link (or if L is a homology boundary 1-link whose longitudes lie in $(\pi_\omega)'$) there is an isomorphism $\overline{E^1(\pi_\omega/(\pi_\omega)')} \cong \pi_\omega/(\pi_\omega)'$. (This could also be obtained from the symmetry of linking coefficients in S^3 and the Mayer-Vietoris resolution

$$0 \to (\tilde{\Gamma})^\nu \to (\tilde{\Gamma})^\nu \to H_1(X;\tilde{\Gamma}) \to 0.$$

In the knot theoretic case this condition implies the symmetry of the Alexander polynomials.)

THEOREM 11.12. [**Br81**] *Let L be a 2-component homology boundary 1-link. Then L is a boundary link if and only if $H_2(X;\tilde{\Gamma}) \neq 0$.*

PROOF. (Sketch.) This condition is clearly necessary.

Assume that $H_2(X;\tilde{\Gamma}) \neq 0$, and let F be a closed connected surface in X^ω representing a generator of this module. Since $H_1(X;\tilde{\Gamma})^*$ is 0 every closed 2-sided surface in X^ω separates. Therefore the complement $X^\omega \setminus F$ must have at least two components. The complementary components cannot be compact, since $[F] \neq 0$. Therefore F separates two ends. Now X^ω has infinitely many ends. Hence we may assume that F is disjoint from its translates, and so projects to a non-singular, two-sided surface in X, by Theorem A of [**Br81**]. Since $H_2(X;\tilde{\Gamma})$ maps onto $H_2(X;\mathbb{Z})$ this surface is homologically nontrivial, and so it separates the components of L. Since this separating surface lifts to the maximal abelian cover X' the link is a boundary link, by duality. □

If L is the homology boundary link of Figure 1.5 (and §7 of Chapter 8) then $Tor_1(\mathbb{Q}, \pi_\omega/(\pi_\omega)') \neq 0$, and so $H_2(X;\tilde{\Gamma}) = 0$.

A *good boundary link* is a boundary 1-link L such that $(\pi L)_\omega$ is perfect. Replacing each component of a link by an (untwisted) Whitehead double gives a good boundary link if and only if all the linking numbers of the original link are 0. It is known that the question of whether 4-dimensional TOP surgery techniques are valid without restriction on fundamental groups is equivalent to showing that certain infinite families of such Whitehead doubles of iterated Bing doubles of the Hopf link are "freely slice", i.e., bound disjoint locally flat discs in D^4 such that the fundamental group of the complement is free. The existence of such slicing discs has been established for ∂^2-links, which form a somewhat more restricted class [**Fr93, FT95**].

If T is a torus in the exterior of a boundary 1-link, with core in π_ω, then ± 1-framed surgery on T gives a new boundary link. Can all boundary 1-links be obtained from the trivial link by a sequence of such surgeries?

11.7. The case $n = 2$

When $n = 2$ the UCSS gives $H^0(X; \tilde{\Gamma}) = 0$, $H^1(X; \tilde{\Gamma}) \cong E^1\mathbb{Z}$, $H^2(X; \tilde{\Gamma}) \cong E^1(\pi_\omega/(\pi_\omega)')$ and an exact sequence

$$0 \to J_{2,1} \to H^3(X; \tilde{\Gamma}) \to H_3(X; \tilde{\Gamma})^* \to E^2 H_2(X; \tilde{\Gamma}) \to 0,$$

where $J_{2,1}$ is an extension of $E^1 H_2(X; \tilde{\Gamma})$ by $E^2(\pi_\omega/(\pi_\omega)')$ Hence $E^2(\pi_\omega/(\pi_\omega)') \cong E^2 E^2 J_{2,1} \cong E^2 E^2 H^3(X; \tilde{\Gamma})$. Moreover, $\overline{H^3(X; \tilde{\Gamma})} \cong H_1(X, \partial X; \tilde{\Gamma})$ and there is an exact sequence

$$0 \to \pi_\omega/(\pi_\omega)' \to H_1(X, \partial X; \tilde{\Gamma}) \to \oplus(\tilde{\Gamma}/\tilde{\Gamma}(x_i - 1)) \to \mathbb{Z} \to 0;$$

it follows that $E^2 E^2(\pi_\omega/(\pi_\omega)') \cong E^2 E^2 H_1(X, \partial X; \tilde{\Gamma})$. Hence there is an isomorphism $\overline{E^2(\pi_\omega/(\pi_\omega)')} \cong E^2 E^2(\pi_\omega/(\pi_\omega)')$. (This isomorphism could have been obtained by considering π as the fundamental group of the closed 4-manifold obtained by surgery on the link.)

In the knot theoretic case this isomorphism gives the Farber-Levine duality pairing, since $\pi_\omega = \pi'$ when $\mu = 1$. This was used independently by Farber and Levine to show that there are high dimensional knot groups which are not 2-knot groups [**Fa77, Le77**].

If L is a boundary 2-link then $H_2(X; \tilde{\Gamma}) \cong H_2(X, \partial X; \tilde{\Gamma})$ and so $H_2(\pi_\omega; \mathbb{Z})$ is a quotient of $\overline{E^1(\pi_\omega/(\pi_\omega)')}$, by Poincaré duality and Theorem 11.11.

11.8. An unlinking theorem

A 1-link is trivial if and only if its group is free, by Theorem 1.1. Gutiérrez showed that if $n \geq 4$ an n-link is trivial if and only if its group is freely generated by meridians and the higher homotopy groups of the exterior are 0, up to the middle dimension. The following proof uses the s-cobordism theorem and is valid also for $n = 3$. While the fundamental group condition is necessary when $n = 2$, we do not yet know whether it is a complete criterion for triviality of 2-links with more than one component, and we can only show that such a 2-link is s-concordant to a trivial 2-link.

THEOREM 11.13. *Let L be a μ-component n-link with $n \geq 3$. Then L is trivial if and only if $\pi L \cong F(\mu)$, with basis a set of meridians, and $\pi_j(X(L)) = 0$ for $1 < j \leq [(n+1)/2]$.*

PROOF. Let U be a trivial μ-component n-link. Then $X(U) = \natural^\mu(D^{n+1} \times S^1)$, where the summands correspond to the components of U. Hence πU is freely generated by meridians and $\pi_j(X(U)) = 0$ for $1 < j \leq n$, by general position. Thus the conditions are necessary.

Assume that they hold, and choose a basepoint $* \in X = X(L)$ and arcs α_i from $*$ to $\partial X(L_i)$ which meet only at $*$. Let N be a closed regular neighbourhood of $\cup \alpha_i$ in X and let $W = \overline{X \setminus N}$. Then $\pi_j(W) = \pi_j(X)$ for all $j < n$, by general position. Let V be a closed regular neighbourhood of a wedge of loops $\vee^\mu S^1 \subset int\, W$ representing a set of meridians.

Poincaré duality gives $H_q(X; \Gamma_\mu) = 0$ for each $[(n+1)/2] \leq q \leq n$. If moreover $n = 2r$ is even then $H_{r+1}(X; \Gamma_\mu) \cong H^{r+1}(X; \Gamma_\mu) \cong H_{r+1}(X; \Gamma_\mu)^*$, and so $H_{r+1}(X; \Gamma_\mu)$ is free. Hence it must also be 0, by the considerations of §5. The Hurewicz Theorem now implies that $\pi_j(X) = 0$ for $1 < j \leq n$. Therefore $\pi_j(W) = 0$ for $1 < j < n$. Hence the inclusion of ∂W into W is $(n-1)$-connected and so $H_q(W, \partial W; \Gamma_\mu) = 0$ for $0 \leq q < n$. Poincaré duality now implies that $H^q(W; \Gamma_\mu) = 0$ for all $q > 2$. Hence $H_n(W; \Gamma_\mu) = H_{n+1}(W; \Gamma_\mu) = 0$, and so $\widetilde{W}$ is contractible. Thus the inclusion $V \subset W$ is a homotopy equivalence. The region $\overline{W \setminus V}$ is an h-cobordism, by Van Kampen's Theorem and the Mayer-Vietoris sequence in the universal covers. Since $Wh(F(\mu)) = 0$ this region is an s-cobordism and so $W \cong V \cong \natural^\mu(D^{n+1} \times S^1)$. Hence $M(L) = W \cup \natural^\mu(D^{n+1} \times S^1) \cong \natural^\mu(S^{n+1} \times S^1) \cong M(U)$. (Note that every self homeomorphism of ∂P extends to a self homeomorphism of P.)

The elements of the meridianal basis for πL are freely homotopic to loops representing the standard basis for $\pi_1(M(U))$. We may realise such homotopies by μ disjoint embeddings of annuli in the product $M(U) \times [0,1]$ running from meridians for L in $M(L) \cong M(U) \times \{0\}$ to such standard loops in $M(\mu)$. Surgery on these annuli (i.e., replacing $D^{n+1} \times S^1 \times [0,1]$ by $S^n \times D^2 \times [0,1]$) then gives an s-concordance from L to the trivial μ-component n-link. Hence L is trivial, by the s-cobordism theorem. $\qquad\square$

ADDENDUM. *A μ-component 2-link L is s-concordant to a trivial 2-link if and only if $\pi L \cong F(\mu)$, with basis a set of meridians.*

PROOF. The homotopical part of the above argument applies without change. We may then use Proposition 11.6A of [**FQ**] to conclude that W is s-cobordant *rel $\partial W = \natural^\mu(S^1 \times S^2)$* to P. Hence $M(L)$ is s-cobordant to $M(U)$, and we may surger embedded annuli as before to get an s-concordance from L to U. $\qquad\square$

The hypothesis that πL be freely generated by meridians cannot be dropped entirely [**Po71**]. (See also §7 of Chapter 8 above.) On the other hand, if L is a 2-link whose longitudinal 2-spheres are all null homotopic then $(X(L), \partial X(L))$ is homotopy equivalent to the model $(X(U), \partial X(U))$ [**Sw77**], and hence the Addendum applies.

11.9. Patterns and calibrations

If L is a μ-component homology boundary n-link the images of a set of meridians under an epimorphism from $\pi = \pi L$ to $F(\mu)$ is an ordered μ-tuple of words which normally generates $F(\mu)$. Although this μ-tuple depends on various choices, we may obtain an invariant for L as follows. Let $\theta : Aut(F(\mu)) \to Aut(F(\mu)^\mu)$ be the diagonal homomorphism, and let $G = F(\mu)^\mu \times_\theta Aut(F(\mu))$. The group G acts on $F(\mu)^\mu$ by $g.(f_1, \ldots, f_\mu) = (g_1\alpha(f_1)g_1^{-1}, \ldots, g_\mu\alpha(f_\mu)g_\mu^{-1})$, for $g = (g_1, \ldots, g_\mu, \alpha) \in G$ and $(f_1, \ldots, f_\mu) \in F(\mu)^\mu$. The *pattern* of L is the G-orbit of the μ-tuple determined by a set of meridians and an epimorphism from π onto $F(\mu)$. This notion is clearly well defined. A homology boundary link is a boundary link if and only if its pattern is represented by $(x_1, \ldots, x_\mu)$. Every μ-tuple of elements which normally generate $F(\mu)$ represents the pattern of some ribbon homology boundary link [**CL91**]. Is every homology boundary link group the group of a boundary link?

The notion of pattern has been extended to a wider class of links [**Le89**]. A group G is a *(finite) E-group* if it is the fundamental group of a (finite) 2-complex K with $H_1(K;\mathbb{Z})$ torsion free and $H_2(K;\mathbb{Z}) = 0$. It is a finite E-group if and only if it is finitely presentable and $G/G' \cong \mathbb{Z}^d$, where $d = \operatorname{def}(G)$ is the deficiency of G. A *calibration* for a finitely generated E-group G is a set of $\operatorname{def}(G)$ elements whose normal closure is G. (Thus a finite E-group admits a calibration if and only if its weight is equal to its deficiency. A calibration for a free group determines a pattern.) A μ-component link L is a *(finite) E-link* if there is a normally surjective homomorphism ψ from πL to a (finite) E-group G of weight μ. In the classical case we require also that the longitudes of L are in $\operatorname{Ker}(\psi)$. It follows easily from the argument of Theorem 1.14 that every SHB link is a finite E-link. Every pair (finite E-group, calibration) is realized by some ribbon n-link, for all $n \geq 1$ [**Be98**].

Kaiser has computed the patterns of links obtained by strong fusion of homology boundary links, and has given thereby a more systematic treatment of some of the examples of Chapter 8. If L is an n-link a choice of arcs α_i from a basepoint to the components $\partial X(L_i)$ of $\partial X(L)$ determines a basing $m : F(\mu) \to \pi L$. If $\beta = (b, u)$ is a fusion band from L_1 to L_2 (as in §7 of Chapter 1) and $n > 1$ then the isotopy type of the strong fusion of L along β is determined by πL and the double coset of the loop $\alpha_1.b.\bar{\alpha}_2$ with respect to the images of $\pi_1(\partial X(L_1))$ on the left and $\pi_1(\partial X(L_2))$ on the right. If L is a homology boundary link and $\phi : \pi L \to F(\mu)$ is an epimorphism then the strong fusion of L along β is a homology boundary link if and only if the 1-relator group $\langle x_1, \ldots, x_\mu, z \mid \phi(m_1)z = z\phi(m_2)\rangle$ is free. In particular, if $\phi(m_1) = x_1$ and $\phi(m_2) = x_2$ then this is so. (See [**Ka92**] for more details.)

11.10. Concordance

The high dimensional knot concordance groups have been computed in various ways – using Seifert forms, linking pairings and homological surgery – and each of these approaches has been applied to boundary links [**CS80, Ko87, Mi87**]. In even dimensions boundary links are null concordant [**Gu72, De81, Co84**], but in odd dimensions they need not be concordant to split links [**CS80, Kw80**]. (See also §3 of Chapter 8 above for the classical case.) Thus the problem of link concordance does not reduce to concordance of the components. It is perhaps best approached through the notion of disc link. (This is outlined in Chapter 14 below.)

An $F(\mu)$-(n)-link (L, θ) is a μ-component n-link L with a conjugacy class of epimorphisms $\theta : \pi L \to F(\mu)$ which carry some set of meridians to the standard basis. (Thus it is a boundary link with additional structure refining the pattern.) Let $U^{n,\mu}$ be the μ-component trivial n-link. Then $X(U^{n,\mu}) \cong \natural^\mu (S^1 \times D^{n+1})$, and any such epimorphism θ may be realized by a degree 1 map $f : (X(L), \partial X(L)) \to (\natural^\mu (S^1 \times D^{n+1}), \mu(S^1 \times S^n))$ such that $f|_{\partial X(L)}$ is a homeomorphism. Moreover two such maps are homotopic if and only if the induced epimorphisms are conjugate.

Any sum of two $F(\mu)$-n-links (as defined in §9 of Chapter 2) is again an $F(\mu)$-n-link. Although the sum is not well-defined on link types, if $n > 1$ it induces a well-defined operation making the set of $F(\mu)$-concordance classes of $F(\mu)$-n-links into an abelian group $C_n(F(\mu))$. Moreover every $F(\mu)$-n-link is $F(\mu)$-concordant to a trivial link if n is even and to a simple $F(\mu)$-n-link if n is odd. If $n = 2q - 1$ and $q \geq 3$ the group $C_n(F(\mu))$ may be identified with a homology surgery obstruction group [**CS80**]. Using a different formulation of the surgery approach, Duval identified $C_n(F(\mu))$ with $W_\varepsilon(\Gamma_{\mu\Psi}, \Gamma_\mu, -)$, the Witt group of $(-1)^{q+1}$-linking pairings over Γ_μ. The most substantial part of his argument is a realization theorem, constructing links with given linking pairings. We shall verify only that the Witt class is an additive invariant of $F(\mu)$-concordance. Let $b^{F(\mu)}(L, \theta)$ be the $(-1)^{q+1}$-linking pairing over Γ_μ corresponding to (L, θ) and let $B^{F(\mu)}(L, \theta)$ its Witt class.

THEOREM 11.14. [**Du86**] *Let* (L_-, θ_-) *and* (L_+, θ_+) *be* $F(\mu)$-*n*-*links, where* $n = 2q - 1 \geq 3$.

(1) *If* (L_-, θ_-) *and* (L_+, θ_+) *are* $F(\mu)$-*concordant then*
$$B^{F(\mu)}(L_-, \theta_-) = B^{F(\mu)}(L_+, \theta_+).$$
(2) $B^{F(\mu)}(L_-\natural L_+, \theta_-\natural\theta_+) = B^{F(\mu)}(L_-, \theta_-) + B^{F(\mu)}(L_+, \theta_+).$

PROOF. Let $(\mathcal{L}, \Theta)$ be a concordance from (L_-, θ_-) to (L_+, θ_+). It is easily seen that if α is an inner automorphism of $F(\mu)$ then $b^{F(\mu)}(L, \theta) \cong b^{F(\mu)}(L, \alpha\theta)$. Hence we may assume that the actions of $F(\mu)$ on the free covers of $X_- = X(L_-)$, $X_+ = X(L_+)$ and $\mathcal{X} = X(\mathcal{L})$ are compatible. Let $B_j(X) = H_j(X; \Gamma_\mu)/E^2 E^2 H_j(X; \Gamma_\mu)$, for $j \geq 0$ and let $\phi = b^{F(\mu)}(L_-, \theta_-) \oplus -b^{F(\mu)}(L_+, \theta_+)$ be the linking form on $B_q(X_-) \oplus B_q(X_+)$. Since $\partial\mathcal{X} \cong X_- \cup X_+$, identified along copies of $\mu(S^{2q-1} \times S^1)$, the exact sequence of homology for the pair $(\mathcal{X}, \partial) = (\mathcal{X}, \partial\mathcal{X})$ and Poincaré duality give a commutative diagram

$$
\begin{array}{ccccc}
B_{q+1}(\mathcal{X}, \partial) & \xrightarrow{\partial_*} & B_q(X_-) \oplus B_q(X_+) & \longrightarrow & B_q(\mathcal{X}) \\
\downarrow & & \phi\downarrow & & \downarrow \\
\overline{E^1 B_q(\mathcal{X})} & \longrightarrow & \overline{E^1 B_q(X_-)} \oplus \overline{E^1 B_q(X_+)} & \longrightarrow & \overline{E^1 B_{q+1}(\mathcal{X}, \partial)}
\end{array}
$$

in which the rows are exact ımod $\mathbb{Z}$-torsion. It is easily verified that $J = \mathrm{Im}(\partial_*)^\perp$ is of type L and that $J^\perp = J$. Hence ϕ is neutral and so $B^{F(\mu)}(L_-, \theta_-) = B^{F(\mu)}(L_+, \theta_+)$.

Since $B^{F(\mu)}(L, \theta)$ is an invariant of $F(\mu)$-concordance, by (1), we may assume that L_- and L_+ are simple $F(\mu)$-links such that $L_\pm \subset D_\pm^{2q+1}$ and $L_- \cap S^{2q} = L_+ \cap S^{2q} \cong \mu D^{2q-1}$. Additivity of the linking pairing then follows by a standard Mayer-Vietoris argument. $\square$

The addition may also be defined in terms of boundary connected sums of spanning hypersurfaces corresponding to the splitting, and $C_n(F(\mu))$ may be identified with a cobordism group of "Seifert" matrices $G(\mu, (-1)^q)$ [**Ko87**]. These descriptions extend to the case $n = 3$, provided that we either work in the TOP locally flat category or pass to subgroups of index 2^μ determined by (Rochlin) signature parity condition. Sheiham has recently shown that the torsion subgroup of $W_\varepsilon(\Gamma_{\mu\Psi}, \Gamma_\mu, -)$ has exponent 8, and the quotient is a direct sum of copies of $\mathbb{Z}$, with one summand for each isomorphism class of simple self dual complex representations of Farber's ring P_μ [**She**].

The group $\mathcal{H}(\mu)$ of "special" automorphisms of $F(\mu)$, which send each basis element to a conjugate, acts on $C_n(F(\mu))$. The boundary concordance set of boundary links is then obtained by factoring out the action of $\mathcal{H}(\mu)$.

In [**CO94**] the notion of pattern is refined to fix the combinatorial "scheme" in which the singular Seifert hypersurfaces of a homology boundary link meet the components of ∂X, and the linking pairing is extended to odd-dimensional homology boundary links. It is shown that any scheme and any Witt class of linking pairings is realized by some homology boundary $(2q - 1)$-link, for $q \geq 2$. (When $q = 2$ there are again signature parity constraints.) The set of homology boundary concordance classes of homology boundary $(2q - 1)$-links with fixed pattern is then described in terms of quotients of homology surgery obstruction groups modulo the action of a group of pattern-preserving automorphisms of $F(\mu)$.

It is not known whether concordant boundary links need be boundary concordant, and in high dimensions there are homology boundary links which are not concordant to boundary links [**CO93, GL92**]. The examples of [**CO93**] are constructed from the untwisted double of a knot with nontrivial signature invariants by taking the strong fusion with respect to a suitable band. Let BL, HBL and SHB denote the sets of boundary links, homology boundary links and sublinks of homology boundary links, respectively, and let cBL, $cHBL$ and $cSHB$ denote the set of links conconcordant to boundary links, etc. We may display schematically the inclusions between these sets as in the following diagram:

$$
\begin{array}{ccccc}
BL & \xrightarrow{\neq} & HBL & \xrightarrow{\neq} & SHB \\
\neq\downarrow & & \neq\downarrow & & \downarrow \\
cBL & \xrightarrow{\neq} & cHBL & \longrightarrow & cSHB
\end{array}
$$

Is $SHB = cHBL$? Is every high dimensional link concordant to (a sublink of) a homology boundary link?

CHAPTER 12

Nilpotent Quotients

The theorem of Stallings (Theorem 1.3) demonstrates a close connection between the homology of a group and its lower central series. To what extent can we derive the nilpotent quotients homologically? The answer is that, roughly speaking, the Massey product structure on $H^1(G; \mathbb{Q})$ determines the rational Mal'cev completion of G (defined below).

The importance of the lower central series for links is that the quotients are the primary invariants of I-equivalence. Duality then gives rise to other invariants (such as Witt classes) derived from the homology of coverings associated to canonical subgroups of the link group. The Massey products in the cohomology of a 1-link group are essentially equivalent to the Milnor invariants, which are defined in terms of Magnus expansions of words representing longitudes.

12.1. Massey products

Let G be a group and R a commutative ring. The cohomology of G with coefficients in the trivial G-module R may be computed from the complex of inhomogeneous cochains $C^*(G; R)$. The module $C^q(G; R)$ is the module of functions from the q-fold product G^q to R. In low degrees the differentials are given by

$$\delta^0 = 0, \quad \delta^1(f)(g, h) = f(g) + f(h) - f(gh) \quad \text{and}$$

$$\delta^2(F)(g, h, j) = F(h, j) - F(gh, j) + F(g, hj) - F(g, h),$$

for g, h, $j \in G$. Therefore $H^0(G; R) = R$, $H^1(G; R) = Hom(G, R)$ and $H^2(G; R) = \{F : G^2 \to R \mid \delta^2(F) = 0\}/\mathrm{Im}(\delta^1)$. The cup product of f_1, $f_2 \in H^1(G; R)$ is represented by the function F defined by

273

$F(g, h) = f_1(g)f_2(h)$, for all g, $h \in G$. Cup product is anticommutative, since

$$f_1(g)f_2(h) + f_2(g)f_1(h) = -\delta^1(f_1f_2)(g, h), \quad \text{for all } g, h \in G,$$

if f_1 and f_2 are 1-cocycles. Therefore it induces a "cup product" homomorphism from $H^1(G; R) \wedge H^1(G; R)$ to $H^2(G; R)$, if 2 is invertible in R. This is also the case if $R = \mathbb{Z}$. For if $f : G \to \mathbb{Z}$ is a homomorphism and F is defined by $F(h) = f(h)(1 - f(h))/2$, for all $h \in G$, then $f(g)f(h) = \delta^1 F(g, h)$, for all g, $h \in G$, and so $f \cup f = 0$. If $2 = 0$ in R there is a cup product homomorphism from the symmetric product $H^1(G; R) \odot H^1(G; R)$ to $H^2(G; R)$ instead.

Massey products were introduced as "higher linking invariants" in [**Ma68**]. The nth order Massey product of $f_1, \ldots, f_n \in H^1(G; R)$ is defined if there is a strictly upper triangular $(n + 1) \times (n + 1)$ matrix M with entries in $C^1(G; R)$ such that $m_{i,i+1} = f_i$ for $1 \leq i \leq n$, $m_{1,n+1} = 0$ and $\delta^1 m_{i,j}(g, h) = \Sigma_{k=i+1}^{k=j-1} m_{i,k}(g)m_{k,j}(h)$ for all $(i, j) \neq (1, n + 1)$. Its value relative to this *defining system* M is the class of the cocycle $\Sigma_{k=2}^{k=n} m_{1,k} \cup m_{k,n+1}$ in $H^2(G; R)$, and is denoted $\langle f_1, \ldots, f_n \rangle_M$. The Massey product is the subset $\langle f_1, \ldots, f_n \rangle$ of $H^2(G; R)$ consisting of all elements obtained in this way from some defining system. It is *essential* if it is defined and $\langle f_1, \ldots, f_n \rangle_M \neq 0$ for all defining systems M. It is inessential, or said to contain 0 if $\langle f_1, \ldots, f_n \rangle_M = 0$ for some choice of defining system. The *indeterminacy* of $\langle f_1, \ldots, f_n \rangle$ is the subset $\{a - b \mid a, b \in \langle f_1, \ldots, f_n \rangle\}$. When $n = 2$ the Massey product is just the cup product, and there is no indeterminacy.

The submatrices determined by the ith rows and j^{th} columns for $i' \leq i \leq j \leq j'$ (for some fixed $i' < j'$) define Massey products of order $j' - i'$ called subproducts. If $\langle f_1, \ldots, f_n \rangle$ is defined each proper subproduct contains 0. The Massey product is *strictly defined* if all the proper subproducts are 0. If $\langle f_1, \ldots, f_n \rangle$ is defined and $h : H \to G$ is a homomorphism $\langle h^* f_1, \ldots, h^* f_n \rangle$ is defined and $h^* \langle f_1, \ldots, f_n \rangle \subseteq \langle h^* f_1, \ldots, h^* f_n \rangle$.

The nth order Massey products in $H^2(G; R)$ may be identified with the cohomology classes of the central extensions of G by R

induced from the canonical extension of the group of upper triangular $(n+1) \times (n+1)$ matrices over R (with 1s on the diagonal), modulo its centre [**Dw75**].

Let $G = \pi Bo$ be the group of the Borromean rings Bo, and let f_1, f_2 and f_3 be the generators of $H^1(G; R)$ which are *Kronecker dual* to the meridians $\{x_1, x_2, x_3\}$ (i.e., $f_i(x_j) = \delta_{ij}$). Then all cup products $f_i \cup f_j$ are 0. The Massey triple product $\langle f_1, f_2, f_3 \rangle$ is defined, with 0 indeterminacy, and is a generator for $H^2(G; R) \cong R$. See [**O'N79**] for connections between 4th order Massey products and Steenrod functional products, with exemplary calculations for 2-component links.

12.2. Products, the Dwyer filtration and gropes

Lot $G(R) = \cap \mathrm{Ker}(\lambda)$, where the intersection is taken over all homomorphisms $\lambda : G \to R$. Then the epimorphism from G to $\overline{G} = G/G(R)$ induces an isomorphism $H^1(\overline{G}; R) \to H^1(G; R)$. Moreover if G is finitely generated and $R = \mathbb{Z}$ or $\mathbb{F}_p$ (for some prime p) then $\overline{G} \cong R^b$ for some b. The 5-term exact sequence of low degree for the LHS cohomology spectral sequence for G as an extension of $\overline{G}$ by $G(R)$ then gives an exact sequence

$$0 \to H^0(\overline{G}; H^1(G(R); R)) \overset{\tau}{\longrightarrow} H^2(\overline{G}; R) \to H^2(G; R),$$

where τ is the transgression homomorphism. This leads to an exact sequence

$$\mathrm{Ker}(\cup_{\overline{G}}) \to \mathrm{Ker}(\cup_G) \to H^0(\overline{G}; H^1(G(R); R)) \to$$

$$\mathrm{Coker}(\cup_{\overline{G}}) \to \mathrm{Coker}(\cup_G).$$

Note that $H^0(\overline{G}; H^1(G(R); R)) \cong Hom(G(R)/[G, G(R)], R)$.

If $R = \mathbb{Z}$ then $\overline{G}$ is free abelian and $H^*(\overline{G}; \mathbb{Z})$ is the exterior algebra generated by $Hom(G, \mathbb{Z})$. In particular, $\cup_{\overline{G}}$ is an isomorphism and so $\mathrm{Ker}(\cup_G) = Hom(G(\mathbb{Z})/[G, G(\mathbb{Z})], \mathbb{Z})$. The inclusion of G_2 into $G(\mathbb{Z})$ induces a homomorphism $G_2/G_3 \to G(\mathbb{Z})/[G, G(\mathbb{Z})]$ which has finite kernel and cokernel. Since localization is exact we may then identify the kernel of cup product with coefficients $\mathbb{Q}$ with $Hom(G_2/G_3, \mathbb{Q})$. In particular, cup product with coefficients $\mathbb{Z}$ or $\mathbb{Q}$ is injective if and only if G_2/G_3 is finite.

On the other hand, let $\theta : F(b) \to G$ induce an isomorphism $H^1(\theta; \mathbb{Z})$. Then cup product with coefficients $R = \mathbb{Z}$ or $\mathbb{Q}$ is identically 0 if and only if the induced map from $F(b)/F(b)_3$ to G/G_3 is a monomorphism.

Dwyer gave an analogous criterion for the relative freedom of nilpotent quotients of a group in terms of the vanishing of Massey products, involving an extension of Stallings' Theorem 1.3. If Y is a topological space with $\pi_1(Y) = \pi$ let $\Phi_k(Y)$ be the kernel of the composite $H_2(Y; \mathbb{Z}) \to H_2(\pi; \mathbb{Z}) \to H_2(\pi/\pi_{k-1}; \mathbb{Z})$, and let $\Phi_k(G) = \Phi_k(K(G, 1))$. The image of $\pi_2(Y)$ under the Hurewicz homomorphism is contained in $\Phi_\omega(Y) \leq \cap_{k \geq 2} \Phi_k(Y)$.

THEOREM. [**Dw75**] *Let* $f : G \to H$ *be a homomorphism of groups such that* $H_1(f; \mathbb{Z})$ *is an isomorphism, and let* $k \geq 2$. *Let* $\phi_k(f) : H_2(G; \mathbb{Z})/\Phi_k(G) \to H_2(H; \mathbb{Z})/\Phi_k(H)$ *be the homomorphism induced by* f. *Then the following conditions are equivalent:*

(1) $\phi_k(f)$ *is an epimorphism;*
(2) $\phi_k(f)$ *is an isomorphism and* $\phi_{k+1}(f)$ *is a monomorphism;*
(3) f *induces an isomorphism* $G/G_k \to H/H_k$.   □

The proof is similar to that of Theorem 1.3, being a finite induction based on the exact sequence of low degree from the homology LHS spectral sequence and the Five-Lemma.

If these conditions hold then

(4) f *induces an isomorphism* $\Phi^{k-1}(H; A) \cong \Phi^{k-1}(G; A)$ *and an epimorphism from* $\Phi^k(H; A)$ *to* $\Phi^k(G; A)$, *where* $\Phi^n(G; A)$ *is the subgroup of* $H^2(G; A)$ *determined by (a generalized form of)* n^{th} *order Massey products, with coefficients* A;

and

(5) *if* $f : F(b) \to G$ *induces an isomorphism* $H^1(f; \mathbb{Z})$ *and all Massey products of order* $\leq k$ *contain 0 then* $\mathrm{Ker}(f) \leq F(b)_{k+1}$, *and so* f *is a monomorphism if all Massey products contain 0.*

The *Dwyer filtration* $\{\Phi_k(Y)\}_{k \geq 1}$ has a geometrical interpretation in terms of singular *gropes* in Y. This notion is defined inductively. The pair (S^1, S^1) is the standard grope of class 1. A grope of class 2 is a pair $(S, \partial S)$, where S is a compact connected oriented surface with a single boundary component. Let $\{\alpha_i, \beta_i \mid 1 \le i \le g\}$ be embedded circles in S representing a symplectic basis for $H_1(S; \mathbb{Z})$. A grope of class k is a pair (X, S^1), where X is a 2-complex constructed from such a grope of class 2 by attaching gropes of class $p_i \ge 1$ to each circle α_i and gropes of class $q_j \ge 1$ to each circle β_j and such that $p_i + q_i \ge k$ for at least one index i. An ∞-grope is a pair (X, S^1) where $X = \cup_{k \ge 1} X_k$ is an increasing union of finite subcomplexes such that (X_k, S^1) is a k-grope, for all $k \ge 1$.

LEMMA. [**FT95**] *An element $g \in \pi = \pi_1(Y)$ is in π_k if and only if it bounds a map of a k-grope into Y, for $1 \le k < \infty$. It is in π_ω if it bounds a map of an ∞-grope into Y.* $\square$

The *spherical k-grope* determined by (X, S^1) is the 2-complex $\hat{X} = X \cup D^2$. The notions of annular grope, etc., are defined similarly.

LEMMA. [**FT95**] *Let Y be a cell complex. Then $\Phi_k(Y)$ is the subset of $H_2(Y; \mathbb{Z})$ represented by maps of spherical k-gropes into Y.* $\square$

12.3. Mod-p analogues

Let $X^n(H)$ denote the verbal subgroup of H generated by all n^{th} powers. If p is a prime the restricted p-lower central series is given by $\{\Gamma_k(G; p)\}_{k \ge 1}$, where $\Gamma_k(G; p)$ is the subgroup generated by $\cup_{p^m n \ge k} X^{p^m}(G_n)$.

If $p = 2$ then $G(\mathbb{F}_2) = X^2(G) \ge G_2$ and $\overline{G} = G/G(\mathbb{F}_2) \cong (\mathbb{Z}/2\mathbb{Z})^b$, say, where $X^p(G)$ is the verbal subgroup of G generated by all p^{th} powers. The cohomology ring $H^*(\overline{G}; \mathbb{F}_2)$ is now the polynomial algebra over $\mathbb{F}_2$ generated in degree 1 by $H^1(G; \mathbb{F}_2)$. Since $\cup_{\overline{G}}$ is again an isomorphism $\mathrm{Ker}(\cup_G) = Hom(X^2(G)/[G, X^2(G)], \mathbb{F}_2) \cong Hom(X^2(G)/[G, X^2(G)]X^4(G), \mathbb{F}_2)$. Hence cup product is injective if and only if G does not map onto $Z/4Z$ and G_2/G_3 is finite of odd order. If G is abelian the kernel of cup product with coefficients $\mathbb{F}_2$ is isomorphic to the image of reduction *mod* (2) from $H^1(G; \mathbb{Z}/4\mathbb{Z})$ to $H^1(G; \mathbb{F}_2)$, which is also the kernel of the *mod*-2 Bockstein homomorphism $\beta_2 : H^1(G; \mathbb{F}_2) \to H^2(G; \mathbb{F}_2)$. On the other hand, cup product is identically 0 if and only if there is a homomorphism from a free group F to G which induces an isomorphism from $F/[F, X^2(F)]X^4(F)$ to $G/[G, X^2(G)]X^4(G)$ [**Hi87**].

If p is odd then $G(\mathbb{F}_p) = G_2(p) = G_2 X^p(G)$ and $\overline{G} = G/G(\mathbb{F}_p) \cong (Z/pZ)^b$, say. The cohomology ring $H^*(\overline{G}; \mathbb{F}_p)$ is the tensor product of an exterior algebra generated in degree 1 by $Hom(G, \mathbb{F}_p)$ with a polynomial algebra generated in degree 2 by $\beta_p(H^1(\overline{G}; \mathbb{F}_p))$, where β_p is the *mod-p* Bockstein homomorphism. In this case we have $\mathrm{Ker}(\cup_G) \cong Hom(G_2(p)/[G, G_2(p)], \mathbb{F}_p)$. Cup product is identically 0 if and only if there is a homomorphism from a free group F to G which induces an isomorphism from $F/\Gamma_3(F; p)$ to $G/\Gamma_3(G; p)$, where $\Gamma_3(G; p)$ is the subgroup generated by $X^p(G) \cup (G_3)$ [**Hi85'**].

If $f : G \to H$ is a homomorphism such that $H^1(f; \mathbb{F}_p)$ is an isomorphism then there are Massey product criteria (analogous to those cited in §2 above) for f to induce an isomorphism $G/\Gamma_k(G; p) \cong H/\Gamma_k(H; p)$ for $k > 3$ [**Dw75**].

12.4. The graded Lie algebra of a group

Let G be a group. In any nilpotent group the subset of elements of finite order forms a normal subgroup. If the group is finitely generated this subgroup is finite. Let $G_1^{\mathbb{Q}} = G$ and let $G_{k+1}^{\mathbb{Q}}$ be the preimage in G of the torsion subgroup of $G/[G, G_k^{\mathbb{Q}}]$. Then $G/G_k^{\mathbb{Q}}$ is a torsion free nilpotent group, and $\{G_k^{\mathbb{Q}}\}_{k \geq 1}$ is the most rapidly descending series of subgroups of G with this property. (See [**Rob**].)

If $g \in G_m$ and $h \in G_n$ then $[g, h] = ghg^{-1}h^{-1} \in G_{m+n}$, while the Hall-Witt identity $[yxy^{-1}, [z, y]][zyz^{-1}, [x, z]][xzx^{-1}, [y, x]] = 1$ holds for any three elements x, y, z of G. The *graded Lie algebra* associated to G is $\mathcal{L}(G) = \oplus_{k \geq 1}(G_k/G_{k+1})$, with the Lie bracket determined by the commutator. (The Jacobi relation follows from the Hall-Witt identity.) Since $\mathbb{Q} \otimes_{\mathbb{Z}} G_k/G_{k+1} \cong \mathbb{Q} \otimes_{\mathbb{Z}} G_k^{\mathbb{Q}}/G_{k+1}^{\mathbb{Q}}$, the tensor product $\mathbb{Q} \otimes_{\mathbb{Z}} \mathcal{L}(G)$ is determined by the rational lower central series $\{G_k^{\mathbb{Q}}\}$. (The restricted p-lower central series determines an analogous p-Lie algebra $\mathcal{L}(G; \mathbb{F}_p) = \oplus_{k \geq 1}(\Gamma_k(G; p)/\Gamma_{k+1}(G; p))$, which inherits a pth power operation from G.)

THEOREM. [**Ma49**] *Let $\mathcal{L}$ be a nilpotent graded Lie algebra which is generated in degree 1. Let $G_{\mathcal{L}}$ be the group whose underlying set is $\mathbb{Q} \otimes_{\mathbb{Z}} \mathcal{L}$ and whose multiplication is given by the Campbell-Baker-Hausdorff formula. Then $G_{\mathcal{L}}$ is nilpotent and $\mathcal{L}(G_{\mathcal{L}}) \cong \mathbb{Q} \otimes_{\mathbb{Z}} \mathcal{L}$. If moreover $\mathcal{L} \cong \mathcal{L}(G)$ for some finitely generated nilpotent group G then there is a homomorphism $f : G \to G_{\mathcal{L}(G)}$ such that $\mathbb{Q} \otimes_{\mathbb{Z}} \mathcal{L}(f)$ is an isomorphism.* □

The tower of groups constructed in this way from the truncations of $\mathbb{Q} \otimes \mathcal{L}(G)$ is called the $\mathbb{Q}$-*Mal'cev completion* of G. The denominators in the CBH formulae for a nilpotent Lie algebra of nilpotency class k divide $k!$, and Mal'cev's Theorem can be refined to show that the Lie algebra determines G/G_k mod p-torsion for primes $p \leq k$ [**CP84**].

Let $Gr(\mathbb{Z}[G]) = \oplus_{n \geq 0}(I_G^n/I_G^{n+1})$ be the graded algebra determined by the I_G-adic filtration of $\mathbb{Z}[G]$. Let $L(A)$ and $T_{\otimes}(A)$ be the free Lie algebra and free tensor algebra generated by the abelian group A, and let U be the universal envelope functor from graded Lie algebras to graded associative algebras. There are canonical homomorphisms $\alpha : L(H_1(G; \mathbb{Z})) \to \mathcal{L}(G)$, $\beta : T_{\otimes}(H_1(G; \mathbb{Z})) \to Gr(\mathbb{Z}[G])$ and $\gamma : U\mathcal{L}(G) \to Gr(\mathbb{Z}[G])$. If $G \cong F(\mu)$ these homomorphisms are isomorphisms, by results of Witt and Magnus; in general they are not bijective.

Grünenfelder has constructed (for any subring R of $\mathbb{Q}$) two spectral sequences $E(R, G)$ and $\widetilde{E}(R, G)$, with E^1 page determined by $H_*(G; R)$, and a homomorphism $\kappa : E(R, G) \to \widetilde{E}(R, G)$, which converge (in an appropriate sense) to $R \otimes_{\mathbb{Z}} \mathcal{L}(G)$, $Gr(R[G])$ and $\gamma|_{R \otimes \mathcal{L}(G)}$, respectively [**Gr80**]. (See also [**Wa80**].)

12.5. DGAs and minimal models

The rational de Rham algebra of a finite simplicial complex P is the algebra $\mathcal{A}(P)$ generated by forms in the barycentric coordinates, with polynomial coefficients. The cohomology of this algebra is isomorphic to the rational cohomology of P, but the algebra carries also deeper information about the rational homotopy type of P.

A *connected differential graded algebra* $\mathcal{A} = (A, d)$ over $\mathbb{Q}$ (or *DGA*) is an associative $\mathbb{Q}$-algebra $A = \oplus_{n \geq 0} A^n$ with multiplication $\wedge$ and differential d of degree $+1$ such that $A^0 = \mathbb{Q}$, $x \wedge y = (-1)^{pq} y \wedge x$, $d^2 = 0$ and $d(x \wedge y) = (dx) \wedge y + (-1)^p x \wedge dy$, for all $x \in A^p$ and $y \in A^q$. The augmentation ideal is $A^+ = \oplus_{n > 0} A^n$ and the *space of indecomposables* is $I(A) = A^+ / A^+ \wedge A^+$. The *DGA* is *minimal* if dx is a sum of products of terms of degree $\leq p$, for all $x \in A^p$. (If A is generated as an algebra by A^1 then $\mathcal{A}$ is minimal.) We may define Massey products in the cohomology of $\mathcal{A}$ just as before.

A *Hirsch extension* of degree 1 of a *DGA* $\mathcal{A}$ is a *DGA* $\mathcal{B}$ such that the underlying algebra is a graded tensor product $B = A^* \otimes_{\mathbb{Q}} \wedge^* V$, for some vector space V, and such that $dV \leq A^1$. (In particular, $B^1 = A^1 \oplus V$.) The extension is *finite* if V is finite dimensional.

If $\mathcal{A}$ is a *DGA* such that $A^2 = A^1 \wedge A^1$ the dual of $d : A^1 \to A^2$ induces a Lie bracket on $L = Hom_{\mathbb{Q}}(A^1, \mathbb{Q})$. Conversely, if L is a finite dimensional $\mathbb{Q}$-Lie algebra the dual of the bracket defines a graded differential on $L^* = \wedge_k Hom_{\mathbb{Q}}(L, \mathbb{Q})$, such that $d\lambda(m, n) = \lambda([m, n])$ for $\lambda \in L^1$ and $m, n \in L$. (In each case the Jacobi identity and the property "$d^2 = 0$" are dual.) The Lie algebra is nilpotent if and only if $\mathcal{A}$ may be obtained by iterated finite Hirsch extensions from the trivial algebra $\mathbb{Q}$.

A 1-*minimal model* for $\mathcal{A}$ is a *DGA*-morphism $\rho : \mathcal{M} \to \mathcal{A}$ such that $\mathcal{M}$ is generated in degree 1, $H^0(\rho)$ and $H^1(\rho)$ are isomorphisms, and $H^2(\rho)$ is injective. We may construct the 1-minimal model for $\mathcal{A}$ as an increasing union of Hirsch extensions of degree 1. Let $M_1 = \wedge^* V$, where $V = H^1(\mathcal{A})$, with differential 0. Choose a splitting $: V \to \text{Ker}(d|_{A^1})$, and let $\rho_1 : M_1 \to \mathcal{A}$ be the homomorphism extending this splitting. Assuming $\mathcal{M}_k$ and ρ_k have been defined, let $V_{k+1} = \text{Ker}(H^2(\rho_k))$. Let $M_{k+1} = M_k \otimes \wedge^* V_{k+1}$ and choose a linear homomorphism $d : V_{k+1} \to Z^2(\mathcal{M}_k)$ lifting the inclusion $V_{k+1} \leq H^2(\mathcal{M}_k)$, to give $\mathcal{M}_{k+1}$. We then choose a linear homomorphism $\rho_{k+1} : V_{k+1} \to A^1$ such that $d\rho_{k+1}(v) = \rho_k(dv)$, for all $v \in V_{k+1}$, and extend as a *DGA*-morphism from $\mathcal{M}_{k+1}$ to $\mathcal{A}$. The 1-minimal model of a finitely presentable group G is the 1-minimal model of $\mathcal{A}(P)$, where P is a finite simplicial complex with $\pi_1(P) \cong G$.

THEOREM. [**Su77**] *The 1-minimal model of a connected DGA is well-defined up to isomorphism. If G/G' is finitely generated and the quotients G/G_k are all torsion free then $\mathbb{Q} \otimes \mathcal{L}(G)$ is isomorphic to the Lie algebra of the 1-minimal model of G.*   □

The 1-minimal model of a group G determines $H^1(G; \mathbb{Q})$ and its Massey product structure; to what extent is the converse true? Suppose that $\mathcal{M}$ and $\mathcal{N}$ are DGAs which are generated in degree 1 and $h^s : H^s(\mathcal{N}) \to H^s(\mathcal{M})$ are homomorphisms for $s = 1$, 2 such that h^1 is an isomorphism, $\langle f_1, \ldots, f_n \rangle$ is defined if and only if $\langle h^1(f_1), \ldots, h^1(f_n) \rangle$ is defined and $h^2(\langle f_1, \ldots, f_n \rangle)$ is a subset of $\langle h^1(f_1), \ldots, h^1(f_n) \rangle$. Are h^1 and h^2 induced by a DGA-morphism $h : \mathcal{N} \to \mathcal{M}$?

A space X is *formal* if there is a DGA-morphism from the minimal model of X to $H^*(X; \mathbb{Q})$ (considered as a DGA with trivial differential) which induces an isomorphism on cohomology. (Thus its rational homotopy type is determined by its cohomology ring.) A connected, graded commutative $\mathbb{Q}$-algebra $\mathcal{H}^*$ is *intrinsically formal* if any space X such that $H^*(X; \mathbb{Q}) \cong \mathcal{H}^*$ is formal. For the study of classical links a variation is more useful. The algebra $\mathcal{H}^*$ is *1-intrinsically formal* if the existence of a 2-connected homomorphism from $\mathcal{H}^*$ to $H^*(G; \mathbb{Q})$ determines $\mathbb{Q} \otimes_{\mathbb{Z}} \mathcal{L}(G)$ up to isomorphism.

Let L be a μ-component 1-link with group $\pi = \pi L$ and let Γ be the $\mathbb{Z}$-weighted graph with μ vertices and with an edge of weight ℓ_{ij} joining the ith and j^{th} vertices if and only if $\ell_{ij} \neq 0$. Then $H^*(X(L); \mathbb{Q})$ is 1-intrinsically formal if and only if Γ is connected [**MP92**]. (In particular, the complements of algebraic links in S^3 are formal.) The integral cohomology ring determines $\mathcal{L}(\pi)$ if the graphs Γ_p obtained by reducing the weights *mod* (p) are connected for all primes p. (See also [**BP94**].) In this case $\mathcal{L}(\pi) \cong L^*(\pi/\pi')/\text{Im}(\partial)$, where $L^*(\pi/\pi')$ is the free graded Lie algebra generated in degree 1 by $\pi/\pi' = H_1(\pi; \mathbb{Z})$ and ∂ is the dual of the cup product from $H^1(\pi; \mathbb{Z}) \wedge H^1(\pi; \mathbb{Z})$ to $H^2(\pi; \mathbb{Z})$.

The 1-minimal model approach can be extended to the derived series in terms of twisted Hirsch extensions [**Su77**].

12.6. Free derivatives

As observed in §1 of Chapter 11, the augmentation ideal of the group ring $\Gamma_\mu = \mathbb{Z}[F(\mu)]$ is freely generated by $\{x_1 - 1, \ldots, x_\mu - 1\}$ as a left Γ_μ-module. Thus if $w \in F(\mu)$ we may write

$$w - 1 = \Sigma_{1 \le i \le \mu} \delta_i(w)(x_i - 1),$$

where the coefficients $\delta_i(w) \in \Gamma_\mu$ are determined by w. Clearly $\delta_i(1) = 0$ for $1 \le i \le \mu$. Moreover, since $vw - 1 = v - 1 + v(w - 1)$, for all $v, w \in F(\mu)$, the Leibniz conditions

$$\delta_i(vw) = \delta_i(v) + v\delta_i(w)$$

hold for all $v, w \in F(\mu)$ and $1 \le i \le \mu$. Hence $\delta_i(w^{-1}) = -w^{-1}\delta_i(w)$. We may extend these functions linearly to "derivations" $\delta_i : \Gamma_\mu \to \Gamma_\mu$. The notation $\frac{\partial w}{\partial x_i} = \delta_i(w)$ is also used.

Now let π be a group with a finite presentation

$$\langle x_1, \ldots, x_g | w_1, \ldots, w_r \rangle^\phi,$$

where $\phi : F(g) \to \pi$ is an epimorphism with kernel normally generated by $\{w_1, \ldots, w_r\}$. Extend ϕ linearly to the group rings. Let X be the 2-complex corresponding to this presentation. Then $C_q(\widetilde{X})$ is a free left $\mathbb{Z}[\pi]$-module with basis given by a fixed choice of lifts of the q-cells of X. It is not hard to see that the differentials are given by $\partial_1(c_1^{(i)}) = (\phi(x_i) - 1)c_0$ and $\partial_2(c_2^{(j)}) = \Sigma_{1 \le i \le g}\delta_i(w_j)c_1^{(i)}$. (See Chapter 9 of [**BZ**].) We may identify the module of 1-cycles $Z_1(\widetilde{X})$ with the left $\mathbb{Z}[\pi]$-module $I(\pi)$. Thus if we view the free modules $C_q(\widetilde{X})$ as modules of column vectors then $I(\pi)$ has a $g \times r$ presentation matrix with (i, j)th entry $\phi(\frac{\partial w_j}{\partial x_i})$. On composing ϕ with the abelianization, we obtain a presentation for $A(\pi)$ as a $\mathbb{Z}[\pi/\pi']$-module.

Augmentation ideals are also right ideals, and, similarly,

$$w - 1 = \Sigma_{1 \le i \le \mu}(x_i - 1)\delta_i'(w),$$

where $\delta_i'(vw) = \delta_i'(v)w + \delta_i'(w)$, for all $v, w \in F(\mu)$ and $1 \le i \le \mu$. Using the involution of the group ring to switch from right to left, we find that $\delta_i'(w) = -x_i^{-1}\delta_i(w^{-1}) = x_i^{-1}w\delta_i(w)$. These lead to a presentation matrix for $I(\pi)$ as a right module.

12.7. Milnor invariants

Let L be a 2-component 1-link with linking number ℓ, and let $X_i = X(L_i)$, for $i = 1$, 2. The inclusion $j_1 : \partial X_1 \to X$ induces an isomorphism $H_2(j_1; \mathbb{Z})$, since $H_2(X, \partial X_1; \mathbb{Z}) \cong H^1(X, \partial X_2; \mathbb{Z}) = 0$. Let $ab : X \to T = K(\mathbb{Z}^2, 1)$ be the map corresponding to the abelianization homomorphism. Since the composition $ab \circ j_1$ has degree ℓ, it follows that if $\{z_1, z_2\}$ is the basis for $H^1(X; \mathbb{Z})$ which is Kronecker dual to the meridians $\{x_1, x_2\}$ and ξ is the generator of $H^2(X; \mathbb{Z})$ which is Lefschetz dual to the class of a path from L_1 to L_2 in $H_1(S^3, L; \mathbb{Z})$ then $z_1 \cup z_2 = \ell \xi$.

This connection between cup product and linking numbers extends to a relationship between Massey products and Milnor invariants, which are "higher linking numbers" defined in terms of the coefficients of Magnus expansions of words representing the longitudes. The main results of this section are from [**Mi57**].

If $I = (i_1, \ldots i_r)$ is a multi-index of length $|I| = r \geq 1$, with $1 \leq i_j \leq \mu$ for all $1 \leq j \leq r$, and $u \in F(\mu)$, let

$$X_I = X_{i_1} \ldots X_{i_r} \quad \text{and} \quad \varepsilon_I(u) = \varepsilon(\delta_{i_1} \ldots \delta_{i_r}(u)),$$

and let $X_\emptyset = 1$ and $\varepsilon_\emptyset(u) = 0$ (rather than 1). Then the image of u under the Magnus imbedding is

$$M(u) = 1 + \Sigma_{|I| \geq 1} \varepsilon_I(u) X_I.$$

Let $\widehat{I_{F(\mu)}}$ be the 2-sided ideal generated by $\{X_1, \ldots, X_\mu\}$ in $\widehat{\Gamma_\mu}$.

LEMMA 12.1. *If $u \in F(\mu)$ then $u \in F(\mu)_k$ if and only if $M(u) \equiv 1$* $mod \ \widehat{I_{F(\mu)}}^k$.

PROOF. An easy induction on k shows that if $u \in F(\mu)_k$ then $M(u) \equiv 1 \ mod \ (I_\mu)^k$. The other implication is more difficult. See Chapter V of [**MKS**]. $\square$

Suppose now that L is a μ-component 1-link, and let $\theta : F(\mu) \to \pi = \pi L$ be a basing for L. Let $k \geq 1$ and let $\{w_1, \ldots, w_\mu\}$ be words in $F(\mu)$ representing the images of the longitudes in π/π_k. If $I = (i_1, \ldots, i_r)$ has length $r \leq k$ let $I' = (i_1, \ldots, i_{r-1})$, and let $\mu(I) = \varepsilon_{I'}(w_{i_r})$. (Thus $w_j = 1 + \Sigma_{|I|>0} \mu(Ij) X_I$, for $1 \leq j \leq \mu$.) Let

$\Delta(I) \leq \mathbb{Z}$ be the ideal generated by $\{\mu(J)\}$, where J runs over the cyclic permutations of proper subsequences of I. The image of $\mu(I)$ in $\mathbb{Z}/\Delta(I)$ is the *Milnor invariant*

$$\bar{\mu}(i_1, \ldots, i_r) = \bar{\mu}(I).$$

In particular, $\bar{\mu}(i, j) = \mu(i, j) = \ell_{ij}$ for all $1 \leq i, j \leq \mu$.

THEOREM 12.2. *Let L be a μ-component 1-link and $I = (i_1, \ldots, i_r)$ be a multi-index with $1 \leq i_j \leq \mu$ for all $1 \leq j \leq r$. Then*

(1) *$\bar{\mu}(I)$ is an invariant of the I-equivalence class of L;*
(2) *reflecting L (changing the ambient orientation) multiplies $\bar{\mu}(I)$ by $(-1)^{r-1}$, while reversing the component L_j multiplies $\bar{\mu}(I)$ by $(-1)^s$, where s is the number of times j occurs in the sequence I.*

PROOF. (1). Comparing the coefficients of monomials on either side of the equation $M(x_i w x_i^{-1}) = (1 + X_i)M(w)\Sigma(-1)^m X_i^m$ gives $\varepsilon_I(x_i w x_i^{-1}) \equiv \varepsilon_I(w)$ modulo the ideal generated by the $\varepsilon_J(w)$, for certain proper subsequences J of I. If $u \in F(\mu)_{|I|+1}$ then $\varepsilon_I(uw) = \varepsilon_I(w)$, since $M(u) - 1 \in \widehat{I_{F(\mu)}}^{|I|+1}$, by Lemma 12.1.

Suppose $J = (i_1, \ldots, i_s)$ for some $1 \leq s < r$ is an initial segment of I'. Then $\varepsilon_J(w_k x_k) = \varepsilon_J(w_k)$ if $k \neq i_s$ and $\varepsilon_J(w_{i_s} x_{i_s}) = \varepsilon_J(w_{i_s}) + \mu(J) = \mu(Ji_s) + \mu(J)$. Similarly $\varepsilon_J(x_k w_k) = \varepsilon_J(w_k)$ if $k \neq i_1$ and $\varepsilon_J(x_{i_1} w_{i_1}) = \varepsilon_J(w_{i_1}) + \mu(\tilde{J})$, where $\tilde{J} = (i_2, \ldots, i_s, i_1)$. Let $v = [w_k, x_k] = w_k x_k w_k^{-1} x_k^{-1}$. Since

$$M(v) = 1 + (M(w_k x_k) - M(x_k w_k))M(w_k^{-1})M(x_k^{-1})$$

it follows that $\varepsilon_J(v)$ is in $\Delta(I)$ for all initial segments J of I'. (It is here that cyclic permutations are needed.) Now $\varepsilon_{I'}(v w_{i_r}) = \varepsilon_{I'}(w_{i_r}) + \Sigma \varepsilon_J(v)\mu(K i_r) + \varepsilon_{I'}(v)$, where the sum is over ordered partitions $I' = (J, K)$ with J and K nonempty. Hence $\varepsilon_{I'}(v w_{i_r}) \equiv \varepsilon_{I'}(w_{i_r})$ *mod* $\Delta(I)$.

Hence $\bar{\mu}(I)$ is unchanged if w_{i_r} is conjugated or multiplied by an element of $F(\mu)$ representing 1 in π/π_r. In particular, it does not depend on the value of $k \geq r$. If $\mathcal{L}$ is an I-equivalence from L to L' then the inclusions of $X(L)$ and $X(L')$ into $X(\mathcal{L})$ satisfy the hypotheses of Theorem 1.3, and the ith longitudes of L and L' are conjugate in $\pi\mathcal{L} = \pi_1(X(\mathcal{L}))$. Hence the $\bar{\mu}$ invariants of L and L' are equal.

(2). Since reflection inverts the meridians of L, while reversing L_j inverts the jth meridian and the jth longitude, this follows easily from the fact that $M(x_i^{-1}) = 1 - X_i$ modulo higher order terms. $\square$

Lemma 12.1 and Theorem 1.4 together imply the following result.

THEOREM 12.3. *Let L be a μ-component 1-link with group $\pi = \pi L$ and let $\theta : F(\mu) \to \pi$ be the homomorphism determined by a set of meridians. All Milnor invariants of L of length $< r$ are 0 if and only if θ induces an isomorphism $F(\mu)/F(\mu)_r \cong \pi/\pi_r$.* $\square$

Cochran and Orr have conjectured that if all Milnor invariants of L of length $< r$ are 0 then all Milnor invariants of length up to $2r$ may be defined with zero indeterminacy.

A *proper shuffle* of two sequences I and J of lengths r and s, respectively, is one of the $\binom{r+s}{r}$ sequences obtained by intermeshing I with J. Let $Sh(IJ)$ be the set of such proper shuffles.

THEOREM 12.4. *Let L be a μ-component 1-link. Then*

(1) $\bar{\mu}(i_1, \ldots, i_r) = \bar{\mu}(i_2, \ldots, i_r, i_1) = \cdots = \bar{\mu}(i_r, i_1, \ldots, i_{r-1})$;
(2) *if I and J are given multi-indices of length > 0 then*
$$\Sigma_{H\in Sh(IJ)}\bar{\mu}(Hk) \equiv 0 \ mod \ h.c.f.\{\Delta(Hk) \mid H \in Sh(IJ)\};$$
(3) $\bar{\mu}(i_1, \ldots, i_r)$ *is determined by the sublink $\cup_{1\leq k\leq r}L_{i_k}$;*
(4) *if the indices $i_1, \ldots, i_r$ are all distinct then $\bar{\mu}(i_1, \ldots, i_r)$ depends only on the link homotopy class of L.*

PROOF. Fix $q > r$. The quotient π/π_q has a presentation $\langle x_i, 1 \leq i \leq \mu \mid [w_i, x_i] = 1, 1 \leq i \leq \mu, F(\mu)_q \rangle$, where the w_i represent the longitudes, and there are words $y_i \in F(\mu)$ such that $\Pi y_i[w_i, x_i]y_i^{-1}$ is in $F(\mu)_q$, by Theorem 1.4. Let D be the ideal in $\mathbb{Z}\langle\langle X_1, \ldots, X_\mu\rangle\rangle$ of elements $\Sigma d_I X_I$ such that $\Delta(I)$ divides d_I for all $|I| \leq q$. Write $M(w_i) = 1 + v_i$, for $1 \leq i \leq \mu$. Then $X_j X_i v_i$, $X_j v_i X_i$, $X_i v_i X_j$ and $v_i X_i X_j$ are all in D, since Ii is a cyclic permutation of a proper subsequence of each of jiI, jIi, iIj and Iij, respectively. It follows that $M(y_i[w_i, x_i]y_i^{-1}) \equiv 1 + X_i v_i - v_i X_i \ mod \ D$, for $1 \leq i \leq \mu$, and so

$1 + \Sigma(X_i v_i - v_i X_i)$ is in D. Now the coefficient of X_{iJ} in this sum is $\mu(Ji) - \mu(iJ)$, and so $\mu(Ji) \equiv \mu(iJ) \bmod \Delta(iJ)$, provided $|iJ| \leq q$. This implies (1).

Property (2) follows from general facts about the coefficients of the Magnus expansion of a word in $F(\mu)$ [**CFL58**], while (3) is clear.

To establish homotopy invariance it shall be sufficient to consider an elementary homotopy affecting only one component. We may assume that $\{i_1, \ldots, i_r\} = \{1, \ldots \mu\}$ (so $r = \mu$), by (3), and that the homotopy changes only the last component and $i_\mu = \mu$, by (1). In the construction of Theorem 1.4 we may choose the arcs α_i with $i < \mu$ so that the homotopy takes place in $W(\hat{\mu}) = X(L(\hat{\mu})) \setminus \cup_{i<\mu} \alpha_i$. Since $\bar{\mu}(I)$ depends only on the image of the μth longitude in $\pi_1(W(\hat{\mu}))$ it is unchanged under homotopy of L_μ. $\square$

These cyclic symmetries and shuffle relations are a complete set of relations [**HL98**]. The "indeterminacy" $\Delta(I)$ may be reduced under additional hypotheses. (See Chapter 14.) On the other hand, part (4) of Theorem 12.4 may be extended somewhat at the cost of coarsening the indeterminacy. Let $\Delta^*(I)$ be the ideal generated by $\{\mu(J)\}$, where J runs over *all* permutations of proper subsequences of I, and let $\mu^*(I)$ be the image of $\mu(I)$ in $\mathbb{Z}/\Delta^*(I)$. Then $\Delta(I) \leq \Delta^*(I)$ and $\mu^*(I)$ is a homotopy invariant for all multi-indices I with distinct indices. Moreover $\mu^*(i_r, \ldots, i_1) = (-1)^r \mu^*(i_1, \ldots, i_r)$. (See [**Mi57**].)

COROLLARY 12.4.1. *Let* $I = (i_1, \ldots, i_r)$ *where* $r > 2$ *and* $i_k = i$ *for at least* $r - 1$ *values of* k, *while the remaining index is* j. *Then* $\binom{r-1}{s}\bar{\mu}(I) = 0$ *for all* $0 < s < r$. *If moreover* $\ell_{ij} = 0$ *then* $\bar{\mu}(I) = 0$. $\square$

Turaev showed that the choices of defining systems for Massey products in link complements could be normalized in a way that reduced the indeterminacy and hence that the Milnor invariants are essentially equivalent to Massey products [**Tu76**]. Let $x_1, \ldots, x_\mu$ be a set of meridians for a μ-component 1-link L. A *normalized Massey system* for $(X(L); x_1, \ldots, x_\mu)$ is a system of ideals $\nu(I)$ and cochains $w(I) \in C^1(X(L); \mathbb{Z}/\nu(I))$, where I varies over multi-indices of length

$r \geq 1$, which are required to satisfy the following conditions. Firstly, $\nu(I)$ is the smallest ideal containing $\cup\{\nu(I') \mid I' < I\}$ (taken over all proper subsequences I') and such that $\Sigma_{I=JK}w(J) \cup w(K)$ is a coboundary in $C^2(X(L); \mathbb{Z}/\nu(I))$ (where the summation is taken over all nonempty proper initial segments J). Secondly, this sum is the coboundary of $w(I)$. (In [**Tu76**] it is the coboundary of $-w(I)$.) Thirdly, $w(i)(x_i) = 1$ for $1 \leq i \leq \mu$ and $w(I)(x_j) = 0$ in all other cases. (We have omitted the reduction of coefficents homomorphisms $\mathbb{Z}/\nu(J) \to \mathbb{Z}/\nu(I)$ from the notation.)

It is easily shown by induction on $|I|$ that normalized Massey systems exist. The class $m(I)$ of $\Sigma_{I=JK}w(J)\cup w(K)$ in $H^2(X; \mathbb{Z}/\Delta(I))$ is the corresponding *normalized Massey product* $\langle z_{i_1}, \ldots, z_{i_r}\rangle$, where $\{z_1, \ldots, z_\mu\}$ is the basis for $H^1(X; \mathbb{Z})$ Kronecker dual to the meridians. (This differs in sign from the definition in [**Tu76**].)

THEOREM. [**Tu76**] *Let L be a μ-component 1-link and let I be a multi-index of length r. Let η_i be the image of the fundamental class of $\partial X(L_i)$ in $H_2(X(L); \mathbb{Z}/\Delta(I))$. Then*

(1) *the ideals $\nu(I)$ are independent of the choice of meridians for L;*

(2) *the ideal generated by all $\nu(I')$ with I' a proper subsequence of I is $\Delta(I)$;*

(3) *if $i_1 \neq i_r$ then $m(I)(\eta_{i_r}) = -m(I)(\eta_{i_1}) = (-1)^r\bar{\mu}(I)$, and $m(I)(\eta_j) = 0$ otherwise.* □

Turaev also proves a corresponding result for links in Z/mZ-homology 3-spheres, for any $m \geq 0$. (He defines analogues of the $\bar{\mu}$ invariants after extending Milnor's Theorem 1.4 to give presentations for the quotients of the most rapidly descending central series whose sections have exponent m. See also [**Po80, St90**], and see [**Ko83**] for an application to 4-manifolds.)

All the Milnor invariants of L of length at most $2k$ vanish if and only if L bounds μ disjoint properly embedded oriented surfaces $V_i \subset D^4$ such that the image of $\pi_1(V_i)$ in $G = \pi_1(D^4 \setminus \cup V_i)$ under the homomorphisms induced by pushing V_i off itself in the normal direction lies in the subgroup G_k, for $1 \leq i \leq \mu$ [IO01]. (Compare also the results of [FT95] quoted above.)

12.8. Link homotopy and the Milnor group

Let G be a group which is normally generated by $\{x_1, \ldots, x_\mu\}$. The *Milnor group* $Mil(G)$ is the largest quotient of G in which each of these generators commutes with all of its conjugates. If A_i is the normal closure of $\langle x_i \rangle$ in $Mil(G)$ then A_i is an abelian normal subgroup, and $Mil(G) = \Pi A_i$. Therefore $Mil(G)$ is nilpotent of class at most μ, by Fitting's Theorem (Proposition 5.2.8 of [Rob]). Hence it is generated by the images of the elements $\{x_1, \ldots, x_\mu\}$ and so is a quotient of the "reduced free group" $RF(\mu) = Mil(F(\mu))$. It is easily verified that the images in $RF(3)$ of $x_1 x_2$ and its conjugate by x_3 do not commute. Thus in general the Milnor group depends on the generating set. (The structure of $RF(\mu)$ is described in [Hu93].)

The Milnor group of a 1-link L is the Milnor group of $\pi = \pi L$ with respect to a set of meridians. It is the largest common quotient of the groups of links which are link homotopic to L. Milnor showed that this group together with a certain "peripheral structure" formed a link homotopy invariant, which determines the $\bar{\mu}(i_1, \ldots, i_r)$ with all indices distinct. Moreover L is homotopically trivial if and only if all such $\bar{\mu}$ invariants are 0 if and only if all the longitudes of L have trivial image in $Mil(\pi)$. In particular, boundary 1-links are homotopically trivial. (See [CF88] for a direct geometric proof.)

A *colouring* of a link is a partition into sublinks. Freedman and Teichner considered coloured homotopies, in which components of the same colour are allowed to cross, and defined the coloured Milnor group, in which all conjugates of meridians of the same colour commute. This is again a nilpotent group, with nilpotency class bounded by the number of colours.

Let N_i be the subgroup of $Mil(\pi)$ generated by $\{[g, [g, x_i]] \mid g \in Mil(\pi)\}$. (Thus $N_i = I(\pi)^2 A_i$.) Levine has conjectured that the link homotopy type of L is determined by the isomorphism class of $(Mil(\pi), ((m_1, n_1), \ldots, (m_\mu, n_\mu)))$, where m_i is the image of an ith

meridian in $Mil(\pi)$ and n_i is the image of the corresponding longitude in $Mil(\pi)/N_i$, and has verified this for $\mu \leq 4$. Hughes has given examples of links distinguished by this invariant but not distinguished by Milnor's original formulation of the peripheral structure (which used instead the images of the n_i in $Mil(\pi)/A_i$).

Habegger and Lin have shown that the set of link homotopy classes of μ-component 1-links is the orbit space of a certain group action on the group of link homotopy classes of μ-component string links, and have thereby given an algorithmic classification of links up to link homotopy. (See Chapter 14.)

Krushkal and Teichner have interpreted link homotopies in terms of gropes.

THEOREM 12.5. [**KT97**] *Two μ-component 1-links $L(0)$ and $L(1)$ are link homotopic if and only if they bound μ disjointly immersed annular gropes of class μ in $S^3 \times I$.*

PROOF. (Sketch.) The condition is clearly necessary, since a link homotopy gives disjointly immersed (level-preserving) annuli.

Let $\{G_i\}_{1 \leq i \leq \mu}$ be a set of disjoint immersions of annular gropes of class μ connecting $L(0)$ and $L(1)$ in $S^3 \times [0,1]$. Choose an arc γ in G_1 from $L(0)_1$ to $L(1)_1$. Then the concatenation $c = L(0)_1.\gamma.L(1)_1.\overline{\gamma}$ is in $\pi_1(S^3 \times [0,1] \setminus \amalg_{2 \leq i \leq \mu} G_i)$. After modifying the immersions $\{G_i\}_{i=2}^{i=\mu}$ by finger moves we may assume that every meridian of $\amalg_{i=2}^{i=\mu} G_i$ commutes with its conjugates in $\pi_1(S^3 \times [0,1] \setminus \amalg_{i=2}^{i=\mu} G_i)$, and so this group is nilpotent of class $\mu - 1$. Therefore c bounds a disc in $S^3 \times [0,1] \setminus \amalg_{i=2}^{i=\mu} G_i$. Hence $L(0)_1$ and $L(1)_1$ jointly bound an immersed annulus A_1 which is disjoint from $\amalg_{i=1}^{i=2\mu} G_i$. Repeating this argument gives μ disjointly immersed annuli connecting $L(0)$ and $L(1)$. The theorem now follows from "(singular) concordance implies link homotopy" as in [**Gi79, Go79**]. □

The surgery equivalence class of a link is determined by its Milnor invariants of length at most 3 (with a sharpened indeterminacy). In particular, two links whose linking numbers are all 0 are surgery equivalent if and only if their corresponding 3-component sublinks are link homotopic [**Le87'**].

12.9. Variants of the Milnor invariants

Cochran uses intersections of surfaces to define and compute invariants which tend to have smaller indeterminacy than the Milnor invariants. Moreover they are particularly well suited to the construction of examples. For instance, the Bing double of the Borromean rings (which is a 6-component link) has trivial Alexander module but is not even homotopic to a boundary link, since $\bar\mu(1,2,3,4,5,6) \neq 0$. Many other examples demonstrate quite convincingly that one cannot expect criteria for concordance to a boundary link in terms of the Alexander module alone. (See [Coc] for a detailed exposition.)

In particular, there are nontrivial μ-component links L such that $\pi = \pi L$ is generated by μ meridians and $\alpha(L) = \mu$, for each $\mu \geq 6$ [Co96]. This answers a question of Birman, who observed that the closure of a pure braid in the kernel of the Gassner representation is of this kind. (See page 130 of [Bir].) Since L is nontrivial π is a proper quotient of $F(\mu)$, and so some Milnor invariant must be nontrivial. Are there such links for $2 \leq \mu \leq 5$? (It is known that the Gassner representation is faithful for $\mu \leq 3$.)

Let L be a μ-component 1-link such that all Milnor invariants of length $\leq r$ are 0. A basing $\theta : F(\mu) \to \pi = \pi L$ determines an isomorphism $F(\mu)/F(\mu)_r \cong \pi/\pi_r$ and hence an epimorphism from π to $F(\mu)/F(\mu)_r$, which may be realized by a map $f : X(L) \to K(F(\mu)/F(\mu)_r, 1)$. Since the longitudes are all in π_r this extends to a map from S^3 to the space $K_r(\mu)$ obtained from $K(F(\mu)/F(\mu)_r, 1)$ by adjoining μ 2-cells along loops representing a basis of $F(\mu)$. Links with all $\bar\mu$ invariants of length $\leq r$ zero and realizing a given element of $\pi_3(K_r(\mu))$ may be constructed by taking the transverse inverse image of the midpoints of the 2-cells. These *Orr invariants* $O_r(L; \theta) \in \pi_3(K_r(\mu))$ are invariants of based link concordance, and are all defined and 0 if and only if all the $\bar\mu$ invariants are 0. (See [Or87, Co87].)

Instead of using the Magnus embedding, the Campbell-Baker-Hausdorff formulae may be used to embed $F(\mu)$ in the "free complete Lie algebra on $x_1, \ldots, x_\mu$", leading to concordance invariants similar to the $\bar{\mu}$ invariants [**Pa97**]. At the primary level these invariants agree with the $\bar{\mu}$ invariants, but in general they are independent.

Another scheme for obtaining presentations of the nilpotent quotients of link groups is based on Chen's theory of iterated path integrals and power series connections [**Ha85**].

The Milnor invariants of L may be interpreted as linking numbers in certain finite nilpotent branched coverings of (S^3, L). More precisely, suppose $I = (Jii')$ is a multi-index of length $k+2$ and that $\bar{\mu}(I')$ has been defined for multi-indices I' of length $\leq k+1$. Let m be a nonzero integer dividing $\Delta(I)$, and for each $i \leq \mu$ let $M_J(i)$ be the $(k+1) \times (k+1)$-matrix over $\mathbb{Z}/m\mathbb{Z}$ such that $M_J(i)_{p,q} = 1$ if $q = p+1$ and $j_p = i$, and is 0 otherwise. Then we may define a representation ϕ_J of $F(\mu)$ in $N(k, m)$, the strictly upper triangular subgroup of $\mathrm{GL}(k + 1, \mathbb{Z}/m\mathbb{Z})$, by setting $\phi_J(x_i) = I + M_J(i)$, for $1 \leq i \leq \mu$. Murasugi shows this determines a representation of πL in $N(k, m)$, and that $\bar{\mu}(I)$ is congruent mod (m) to the linking number of certain unions of preimages of L_i and $L_{i'}$ in the corresponding branched covering space [**HM77, Mu85**]. (See also [**HS66, Lf71, Go78**].)

The p-fold branched cyclic cover of S^{n+2}, branched over an n-knot K, is a $\mathbb{Z}_{(p)}$-homology sphere. A concordance between n-links lifts to a "homology concordance" by branching over the first component of the concordance [**CO93**]. Cochran and Orr have defined mod-(p) and $\mathbb{Z}_{(p)}$ versions of the Milnor invariants for links in such homology spheres. These are homology cobordism invariants, and so are potentially useful invariants of link concordance. The mod-(p) invariants vanish if all the Milnor invariants are 0, and all vanish if L is an SHB link. Thus the $\mathbb{Z}_{(p)}$ invariants may detect links with vanishing Milnor invariants which are not SHB links [**CO99**].

12.10. Solvable quotients and covering spaces

The most useful invariants of knots and links after the link groups are the homology modules and duality pairings of canonical covering spaces, in particular those of the maximal abelian covering space. In the classical case this corresponds to studying the metabelian quotients π/π'', which are rarely nilpotent. The study of Alexander modules and of nilpotent quotients have two natural common generalizations.

Firstly, one may consider the homology of nilpotent covers. If G is a finitely generated torsion free nilpotent group, or more generally if it is torsion free and polycyclic then the group ring $\mathbb{Z}[G]$ is noetherian, of finite global dimension and $\widetilde{K}_0(\mathbb{Z}[G]) = 0$. Hence the homology modules of any corresponding covering space of a finite complex are noetherian and have finite free resolutions. We would hope to derive Witt class invariants of concordance from the homology of such covers. (The case of 2-component $Z/2Z$-homology boundary links and the group C of §5 of Chapter 8 seems one deserving of early consideration.)

Cochran, Orr and Teichner found such invariants for classical knots [**COT03**]. They construct a canonical tower of torsion free solvable groups, to which "most" classical knot groups map nontrivially. The group rings of such groups are Ore domains, and have skew fields of fractions. (However, these solvable groups are not necessarily polycyclic, or even of finite Hirsch length.) They define $\mathbb{R}$-valued concordance invariants in terms of L^2 signatures of suitably localized duality pairings. (The localization ensures that the pairings are non-singular; compare §6 of Chapter 2.)

Secondly, solvable linear groups (such as polyclic groups) have subgroups of finite index with nilpotent commutator subgroup, by a theorem of Mal'cev. (See proposition 15.1.4 of [**Rob**].) This suggests we should consider the lower central series of π'. If π is the group of a fibred 1-knot then π' is a finitely generated free group, and so is residually nilpotent. (The commutator subgroup is also residually nilpotent if the knot is in the class generated by fibred knots and 2-bridge knots under the operations of cabling and forming connected sums [**Go81**].) On the other hand, if L is a knot or link such that $\Delta_1(L) \doteq 1$ then π' is perfect, and so the lower central series takes us no further. Fortunately, this can only happen if $\mu \leq 2$ and L is a TOP slice knot or is TOP concordant to the Hopf link.

CHAPTER 13

Algebraic Closure

This chapter shall summarize work of Cochran, Le Dimet, Levine, Orr and Vogel on universal groups for links whose groups have relatively free nilpotent quotients $(\pi/\pi_k \cong F(\mu)/F(\mu)_k$ for $1 \leq k < \infty)$. This includes all 1-links I-equivalent to sublinks of homology boundary links and all higher dimensional links.

13.1. Homological localization

If $\mathcal{L}$ is an I-equivalence from $L(0)$ to $L(1)$ the inclusion of $X(L(i))$ into $X(\mathcal{L})$ induces a (homologically) 2-connected, normally surjective homomorphism on fundamental groups, for $i = 0$ and 1. Therefore if F is a functor from groups (of finite weight) to groups which takes 2-connected, normally surjective homomorphisms to isomorphisms $F(\pi L)$ is an invariant of I-equivalence. For example, $F(G) = G/G_n$ is such a functor for any $1 \leq n < \infty$, by Stallings' Theorem. The tower of nilpotent quotients gives the nilpotent completion $\widehat{G}$, discussed in §1 below.

Is there a universal such functor through which all others factor? Following [**Ad75**], it may be prudent to require that there be a natural transformation η from the identity functor to F such that η_G is 2-connected and normally surjective and $F(\eta_G) = \eta_{F(G)}$, at least for groups of finite weight. Thus F is naturally idempotent. We may then say that a group G is F-*local* if η_G is an isomorphism. (However, the nilpotent quotient (and nilpotent completion) functors do not satisfy these additional conditions, since $H_2(F(\mu)/F(\mu)_n; \mathbb{Z}) \neq 0$ if $\mu > 1$ and $n \geq 1$, and $H_2(\widehat{F}(\mu); \mathbb{Z}) \neq 0$ [**Bsf**].)

Vogel has shown that there is an essentially unique pair (E, η), where E is a functor from the category of pointed CW-complexes to itself and η is an idempotent natural transformation, which localizes maps of finite complexes with contractible cofibre. (See [**LeD**].) Such maps induce 2-connected, normally surjective homomorphisms of fundamental groups, and so $F(G) = \pi_1(EK(G,1))$ is a candidate for a universal functor satisfying the conditions of the previous two paragraphs. We shall discuss this functor from a more group-theoretic point of view, introduced earlier by Gutiérrez and Levine.

13.2. The nilpotent completion of a group

Let G be a group. The *nilpotent completion* of G is the inverse limit of the tower of quotients of the lower central series, $\widehat{G} = \varprojlim G/G_k$. The canonical homomorphism from G to $\widehat{G}$ is a monomorphism if and only if G is residually nilpotent, i.e., $G_\omega = 1$. If G is finitely generated then $G/G_n \cong \widehat{G}/(\widehat{G})_n$ [**Bsf**]. The group G is *weakly parafree* (of rank μ) if it satisfies the following equivalent conditions:

(1) $G/G_k \cong F(\mu)/F(\mu)_k$ *for* $1 \le k < \infty$;
(2) *there is a homomorphism* $\theta : F(\mu) \to G$ *which induces isomorphisms* $F(\mu)/F(\mu)_k \cong G/G_k$ *for* $1 \le k < \infty$;
(3) $\widehat{F}(\mu) \cong \widehat{G}$.

It is *parafree* if moreover it is residually nilpotent.

Let L be a μ-component n-link with group $\pi = \pi L$. If either $n > 1$ or $n = 1$ and all the Milnor invariants of L are 0 then π is weakly parafree. The link groups $G(i,j)$ introduced in Chapter 1 are parafree [**Ba69**]. Every finitely generated E-group is weakly parafree, by Stallings' Theorem.

The *Parafree Conjecture* is that if G is finitely generated and parafree then $H_2(G;\mathbb{Z}) = 0$. The more optimistic *Strong Parafree Conjecture* is that $c.d.G \le 2$ also. If $H_2(G/G_\omega;\mathbb{Z}) = 0$ then $G_{\omega+1} = G_\omega$, by the 5-term exact sequence of low degree for G as an extension of G/G_ω by G_ω. (This is easily proven group-theoretically if G/G_ω is free.) Cochran and Orr show that, conversely, the Parafree Conjecture is true if $G_{\omega+1} = G_\omega$ for every finitely generated weakly parafree group G [**CO98**].

An automorphism of $\widehat{F}(\mu)$ is *special* if it sends each basis element x_i (for $1 \leq i \leq \mu$) to a conjugate. The set of such automorphisms clearly forms a group, which we shall denote by $\widehat{\mathcal{H}}(\mu)$. If $g_1, \ldots, g_\mu$ are any μ elements of $\widehat{F}(\mu)$ there is a unique automorphism of $\widehat{F}(\mu)$ sending x_i to $g_i x_i g_i^{-1}$, for $1 \leq i \leq \mu$ [**Le89**]. (See Theorem 13.6 below, which proves the analogous result for the algebraic closure $\breve{F}(\mu)$.)

Let $K_\omega(\mu)$ be the cofibre of the map from $\vee^\mu S^1$ to $K(\widehat{F}(\mu), 1)$ induced by the inclusion of $F(\mu)$ into its nilpotent completion. This space is simply-connected, and its higher homology groups agree with those of $\widehat{F}(\mu)$. In particular, $\pi_2(K_\omega(\mu)) \cong H_2(\widehat{F}(\mu); \mathbb{Z})$, which is uncountable [**Bsf**]. As in the definition of the Orr invariants $O_k(L; \theta)$ (in §9 of Chapter 12) there are Orr invariants $O_\omega(L; \theta)$ in $\pi_{n+2}(K_\omega(\mu))$ for based links L such that $\pi = \pi L$ is weakly parafree. A basing $\theta : F(\mu) \to \pi$ determines an isomorphism $\widehat{F}(\mu) \cong \pi$, and the corresponding homomorphism from π to $\widehat{F}(\mu)$ may be realized by a map from $X(L)$ to $K(\widehat{F}(\mu), 1)$. This extends to a map from S^{n+2} to $K_\omega(\mu)$, which defines the homotopy class $O_\omega(L; \theta)$.

A concordance $\mathcal{L}$ between based links $L(0)$ and $L(1)$ is *based* if there is a homomorphism $\theta : F(\mu) \to \pi\mathcal{L}$ which agrees with the given basings up to composition with conjugations by elements of $\pi\mathcal{L}$. The element $O_\omega(L; \theta) \in \pi_{n+2}(K_\omega(\mu))$ is an obstruction to a based link L with weakly parafree group being based-concordant to a boundary link. The special automorphisms of $\widehat{F}(\mu)$ determine self homotopy equivalences of $K(\widehat{F}(\mu), 1)$, and there is an induced homotopy action of $\widehat{\mathcal{H}}(\mu)$ on $K_\omega(\mu)$. Factoring out the action of $\widehat{\mathcal{H}}(\mu)$ gives a concordance invariant of unbased links. It is not known how large these obstruction groups are, nor whether every element is realized by some based link [**Or87**].

13.3. The algebraic closure of a group

The notion of algebraic closure of a group was introduced into link theory by Gutiérrez [**Gu79**]. There are various interpretations of this notion in the literature. (See for instance [**BDH80**].) The particular version that we shall describe is due to Levine, who used it to define a modification of Orr's invariant, which takes values in the homology of a countable group and has the further advantage that all elements of the obstruction group can be realized. (If $n > 1$ the notion of link must be broadened somewhat.) Most of the arguments in this section are from [**Le89'**].

A *contractible system of equations in m variables* $x_i = w_i$ over a group G is determined by a finite set of words $\{w_j \mid 1 \le j \le m\}$ in the free product $G * F(m)$, such that each word has trivial image in $F(m)$. Let $e : G \to G*F(m)/\langle\langle x_i^{-1}w_i \mid 1 \le i \le m\rangle\rangle$ be the canonical homomorphism. (This is a monomorphism if $m = 1$ and G is torsion free [**Kl93**].) If $h : G \to H$ is a homomorphism, a *solution under h* to such a system is a homomorphism

$$\check{h} : G * F(m)/\langle\langle x_i^{-1}w_i \mid 1 \le i \le m\rangle\rangle \to H$$

such that $\check{h} \circ e = h$. The group is *algebraically closed* if every contractible system of equations over G has a unique solution under id_G. (In this case $\check{h}$ is a retraction and we speak of solutions in G.)

LEMMA 13.1. *Let G be a group. Then G is algebraically closed if whenever $f : H \to J$ is a 2-connected, normally surjective homomorphism of finitely presentable groups and $\alpha : H \to G$ is a homomorphism there is an $\alpha' : J \to G$ such that $\alpha'f = \alpha$.*

PROOF. Let $\{w_j \mid 1 \le j \le m\}$ be a finite subset of $G * F(m)$ determining a contractible system of equations. These words involve only finitely many elements of G. Therefore there is a homomorphism $\alpha : F(p) \to G$ and a contractible system of equations $\{W_j \mid 1 \le j \le m\}$ over $F(p)$ whose image is the given system. Let $S = F(p) * F(m)/\langle\langle x_i^{-1}W_i \mid 1 \le i \le m\rangle\rangle$. Then S has deficiency p and weight p, so the natural homomorphism $f : F(p) \to S$ is 2-connected and normally surjective. Hence there is a unique homomorphism $\alpha' : S \to G$ extending α, which clearly induces a solution to the original system of equations. $\qquad\square$

The next lemma is easily verified.

LEMMA 13.2. *The inverse or direct limit of a system of algebraically closed groups is algebraically closed. If G is a central extension of H then G is algebraically closed if and only if H is algebraically closed.* □

It follows easily that nilpotent groups and nilpotent completions of groups are algebraically closed.

An *algebraic closure* of a group G is a homomorphism $\phi : G \to H$ with range an algebraically closed group, and which is universal with respect to such homomorphisms. We shall show next that every finitely presentable group has an algebraic closure. It is convenient to introduce the following notion. A subgroup K of a group G is *invisible* if it is the normal closure in G of a finite subset and $[G, K] = K$. The terminology is motivated by the next lemma.

LEMMA 13.3. *Let G be a group. Every contractible system of equations over G has at most one solution in G if and only if G has no nontrivial invisible subgroup.*

PROOF. Let K be an invisible subgroup which is the normal closure in G of $\{k_1, \ldots, k_m\}$. Then $k_i = w_i(k_1, \ldots, k_m)$, where $w_i \in G*F(m)$ is a product of conjugates of commutators of the form $[x_i, g]$, since $K = [G, K]$. Hence $\{\phi(k_1), \ldots, \phi(k_m)\}$ is a solution set in H for the system of equations $x_i = \tilde{\phi}(w_i(x_1, \ldots, x_m)) \in H * F(m)$, where $\tilde{\phi} = \phi * id_{F(m)}$ is the obvious extension of ϕ to a homomorphism from $G * F(m)$ to $H * F(m)$. But this is clearly a contractible system of equations with the trivial solution $x_i = 1$, for all i. Therefore if every contractible system of equations over G has at most one solution in G the invisible subgroup K must be trivial.

Suppose $S : x_i = w_i$ is a contractible system over G with two solution sets $x_i = g_i$ and $x_i = h_i$. Then $x_i = 1$ and $x_i = g_i h_i^{-1}$ are solution sets for the contractible system $S' : x_i = w_i'$, where $w_i'(x_1, \ldots, x_m) = w_i(x_1 h_1, \ldots, x_m h_m) h_i^{-1}$. Now we see that w_i' is a product of conjugates of commutators $[x_j, a]$, where a can be any monomial. In fact, since $w_i'(1, \ldots, 1) = 1$ we can first write w_i' as a product of conjugates $\Pi_{r=1}^{r=s} a_r x_{i_r}^{\epsilon_r} a_r^{-1}$ where $a_r \in G$. This can be rewritten $w_i' = (\Pi_{r=1}^{r=s} b_r [a_r, x_{i_r}^{\epsilon_r}] b_r^{-1}) b_{s+1}$ where $b_1 = 1$ and $b_{r+1} =$

$b_r x_{i_r}^{\epsilon_r}$. But since w_i' is also contractible over H, $b_{s+1} = 1$. So $\{g_i h_i^{-1}\}$ are seen to be solutions of a system which displays them as normally generating a nontrivial invisible subgroup.   □

In particular, if K is an invisible subgroup of a group G and $\phi : G \to H$ is a homomorphism with algebraically closed codomain H then the normal closure of $\phi(K)$ in H is invisible, and so $\phi(K) = 1$.

THEOREM 13.4. *Let G be a finitely presentable group. There is a universal homomorphism $\phi : G \to \check{G}$ where $\check{G}$ is algebraically closed, through which every other homomorphism to an algebraically closed group factors uniquely. Moreover, $\check{G}$ is countable, and ϕ is 2-connected and normally surjective.*

PROOF. Enumerate the contractible systems of equations over G as $\{S_n\}_{n \geq 1}$. We shall assume that $S_n = \{x_i^{-1} w_i\}$ where the w_i are words in $G * F(X_n)$ with $X_n - \{x_l\}$, and the alphabets X_m and X_n are disjoint if $m \neq n$. Let $Q_0 = G$. Assume that we have defined groups Q_k and 2-connected, normally surjective homomorphisms $i_k : Q_{k-1} \to Q_k$ for $1 \leq k \leq n$. Let $Q_{n+1} = Q_n * F(X_n)/\langle\langle S_n \rangle\rangle$. Let $Q = \varinjlim Q_n$. Then every contractible system of equations over G has a solution in Q, but the solution need not be unique.

Let N be the union of all the invisible subgroups of Q. It is easily seen that the normal closure of the union of two invisible subgroups is again invisible, and so N is an increasing union $N = \cup N_k$ of invisible subgroups. Let $\check{G} = Q/N$, and let ϕ be the composition of the natural homomorphism from G to Q with projection onto Q/N. It is easily verified that $\check{G}$ has no nontrivial invisible subgroups. Hence any contractible system of equations over $\check{G}$ has at most one solution in $\check{G}$, by Lemma 13.3.

Let S be a contractible system over $\check{G}$ and lift S to a system over Q. The finitely many elements of Q involved in the contractible words defining S are themselves solutions of equations $S' : y_j = v_j(y_i, \ldots, y_p)$ over G. We may rewrite the words w_i as words w_i' in $G * F(x_i) * F(y_j)$, and obtain a system $S'' : x_i = w_i', y_j = v_j$ over G, which has a solution $x_i = g_i, y_j = h_j$ in Q. The images of these elements in $\check{G}$ give a solution to the system S. Hence $\check{G}$

is algebraically closed. The universal property follows easily as $\check{G}$ is generated by solutions of contractible systems of equations over G. $\qquad\square$

The construction actually shows that $\check{G} = \varinjlim P_n$ is the direct limit of a sequence of finitely presentable groups such that the connecting homomorphisms are homologically 2-connected and normally surjective. For let $a_1, \ldots, a_p$ normally generate N_k, and write each a_i as a product w_i of conjugates of commutators of the form $[a_j, g]$. We may choose Q_s such that a_j lifts to $a'_j \in Q_s$ and w_i lifts to a product w'_i of conjugates of commutators of the form $[a'_j, g']^{\pm}$ in Q_s. After increasing s, if necessary, we may assume that $a'_i = w'_i$ in Q_s. Let K_k be the normal closure of $\{a'_1, \ldots, a'_p\}$ in Q_s. After enlarging K_k and renumbering a subsequence, if necessary, we may assume that $i_k(K_k) \leq K_{k+1}$ and that $Q_s = Q_k$. Let $P_n = Q_n / K_n$ for all $n \geq 1$. Then P_n is finitely presentable and the natural homomorphism from G to P_n is 2-connected and normally surjective and $\check{G} = \varinjlim P_n$.

Lemma 13.1 and Theorem 13.4 together imply the following corollary.

COROLLARY 13.4.1. *A 2-connected, normally surjective homomorphism of finitely presentable groups* $f : G \to H$ *induces an isomorphism* $\check{f} : \check{G} \to \check{H}$ *of the algebraic closures.* $\qquad\square$

The algebraic closure of a finitely generated perfect group is trivial, but this is not true in general. (Consider G an increasing union of non-abelian free groups, each embedded in the commutator subgroup of the next. Then $G = G'$ and G embeds in $\check{G}$. Moreover $\check{G}$ is not even transfinitely nilpotent.)

We shall show next that this algebraic closure is the group theoretic aspect of the homological localization functor (E, η) of Vogel.

THEOREM 13.5. *If* $G = \pi_1(X)$ *for some finite complex* X *then* $\check{G} \cong \pi_1(EX)$.

PROOF. A contractible system of equations over $\pi_1(EX)$ corresponds to a map $\alpha : W = \vee^p S^1 \to EX$ and a map $f : W \to W' = (W \vee \bigvee^m S^1) \cup mD^2$ which is normally surjective and (so) induces an isomorphism on integral homology. Hence there is a map $\alpha' : W' \to EX$ extending α, by the universal property of Vogel's homology localization functor E. It follows easily that $\pi_1(EX)$ is algebraically closed.

Let $\eta_{X*} : G \to \pi_1(EX)$ be the homomorphism of fundamental groups induced by η_X. Then η_{X*} is normally surjective, and is 2-connected since η_X induces isomorphisms on integral homology. As in the remark following Theorem 13.4, this implies that $\check{G} \cong \pi_1(EX)$. $\qquad\square$

It is not known in general whether $EK(G,1)$ is aspherical.

Let $\check{\mathcal{H}}(\mu)$ denote the group of special automorphisms of $\check{F}(\mu)$ (i.e., those which send each basis element x_i to a conjugate, for $1 \leq i \leq \mu$). The special automorphisms of $\widehat{F}(\mu)$ and $\check{F}(\mu)$ were determined in [**Le89, Le89'**].

THEOREM 13.6. *Let $G = \widehat{F}(\mu)$ or $\check{F}(\mu)$ and let $g_1, \ldots, g_\mu \in G$. Then there is a unique automorphism of G sending x_i to $g_i x_i g_i^{-1}$, for $1 \leq i \leq \mu$.*

PROOF. Let $\phi : F(\mu) \to \check{F}(\mu)$ be the homomorphism defined by $\phi(x_i) = g_i x_i g_i^{-1}$, for $1 \leq i \leq \mu$. Since $\check{F}(\mu)$ is the algebraically closure of $F(\mu)$, this extends uniquely to an endomorphism $\check{\phi}$ of $\check{F}(\mu)$. If $g \in \check{F}(\mu)$ then $F(\mu) \cup \{g, g_1, \ldots, g_\mu\}$ is contained in a finitely generated subgroup H which is normally generated by $F(\mu)$. Now $\phi(F(\mu)) \leq H$ and H is also normally generated by $\phi(F(\mu)) \cup \{g, g_1, \ldots, g_\mu\}$. Therefore H is generated by the solutions of some contractible system S over $\phi(F(\mu))$. We may write $S = \phi(S')$ for some contractible system S' over $F(\mu)$. Since S' has a unique system of solutions in $\check{F}(\mu)$, the solutions of S are in $\check{\phi}(\check{F}(\mu))$, and so g is in $\check{\phi}(\check{F}(\mu))$.

Hence $\check{\phi}$ is surjective. If we now define ψ similarly by $\psi(x_i) = g_i^{-1} x_i g_i$, for $1 \leq i \leq \mu$, we see that $\check{\psi}\check{\phi} = id$, and so $\check{\phi}$ is an automorphism. The case of $\widehat{F}(\mu)$ is similar. $\qquad\square$

The following related result is used at one point in Chapter 14.

LEMMA 13.7. [**Le89**] *The centralizer in* $\widehat{F}(\mu)$ *of a primitive element* $x \in F(\mu)$ *is the cyclic group* $\langle x \rangle$ *generated by* x.

PROOF. Let $Q(1) = 0$ and let $Q(g)$ be the nonzero homogeneous term of lowest degree > 0 in the Magnus expansion of g in $\widehat{\Gamma_\mu}$, if $g \neq 1$. (Thus $Q(g) = 0$ if and only if $g = 1$.) We may assume that $x = x_1$ and $Q(x) = X_1$. An easy calculation shows that $Q([g, h]) = Q(g)Q(h) - Q(h)Q(g)$. Therefore $gx = xg$ implies that $Q(g)X_1 = X_1 Q(g)$, and so $Q(g) = m(X_1)^k$ for some $k > 0$. If $k > 1$ then $g \in \widehat{F}(\mu)'$ and so the image of g in the commutative ring $\widehat{\Lambda_\mu}$ must be 1. Hence $Q(g) = 0$ and so $g = 1$. If $k = 1$ then on applying the same argument to gx^{-m} we find that $gx^{-m} = 1$ and so $g \in \langle x \rangle$. $\square$

13.4. Complements on $\check{F}(\mu)$

Since the algebraic closure $\phi : F(\mu) \to \check{F}(\mu)$ is 2-connected $H_1(\check{F}(\mu); \mathbb{Z}) \cong \mathbb{Z}^\mu$ and $H_2(\check{F}(\mu); \mathbb{Z}) = 0$, and so $F(\mu)/F(\mu)_n \cong \check{F}(\mu)/(\check{F}(\mu))_n$, for $1 \leq n < \infty$, by Stallings' Theorem. Let U_n be the group of units in $\widehat{\Gamma_\mu} = \mathbb{Z}\langle\langle X_1, \dots, X_\mu \rangle\rangle$ which are congruent to 1 *mod* the nth power of the augmentation ideal, and let $U = U_1$. Then U is the inverse limit of the tower of nilpotent quotients $\{U/U_n\}$, and so is algebraically closed. Hence the Magnus embedding of $F(\mu)$ extends to a homomorphism from $\check{F}(\mu)$ to $U < \widehat{\Gamma_\mu}^\times$ [**LeD**]. Is this extension a monomorphism?

Let $\check{\Gamma}_{\mu C}$ be the Cohn localization of $\check{\Gamma}_\mu = \mathbb{Z}[\check{F}(\mu)]$ with respect to the augmentation, as in Chapter 11. The following result of Le Dimet is an analogue for $\check{F}(\mu)$ of the fact that the augmentation ideal of $F(\mu)$ is free of rank μ as a left Γ_μ-module.

THEOREM 13.8. [**LD92**] *Let* $v : (\check{\Gamma}_{\mu C})^\mu \to \check{\Gamma}_{\mu C} \otimes_{\check{\Gamma}} I(\check{F}(\mu))$ *be the homomorphism defined by* $v(\gamma_1, \dots, \gamma_\mu) = \Sigma \gamma_i \otimes (x_i - 1)$. *Then* v *is an isomorphism of left* $\check{\Gamma}_{\mu C}$-*modules.*

PROOF. The universal properties of Cohn localization of group rings and Vogel localization of spaces imply that the inclusion η_W of $W = \vee^\mu S^1$ into $E(\vee^\mu S^1)$ induces an isomorphism of $H_1(F(\mu); \check{\Gamma}_{\mu C})$ with $H_1(\check{F}(\mu); \check{\Gamma}_{\mu C})$. Since η_W also induces a homomorphism from the sequence obtained by tensoring the augmentation sequence for

Γ_μ with $\breve{\Gamma}_{\mu C}$ to the augmentation sequence for $\breve{\Gamma}_\mu$, and since $I(F(\mu))$ is freely generated by the elements $x_i - 1$, the claim now follows from the Five-Lemma.   □

As a consequence, $\breve{F}(\mu)$ admits a free differential calculus with values in $\Lambda_\mu S$, where $S = 1 + I_\mu = \{f \in \Lambda_\mu \mid \varepsilon(f) = 1\}$. (This is the Cohn localization of $\mathbb{Z}[F(\mu)/F(\mu)'] = \mathbb{Z}[\breve{F}(\mu)/\breve{F}(\mu)']$ with respect to the augmentation.) For the corresponding homomorphism from $(\Lambda_\mu S)^\mu$ to $\Lambda_\mu S \otimes_{\breve{\Gamma}} I(\breve{F}(\mu))$ is also an isomorphism, and so we may define derivations ∂_i by $\partial_i(\Sigma \gamma_i \otimes (x_i - 1)) = \gamma_i$. These have the usual properties; in particular, if $g \in \breve{F}(\mu)$ and $\partial_i(g) = 0$ for all i then $g \in \breve{F}(\mu)'$, while if $g \in \breve{F}(\mu)''$ then $\partial_i(g) = 0$ for all i. Le Dimet uses these to extend the Gassner representations for the pure braid groups to representations for string link groups. (See Chapter 14.)

The following result of Vogel implies that every invariant of boundary links defined in terms of unitary representations may be extended to an invariant of all higher-dimensional links. (See also [**Le94**] on links and the Atiyah-Patodi-Singer η-invariant.)

THEOREM 13.9. [**Vo92**] *Every representation of $F(\mu)$ in a compact connected Lie group extends to a representation of $\breve{F}(\mu)$.*

PROOF. (Sketch.) A series $u = \Sigma n_\gamma \gamma \in \widehat{\Gamma_\mu}$ is said to be convergent, of size $\leq c$, if there is an $a > 0$ that $|n_\gamma| \leq ac^m$ for every monomial γ of length m. The set of convergent series of size $\leq c$ is a subalgebra of $\widehat{\Gamma_\mu}$. Substitution of convergent series with augmentation 0 into a convergent series gives a convergent series. It is then shown that if w is a convergent series in $\widehat{\Gamma_{\mu+1}}$ with augmentation 0 and such that the coefficient of $X_{\mu+1}$ in w is 1 then there is a unique series $u \in \widehat{\Gamma_\mu}$ with augmentation 0 such that $w(X_1, \ldots, X_\mu, u) = 0$, and this u is convergent. This analogue of the implicit function theorem is used to show that the image of each element of $\breve{F}(\mu)$ under the Magnus embedding is convergent.

Let $R(G)$ denote the set of representations of the group G in the compact connected Lie group J. If G is finitely generated then $R(G)$ is naturally a compact algebraic variety (in the strong topology). In particular, $R(F(\mu))$ is connected and non-singular. Let $R_0(G)$ be the component containing the trivial representation. As $\breve{F}(\mu)$ is the direct limit of a sequence of 2-connected, normally surjective homomorphisms of finitely presentable groups G_p, by Theorem 13.4, $R(\breve{F}(\mu))$ is the inverse limit of the tower of representation varieties $\{R(G_p)\}_{p\geq 1}$. Since these are compact it shall suffice to show that the restrictions from $R_0(G_p)$ to $R(F(\mu))$ are surjective, for all $p \geq 1$. Convergence of the Magnus expansions is used to show that every representation of $F(\mu)$ in $\mathrm{GL}(m, \mathbb{C})$ which is sufficiently close to the trivial representation extends to a representation of G_p. The rest of the argument uses algebraic geometry to show that the complexification of this map has odd degree and hence that there are (real) points of $R_0(G_p)$ over every point of $R(F(\mu))$.   □

13.5. Other notions of closure

Since the algebraic closure $\phi : G \to \breve{G}$ is 2-connected $G/G_n \cong \breve{G}/(\breve{G})_n$ for all $1 \leq n < \infty$, by Stallings' Theorem, and so G and $\breve{G}$ have isomorphic nilpotent completions. The *residually nilpotent algebraic closure* of G is the image $\overline{G}$ of $\breve{G}$ in $\widehat{G}$. (We could also argue that since $\widehat{G}$ is algebraically closed ϕ extends to a homomorphism from $\breve{G}$ to $\widehat{G}$, with image $\overline{G}$.) Thus $\overline{G} \cong \breve{G}/(\breve{G})_\omega$, is countable and contains G/G_ω. If G is finitely presentable is $\breve{G} \cong \overline{G}$? (The known examples of finitely generated groups G such that $\breve{G}$ is not residually nilpotent are infinitely related.)

In particular, $\overline{F}(\mu)$ is the image of $\breve{F}(\mu)$ under the homomorphism extending the Magnus embedding. The argument for Theorem 13.4 implies that $\overline{F}(\mu)$ is a direct limit of finitely generated parafree groups. Hence if the Parafree Conjecture is true then $H_2(\overline{F}(\mu); \mathbb{Z}) = 0$ and $\breve{F}(\mu)_{\omega+1} = \breve{F}(\mu)_\omega$. If the Strong Parafree Conjecture is true then $\overline{F}(\mu)$ has homological dimension 2 and so $H_3(\overline{F}(\mu); \mathbb{Z}) = 0$ also. Is $\overline{F}(\mu) \cong \breve{F}(\mu)$?

Let $\mathcal{H}(\mu)$ denote the group of special automorphisms of $\overline{F}(\mu)$. It is an immediate consequence of Theorem 13.6 that the special automorphism of $\widehat{F}(\mu)$ determined by μ elements $g_i \in \overline{F}(\mu)$ restricts to an automorphism of $\overline{F}(\mu)$. Similarly, Lemma 13.7 implies that the centralizers of the generators of $F(\mu)$ in $\overline{F}(\mu)$ are cyclic.

If we relax the requirement in the definition of algebraic closure that the words $w_i \in G * F(m)$ determining a system of equations have trivial image in $F(m)$, to requiring only that they lie in the kernel of $G * F(m) \to F(m) \to F(m)/F(m)'$ we obtain the notion of *acyclic closure*. It is not known whether the canonical homomorphism from the algebraic closure to the acyclic closure is an isomorphism. Acyclic closure is closely related to Bousfield's HZ-localization [**DOS89**]. A group G is HZ-*local* if for any 2-connected homomorphism $f : H \to J$ and any homomorphism $\alpha : H \to G$ there is a unique $\alpha' : J \to G$ such that $\alpha' f = \alpha$. (We do *not* require that H and J be finitely presentable and f normally surjective.) Every HZ-local group is transfinitely nilpotent.

The HZ-localization may be constructed as a transfinite tower. Let $\phi_1 : G \to G(1) = G/G'$ be the abelianization. For each ordinal α, suppose that a group $G(\alpha)$ and a homomorphism $\phi_\alpha : G \to G(\alpha)$ have been defined. The epimorphism from $H_2(G(\alpha); \mathbb{Z})$ to $\mathrm{Coker}(H_2(\phi_\alpha; \mathbb{Z}))$ determines a central extension

$$0 \to \mathrm{Coker}(H_2(\phi_\alpha; \mathbb{Z})) \to G(\alpha + 1) \to G(\alpha) \to 1$$

and a lift $\phi_{\alpha+1}$. If α is a limit ordinal let $G(\alpha)$ be the HZ-closure of G in $\varprojlim_{\beta < \alpha} G(\beta)$, and let ϕ_α be the corresponding homomorphism. It can be shown that this transfinite construction eventually stabilizes [**Bsf**]. The groups $G(\alpha)$ are algebraically closed, by transfinite induction (since they are either central extensions or projective limits of algebraically closed groups). Hence HZ-localization factors through the algebraic closure.

Direct limits of HZ-local groups are not usually HZ-local.

13.6. Orr invariants and $cSHB$-links

An $\breve{F}$-*link* is a link L admitting a basing $\theta : F(\mu) \to \pi = \pi L$ such that $\breve{\theta}$ is an isomorphism and (if $n = 1$) the longitudes have trivial image in $\breve{\pi}$. (It is not known whether the latter condition is automatic - this is so if $\breve{F}(\mu)$ is residually nilpotent.) If θ' is another basing for L then $\breve{\theta}' = \breve{\theta}h$ for some special automorphism of $\breve{F}(\mu)$, and so whether a link is an $\breve{F}$-link is independent of the choice of basing. Every link concordant to an $\check{F}$-link is an $\check{F}$-link. Is every link with weakly parafree group an $\check{F}$-link?

The map $f : X(L) \to K(\check{F}(\mu), 1)$ determined by $\check{\theta}$ extends to a map $f_M : M(L) \to K(\check{F}(\mu), 1)$ (using the longitude condition, if $n = 1$). On the other hand, f also extends to a map $f_S : S^{n+2} \to K_{ac}(\mu)$, where $K_{ac}(\mu)$ is the cofibre of the map from $\vee^{\mu}S^1$ to $K(\check{F}(\mu), 1)$ induced by the inclusion of $F(\mu)$ into $\check{F}(\mu)$. The maps f_M and f_S together extend to a map $f_W : W(L) \to K_{ac}(\mu)$, where $W(L)$ is the trace of 0-framed surgery on L, with oriented boundary $\partial W(L) = S^{n+2} \amalg -M(L)$.

Let $O_{ac}(L; \theta)$ be the homotopy class of f_S in $\pi_{n+2}(K_{ac}(\mu))$. This is a lift of the Orr invariant $O_{\omega}(L; \theta)$ (for L an $\check{F}$-link), since nilpotent completion factors through algebraic closure. Let $[O_{ac}(L; \theta)]$ be the image of $O_{\omega}(L; \theta)$ in $H_{n+2}(K_{ac}(\mu); \mathbb{Z})$ under the Hurewicz homomorphism. Since the images of the fundamental classes of the boundary components of $W(L)$ in $H_{n+2}(W(L); \mathbb{Z})$ agree we see that $[O_{ac}(L; \theta)] = f_{S*}[S^{n+2}] = f_{M*}[M(L)]$ in $H_{n+2}(\check{F}(\mu); \mathbb{Z}) = H_{n+2}(K_{ac}(\mu); \mathbb{Z})$. (This modification of the Orr invariant is from [**Le89'**].)

This is invariant under based concordance, and is 0 for boundary links. Factoring out the action of $\check{\mathcal{H}}(\mu)$ gives a concordance invariant of unbased links. It is not known how large these obstruction groups are, nor whether $H_3(\check{F}(\mu); \mathbb{Z}) = 0$, the case relevant for classical links. (The latter is so if $EK(F(\mu), 1) = E(\vee^{\mu}S^1)$ is aspherical.) Every element of $H_3(\check{F}(\mu); \mathbb{Z})$ is realized by some $\check{F}$-1-link.

In [**Le89'**] it is shown that the following conditions are equivalent, for L a μ-component 1-link:

(1) L is $cSHB$ (concordant to a sublink of a homology boundary link);

(2) there is a 2-connected, normally surjective homomorphism $\Phi : \pi L \to G$, where G is finitely presentable, $H_2(G; \mathbb{Z}) = 0$, the longitudes of L are in $\mathrm{Ker}(\Phi)$ and $[M(L)]$ has image 0 in $H_3(G; \mathbb{Z})$;

(3) L is an $\check{F}$-link and $[O_{ac}(L; \theta)] = 0$ for some (and hence any) basing θ.

Is there always such a homomorphism Φ to a group G of cohomological dimension ≤ 2? Every SHB link is a finite E-link, by the argument of Theorem 1.14. The 2-complex corresponding to a presentation of maximal deficiency for a finite E-group is a subcomplex of a contractible 2-complex. If the Whitehead Conjecture is true every such group has cohomological dimension ≤ 2, and hence every finite E-link is $cSHB$ [**Le89**]. Is every $cSHB$ link a finite E-link?

If L is a null-concordant 1-link are all the longitudes of L in the intersection of the terms of the transfinite derived series? (See [**CO98**].)

CHAPTER 14

Disc Links

One serious difficulty in attempting to classify links up to concordance is that the set of concordance classes does not have a natural group structure. Le Dimet, working in higher dimensions, and subsequently Levine and Habegger and Lin, in the classical case, have shown that one should first study disc links (or "string links"). This chapter sketches some of their work.

14.1. Disc links and string links

It shall be convenient here to identify the n-disc D^n with the product $(D^1)^n$. Let $j_\mu : \{1, \ldots, \mu\} \to D^2$ be the embedding given by $j_\mu(k) = \frac{1}{2} e^{2\pi i k/\mu}$, for $1 \le k \le \mu$. If $n \ge 1$ the *standard trivial* (n, μ)-*disc link* is $E_0^{n,\mu} = j_\mu \times id_{D^n} : \mu D^n \to D^{n+2} = D^2 \times D^n$. A (n, μ)-*disc link* is a locally flat embedding $E : \mu D^n \to D^{n+2}$ such that $E|_{\mu S^{n-1}} = E_0^{n,\mu}|_{\mu S^{n-1}}$. In the classical case S^0 has two components and we allow a more general possibility. A μ-component *string link* is a locally flat embedding $E : \mu D^1 \to D^3$ such that $E(-1, k) = (-1, j_\mu(k))$ and $E(1, k) = (1, j_\mu(\sigma(k)))$, for $1 \le k \le \mu$, where $\sigma \in S_\mu$ is a permutation of $\{1, \ldots, \mu\}$. It is *pure* if it is a $(1, \mu)$-disc link, i.e., if $\sigma = id$. A string link such that projection onto the last coordinate is a local homeomorphism is a *braid*. (See [**Bir**].) It is already clear in the 1-component case that, in general, string links are not isotopic to braids.

The notions of exterior and group and the equivalence relations isotopy, concordance, I-equivalence and link-homotopy extend in an obvious manner, with the proviso that all I-equivalences and homotopies (etc.) are constant on the boundary. (We take $(*, 0) \in S^1 \times D^n$ as basepoint, where $* \in S^1 = \partial D^2$ is not a root of unity.) If $n > 1$

307

a fixed choice of meridians for the trivial link $E|_{\mu S^{n-1}}$ determines a canonical set of meridians for E. In the string link case there are two natural meridianal homomorphisms to consider, namely the inclusions $e_\pm$ of $D_0 = X(j_\mu) \subset D^2$ as the top and bottom of $X(E)$. (In the $(1,\mu)$-disc link case we may define meridian-longitude pairs. See §2 below.) The inclusion of $\vee^\mu S^1$ into $X(E)$ determined by the meridians induces an isomorphism on homology, and so these spaces have equivalent homological localizations.

We may define standard spanning surfaces for $E_0^{n,\mu}$ by setting $U_k(t,d) = ((\frac{3+t}{4})e^{2\pi ik/\mu}, d)$, for all $1 \le k \le \mu$ and $(t,d) \in D^{n+1} = D^1 \times D^n$. Let $W_k = S^{n+1} \cap \partial U_k(D^{n+1})$, for all $1 \le k \le \mu$. A (n,μ)-disc link E is *boundary* if it extends to an embedding of μ disjoint orientable hypersurfaces V_k such that $\partial V_k = E_k \cup W_k$, for all $1 \le k \le \mu$, where E_k is the kth component of E. A string link is boundary if it is a boundary $(1,\mu)$-disc link.

Taking the pairwise boundary connected sum of two disc links (stacking with respect to the last coordinate) defines an operation $\natural$ which makes the set of ambient isotopy classes of such links into a monoid with identity represented by the trivial (n,μ)-disc link. When $n = 1$ this monoid contains the group P_μ of pure braids on μ strings as a subgroup of the group of invertible elements. This is in fact the full group of invertible elements. (See §3 below.) Stacking also defines a multiplication on the set of μ-component string links, for which the group of invertible elements is the braid group B_μ. A *colouring* of a string link is a partition of the index set $\{1,\ldots,\mu\}$ such that both ends of each string have the same colour, i.e., are images of points in the same subset. The set of string links with a given colouring is closed under stacking. Sending a string link to the permutation σ of its endpoints defines an epimorphism from the monoid of μ-component string links to the symmetric group S_μ.

The set of concordance classes of (n,μ)-disc links is a group $C_{n,\mu}$, with inverse given by changing the signs of the last coordinate in both domain and range. If $\mu = 1$ this is just the n-dimensional knot concordance group C_n, and so is abelian. A straightforward geometric argument (similar to a well-known argument for the commutativity of the higher homotopy groups of a space) shows that $C_{n,\mu}$ is also abelian if $n > 1$ [**LeD**]. However, the natural homomorphism from P_μ to $C_{1,\mu}$ is injective (see §3 below) and so $C_{1,\mu}$ is not abelian, if $\mu > 2$. Is $C_{1,2}$ abelian? There is a corresponding notion of boundary concordance. The set $B(n, \mu)$ of boundary concordance classes of boundary (n, μ)-disc links is a group, and is abelian if $n \geq 2$ [**LeD**].

Fix $\pm$orientation preserving homeomorphisms $h_{n\pm} : D^n \to D^n_\pm$ onto the upper and lower hemispheres of $S^n = D^n_+ \cup D^n_-$ which agree on S^{n-1}. The *closure* of the (n, μ)-disc link E is the μ-component n-link $L(E) = h_{(n+2)+} \circ E \cup h_{(n+2)-} \circ E^{n,\mu}_0$. Closure maps the group $C_{n,\mu}$ onto the set $C_n(\mu)$ and maps the group $B(n, \mu)$ onto the set $C_n(F(\mu))$. (The latter function is a homomorphism if $n \geq 2$.) Composition with the surjection from $C_1(F(\mu))$ to $G(\mu, -1)$ of [**Ko87**] gives an epimorphism from $B(1, \mu)$ to $G(\mu, -1)$ [**deC99**].

A *d-base* for a μ-component link L is an embedding of D^2 in S^{n+2} such that D meets L_i transversely in the point $j_\mu(i)$ with intersection number $+1$, for $1 \leq i \leq \mu$. The closure of a string link has a canonical d-base. This induces bijections between the isotopy classes and homotopy classes (respectively) of string links and of d-based 1-links. Other, weaker notions of basing have been considered [**Le88"**, **HL90**]. The weakest form of basing is a choice of meridians.

14.2. Longitudes

Let E be an (n, μ)-disc link with exterior $X(E)$ and group $\pi E = \pi_1(X(E))$. Then $H_1(X(E); \mathbb{Z}) \cong \mathbb{Z}^\mu$ and $H_2(X(E); \mathbb{Z}) = 0$, and the inclusion of the meridians determines a 2-connected, normally surjective homomorphism from $F(\mu)$ to πE. The natural homomorphism from πE to $\pi L(E)$ is a surjection, and is an isomorphism if $n > 1$. If E is a string link then πE has deficiency μ, and $\pi L(E) \cong \pi E / \langle\langle e_{+*}(g) = e_{*-}(g), \ \forall g \in F(\mu) \rangle\rangle$, by the Van Kampen Theorem. (Here we connect $e_\pm(*)$ to the basepoint $(*, 0)$ via paths in $\{*\} \times D^1$.)

If E is a $(1, \mu)$-disc link we may define *meridian-longitude* pairs $\{(m_i, l_i)\}$. Let N_i be a regular neighbourhood of the ith component E_i. Let $s_i^\pm$ be the arcs from $(*, 0)$ to each end of ∂N_i in $X(E|_{\mu S^0}) \subset$

$S^2 = \partial(D^2 \times D^1)$ given by the intersection of $X(E|_{\mu S^0})$ with the plane through $(*, 0)$ and $(j_\mu(i), \pm 1)$, for $1 \leq i \leq \mu$. Let $m_i = s_i^- . b_i . (s_i^-)^{-1}$, where b_i is a (suitably oriented) loop around the bottom end of N_i and $l_i = s_i^+ . p_i . (s_i^-)^{-1}$, where p_i is a parallel of E_i (running from top to bottom, and chosen so that l_i is null homologous in $X(E_i)$).

The kernel of the natural homomorphism from πE to $\pi L(E)$ is the normal closure of the subgroup generated by $\{[m_i, l_i] \mid 1 \leq i \leq \mu\}$. (This is straightforward, on observing that $p_i . b_i . (p_i)^{-1}$ is homotopic to a loop around the top end of N_i.)

Let $A_k \subset \partial(D^2 \times D^1)$ be the arc connecting $e_-j_\mu(k)$ to $e_+j_\mu(k)$ formed by radii to $e_\pm(S^1)$ and the arc $\{e^{2\pi ik/\mu}\} \times D^1$, for $1 \leq k \leq \mu$, and let α_i be a (based) loop around the boundary of a regular neighbourhood of A_i in S^2, for $1 \leq i \leq \mu$. Then $\pi_1(S^2 \setminus \cup A_i) \cong F(\mu - 1)$ is generated by such loops, with the single relation $\Pi \alpha_i = 1$. Since $[m_i, l_i]$ is homotopic to α_i in $X(E)$ we obtain the relation $\Pi[m_i, l_i] = 1$ [**Le88"**].

Let $\check{l}_i$ and $\bar{l}_i$ be the images of l_i in $\check{\pi}E \cong \check{F}(\mu)$ and $\widehat{\pi}E \cong \widehat{F}(\mu)$, respectively, for $1 \leq i \leq \mu$. Then $\bar{l}_i$ is the image of $\check{l}_i$ under the natural homomorphism from $\check{\pi}E$ to $\widehat{\pi}E$, and so lies in the subgroup $\bar{\pi} \cong \bar{F}(\mu)$. Levine defines the Milnor invariants of E as the images $\bar{l}_i$ [**Le88"**]. If $L = L(E)$ is the closure of a $(1, \mu)$-disc link E and $\pi = \pi L$ the longitudes of L are in π_q if and only if the meridians determine an isomorphism $F(\mu)/F(\mu)_{q+1}$ if and only if the Milnor invariants $\bar{l}_i$ are all in $\widehat{F}(\mu)_q$. In particular, if E is a pure braid $\pi L(E)$ has a presentation $\langle x_i, 1 \leq i \leq \mu \mid [w_i, x_i], 1 \leq i \leq \mu \rangle$, where w_i represents the ith longitude, for $1 \leq i \leq \mu$.

An (n, μ)-disc link E is a boundary disc link if the canonical meridianal homomorphism from $F(\mu)$ to πE induces an isomorphism $F(\mu) \cong \pi E/(\pi E)_\omega$. When $n = 1$ both inclusions $e_\pm$ of D_0 as the top and bottom of $X(E)$ must induce the same isomorphism. The meridianal homomorphism then has a unique splitting.

14.3. Concordance and the Artin representation

If E is a braid $X(E)$ collapses to D_0, and so the meridianal homomorphisms $e_\pm$ are isomorphisms. The *Artin representation* of

B_μ is the homomorphism $Ar : B_\mu \to Aut(F(\mu))$ defined by $Ar(E) = e_{+*}^{-1}e_{-*}$. The image of the ith standard generator σ_i of B_μ is the automorphism which fixes the basis elements x_j for $j \ne i$ or $i+1$, and sends x_i to x_{i+1} and x_{i+1} to $x_{i+1}^{-1}x_ix_{i+1}$. The Artin representation is faithful, and its image consists of the automorphisms of $F(\mu)$ which preserve the product $x_1 \cdots x_\mu$ and send each basis element x_i to a conjugate of $x_{\sigma(i)}$, for some permutation $\sigma \in S_\mu$ [**Bir**]. The image of the pure braid group P_μ is the subgroup $Aut_0(F(\mu))$ consisting of such automorphisms of $F(\mu)$ for which $\sigma = I$. If $E \in P_\mu$ then $Ar(E)(x_i) = w_i^{-1}x_iw_i$, where w_i represents the ith longitude, for $1 \le i \le \mu$. (The ith word w_i is uniquely determined by the condition that it have exponent sum 0 in the letter x_i.) Hence if E is nontrivial some $w_i \ne 1$ in $F(\mu)$ and so some $\bar\mu$-invariant of $L(E)$ is nonzero.

The pure braid groups P_μ are residually nilpotent, by the following result of Ohkawa.

THEOREM 14.1. [**Oh82**] *The pure braid group P_μ is residually a finite p-group, for all primes p.*

PROOF. Let p be a prime and let $\Gamma_q(F(\mu);p)$ be the qth term of the restricted lower central series for $F(\mu)$. Let $P_{\mu q}^{(p)}$ be the set of pure μ-component braids E such that $Ar(E)(g) \equiv g \bmod \Gamma_q(F(\mu);p)$. Then $P_{\mu q}^{(p)}$ is a normal subgroup of B_μ, and $\cap_{q \ge 1}P_{\mu q}^{(p)} = 1$, since free groups are residually finite p-groups ($\cap_{q \ge 1}\Gamma_q(F(\mu);p) = 1$; see 6.1.9 of [**Rob**]) and Ar is injective. As an easy induction on q shows that $P_\mu/P_{\mu q}^{(p)}$ is a finite p-group for all $q \ge 1$, the result follows.   □

Ohkawa observed that $E \in P_{\mu q}^{(p)}$ if and only if $\bar\mu_I(L(E)) \equiv 0 \bmod (p)$ for all multi-indices I of length $|I| < q$, and used the injectivity of the *mod-(p)* analogue of the Magnus embedding. The intersection over all primes gives the subgroup $P_{\mu q} = \cap_p P_{\mu q}^{(p)}$, which consists of pure braids E such that $\bar\mu_I(L(E)) = 0$ for all multi-indices I of length $|I| < q$.

Analogous representations are central to the work of Habegger and Lin. They define the "universal" Milnor invariant of a $(1, \mu)$-disc link as its image under the Artin representation, and argue that the indeterminacy of the classical Milnor invariants arises in representing the link as the closure of a $(1, \mu)$-disc link.

Let E be a $(1, \mu)$-disc link. The inclusions $e_{\pm} : D_0 \to X(E)$ induce 2-connected, normally surjective homomorphisms from $F(\mu)$ to πE. Let $A_n(E) = e_{+*}^{-1} e_{-*}$ be the corresponding automorphism of $F(\mu)/F(\mu)_{n+1}$. It is immediate that $A_n(E \natural E') = A_n(E) A_n(E')$. If E and $\widetilde{E}$ are concordant $(1, \mu)$-disc links then Stallings' Theorem (applied to the exterior of the concordance) implies that $A_n(E) = A_n(\widetilde{E})$. The *level-n Artin representation* is the induced homomorphism $A_n : C_{1,\mu} \to Aut(F(\mu)/F(\mu)_{n+1})$. Since $A_n(E)$ conjugates each meridian by the corresponding longitude and preserves the element represented by ∂D^2, the image of A_n lies in the subgroup $Aut_0(F(\mu)/F(\mu)_{n+1})$ consisting of the automorphisms which preserve the product $x_1 \cdots x_\mu$ and send each basis element x_i to a conjugate. There is a natural homomorphism from $Aut_0(F(\mu)/F(\mu)_{n+2})$ to $Aut_0(F(\mu)/F(\mu)_{n+1})$, which is easily seen to be surjective, and its kernel $K_n(\mu)$ is central. Thus $Aut_0(F(\mu)/F(\mu)_{n+1})$ is nilpotent of class n.

The level-n Artin representations are surjective, by Theorem 1.1 of [**HL98**]. The proof is by induction on n, and involves a string theoretic analogue of Orr's construction of links realizing given Milnor invariants. (One constructs a map from $D^2 \times D^1$ to the space obtained by adjoining 2-cells to $K(F(\mu)/F(\mu)_{n+1})$ along a meridian basis. The string link is then obtained as the transverse inverse image of the midpoints of the 2-cells.) A further argument using the fact that $H_2(\check{F}(\mu); \mathbb{Z})$ is 0 (see Theorem 13.4) gives a surjective representation of $C_{1,\mu}$ onto $Aut_0(\check{F}(\mu))$. (This was constructed earlier in Chapter I.§3.3 of [**LeD**].) This representation is determined by the images of the longitudes in $\check{F}(\mu)$. Conversely, it determines their images $\bar{l}_i$ in $\overline{F}(\mu)$, by Lemma 13.7, which implies that the centralizers of the generators of $F(\mu)$ in $\overline{F}(\mu)$ are cyclic.

Every automorphism of $F(\mu)$ extends uniquely to an automorphism of $\check{F}(\mu)$, and the representation of $C_{1,\mu}$ onto $Aut_0(\check{F}(\mu))$ extends $Ar|_{P_\mu}$. Hence the natural homomorphism from P_μ to $C_{1,\mu}$ is injective. The image of $Aut_0(F(\mu))$ in $Aut_0(\check{F}(\mu))$ is a proper subgroup (see Theorem 13.6) and so there are pure string links which are not concordant to braids. (Le Dimet gives examples of 2-component pure string links which are not I-equivalent to pure braids.)

We may now sketch the argument of [**HL98**] which identifies the units of the $(1, \mu)$-disc link monoid as the pure braids. Suppose that E and E' are $(1, \mu)$-disc links such that $E \natural E'$ is isotopic to $E_0^{1,\mu}$. It follows easily from Van Kampen's Theorem that $\pi E \cong F(\mu)$. Hence the Artin representation of E is in the image of $Aut_0(F(\mu))$ and so there is a pure braid B such that the Artin representation of $E \natural B^{-1}$ is trivial. But then each longitude of $E \natural B^{-1}$ has trivial image in $\pi(E \natural B^{-1})$ and so bounds a disc in $X(E \natural B^{-1})$. Hence this string link is trivial and so E is isotopic to B.

Habegger and Lin give a general structure theorem (Theorem 2.2 of [**HL98**]) which applies to various equivalence relations coarser than isotopy, including concordance and link-homotopy (but not including isotopy itself, unless $\mu = 1$). Under suitable hypotheses on the equivalence relation $\sim$ the operation $\natural$ induces a group multiplication on the set $SL(\mu)/\sim$ of equivalence classes of $(1, \mu)$-disc links. If we regard a string link as a tangle in the upper (respectively, lower) half ball, stacking defines a left (respectively, right) action of $SL(2\mu)/\sim$ on $SL(\mu)/\sim$. Let $S^R(\mu; \sim)$ be the stabilizer of $E_0^{1,\mu}$ under the *right* action. The set of equivalence classes of μ-component links is then the orbit space of the *left* action of $S^R(\mu; \sim)$ on $SL(\mu)/\sim$.

Their main result for link concordance is as follows.

THEOREM. [**HL98**] *The group $C_{1,2\mu}$ acts on the set $C_{1,\mu}$ (on the left and on the right). The stabilizers of the identity under the left and right actions agree. Let $S(\mu)$ be this common stabilizer. Then closure induces a bijection from the set of left $S(\mu)$ orbits in $C_{1,\mu}$ to $C_1(\mu)$. In particular, a pure string link is null concordant if and only if its closure is null concordant.* □

The concordance classification of 1-links remains far from being settled. Even in the knot theoretic case there are only partial invariants known.

The intersection of $S^R(\mu; \sim)$ with $P_{2\mu}$ acts on the set of boundary string links, and this action passes to an action on the cobordism group $B(1, \mu)$ [**deC99**]. (This paper also gives a criterion for two boundary pure string links to have boundary-concordant closures.)

14.4. Homotopy

The hypotheses of the structure theorem were established for the case of link-homotopy in [**HL90**]. This paper gives an algorithmic approach to describing the set of link-homotopy classes of closed links.

As in §3 above, Stallings' Theorem gives rise to a representation of the group $LH(\mu)$ of link-homotopy classes of $(1, \mu)$-disc links in $Aut_0(RF(\mu))$, the group of automorphisms of the reduced free group $RF(\mu)$ which send each basis element x_i to a conjugate and preserve the product $x_1 \cdots x_\mu$. This representation is an isomorphism, by Theorem 1.7 of [**HL90**]. The first step in the proof is to show that the kernel of the projection of $LH(\mu)$ onto $LH(\mu - 1)$ given by omitting the last string is isomorphic to $RF(\mu - 1)$ and hence that $LH(\mu)$ is a semidirect product $RF(\mu - 1) \rtimes LH(\mu - 1)$. (This is essentially due to Goldsmith, who defined the homotopy braid group and gave a presentation for such groups [**Go74**].) Since there are analogous decompositions $P_\mu \cong F(\mu - 1) \rtimes P_{\mu-1}$ for the pure braid groups an easy induction shows that the natural homomorphism from P_μ to $LH(\mu)$ is an epimorphism. In fact $LH(\mu)$ may be identified with the group of link-homotopy classes of pure braids on the same number of strings.

If $\sigma \in LH(\mu)$ is decomposed as $\sigma = \theta g$ with $\theta \in LH(\mu - 1)$ and $g \in RF(\mu - 1)$ then a *partial conjugate* of σ is an element of the form $\theta h g h^{-1}$, where $h \in RF(\mu - 1)$ also. Conjugate string links determine link-homotopic closed links, since cyclically permuted products of string links yield the same closed link. Since partial conjugation corresponds to conjugating one component of the closed link in the link group of the complementary closed link, partially conjugate string links also determine link-homotopic closed links. Conversely, if the closures of two string links are link-homotopic then the string links are equivalent under the equivalence relation generated by conjugacy and partial conjugacy. The analysis of this equivalence relation in [**HL90**] involves a Markov-type theorem, using moves suggested by Levine, which do not change the number of strings (in contrast to the classical Markov moves).

14.5. Milnor invariants again

Habegger and Lin gave also an algebraic version of their structure theorem. Define homomorphisms $q : F(2\mu) \to F(\mu)$, $i_0 : F(\mu) \to F(2\mu)$ and $i_1 : F(\mu) \to F(2\mu)$ by $q(x_i) = i_0(x_i) = x_i$, $q(x_{2\mu+1-i}) = x_i^{-1}$ and $i_1(x_i) = x_{2\mu+1-i}^{-1}$, for $1 \leq i \leq \mu$. We shall use the same symbols for the corresponding maps between nilpotent quotients. If $a \in Aut_0(F(\mu)/F(\mu)_{n+1})$ let $\tilde{a}(x_i) = a(x_i)$ and $\tilde{a}(x_{\mu+i}) = x_{\mu+i}$, for $1 \leq i \leq \mu$. If $A \in Aut_0(F(2\mu)/F(2\mu)_{n+1})$ then $a_j = q\tilde{a}Ai_j$ is an automorphism of $F(\mu)/F(\mu)_{n+1}$, for $j = 0$ and 1. Then $a.A = a_1^{-1}a_0$ and $A.a = (a^{-1}A^{-1})^{-1}$ determine right and left actions of $Aut_0(F(2\mu)/F(2\mu)_{n+1})$ on the set $Aut_0(F(\mu)/F(\mu)_{n+1})$, which are compatible with the geometrically defined actions of $C_{1,2\mu}$ on $C_{1,\mu}$ via the level-n Artin representation. The results in this section are all from [**HL98**].

LEMMA 14.2. *Let* $j_n = id_G$, *where* $G = F(\mu)/F(\mu)_n$, *and let* $S_n^R(\mu)$ *and* $S_n^L(\mu)$ *be the stabilizers of* j_{n+1} *under the right and left actions (respectively) of* $Aut_0(F(2\mu)/F(2\mu)_{n+1})$. *Then*

(1) $S_n^R(\mu) = S_n^L(\mu)$;
(2) *the image of* $S_n^R(\mu)$ *in* $Aut_0(F(2\mu)/F(2\mu)_n)$ *is* $S_{n-1}^R(\mu)$;
(3) $S_n^R(\mu)$ *acts unipotently on the set* $Aut_0(F(\mu)/F(\mu)_{n+1})$.

PROOF. The proofs of (1) and (2) are straightforward calculations.

If $k \in K_{n-1}(\mu)$ and $A \in S_n^R(\mu)$ then $\tilde{k}A = A\tilde{k}$. Hence $k.A = 1.\tilde{k}A = 1.A\tilde{k} = 1.\tilde{k} = k$, and so the action of $S_n^R(\mu)$ on the subgroup $K_{n-1}(\mu)$ is trivial. It follows that if $a \in Aut_0(F(\mu)/F(\mu)_{n+1})$ induces j_r for some $r < n$ then $a.A$ and a have the same image in $Aut_0(F(\mu)/F(\mu)_{r+1})$. Hence (3) holds also. $\square$

THEOREM 14.3. *Let E be a $(1, \mu)$-disc link with closure $L = L(E)$. Then $A_n(E) = 1$ if and only if all Milnor invariants of L of length $\leq n$ vanish.*

PROOF. Let $\theta : F(\mu)/F(\mu)_{n+1} \cong \pi E/(\pi E)_{n+1} \to \pi L/(\pi L)_{n+1}$ be the homomorphism induced by the inclusion of the meridians. Then each condition is equivalent to θ being an isomorphism.   □

COROLLARY 14.3.1. *If all Milnor invariants of $L(E)$ of length $\leq n$ vanish $A_{n+1}(E) \in K_n(\mu)$ is an invariant of the concordance class of $L(E)$. The group $K_n(\mu)$ is isomorphic to the group of $(1, \mu)$-disc links whose $\bar{\mu}$-invariants of length $\leq n$ vanish modulo sum with links whose $\bar{\mu}$-invariants of length $\leq n+1$ vanish. Hence the number of linearly independent $\bar{\mu}$-invariants of length $n + 1$ is the rank of $K_n(\mu)$.*   □

Let $h_n^\mu : (F(\mu)_n/F(\mu)_{n+1})^\mu \to F(\mu)_{n+1}/F(\mu)_{n+2}$ be the homomorphism which sends $(y_1, \ldots, y_\mu)$ to $\Pi[y_i, x_i]$. Then h_n^μ is onto, and $K_n(\mu) \cong \mathrm{Ker}(h_n^\mu)$. (The isomorphism sends $(y_1, \ldots, y_\mu)$ to the automorphism which conjugates each x_i by y_i, for $1 \leq i \leq \mu$.) Thus $K_n(\mu)$ has rank $\mu N_n(\mu) - N_{n+1}(\mu)$, where $N_n(\mu)$ is the rank of $F(\mu)_n/F(\mu)_{n+1}$. (Let mb be the Möbius function. Then $N_n(\mu) = n^{-1}\Sigma_{d|n} mb(d)\mu^{n/d}$. See page 330 of [MKS].) The number of linearly independent $\bar{\mu}$-invariants of length $n + 1$ was first calculated in [Or89].

The $\bar{\mu}$ invariants of pure string links are all of finite type [Li97].

14.6. The Gassner representation

An automorphism $\phi \in \mathrm{Aut}_0(\check{F}(\mu))$ induces the identity on the abelianization $\check{F}(\mu)/\check{F}(\mu)' \cong \mathbb{Z}^\mu$, and hence on its group ring Λ_μ. Hence it also induces an automorphism of the Λ_μ-modules $A(\check{F}(\mu)) = \Lambda_\mu \otimes_{\check{\Gamma}} I(\check{F}(\mu))$ and $B(\check{F}(\mu)) = \check{F}(\mu)'/\check{F}(\mu)''$. The localisation of $A(\check{F}(\mu))$ with respect to the multiplicative system $S = 1 + I_\mu$ is free of rank μ, and so determines a $\mu \times \mu$-matrix $\rho(\phi)$ over $\Lambda_\mu S$, with entries $[\rho(\phi)]_{ij} = \partial_j(\phi(x_i))$, by the work of Le Dimet [LD92]. (See Theorem 13.8.) The function ρ is a homomorphism (essentially by the chain rule) and composition with the Artin representation gives the *Gassner representation* of the $(1, \mu)$-disc link monoid. Since the automorphisms in $Aut_0(\check{F}(\mu))$ preserve the product $x_1 \cdots x_\mu$ this representation is equivalent to the sum of one of rank $\mu - 1$ with the trivial rank 1 representation. (In the case of pure braids the representation has values in $GL(\mu, \Lambda_\mu)$. The argument of [**Bir**] for this case applies without essential change to string links.)

Our exposition of these representations is taken from [**KLW01**].

LEMMA 14.4. *Let X be the exterior of a $(1, \mu)$-disc link E, and fix a basepoint $*$ in ∂D^2. Then the inclusions $e_\pm : D_0 \to X$ as the top and bottom of X induce isomorphisms*

$$H_1(D_0; \Lambda_{\mu S}) = (\Lambda_{\mu S})^{\mu - 1} \cong H_1(X; \Lambda_{\mu S}) \quad and$$

$$H_1(D_0, *; \Lambda_{\mu S}) = (\Lambda_{\mu S})^\mu \cong H_1(X, \{*\} \times D^1; \Lambda_{\mu S}).$$

PROOF. Let $C_* = C_*(X, D_\pm; \Lambda_\mu) = C_*(X', D'_\pm)$ be the chain complex of the maximal abelian cover of the pair $(X, D_\pm = e_\pm(D_0))$. Then $\mathbb{Z} \otimes_\Lambda C_*$ is acyclic and so $C_{*S} = \Lambda_{\mu S} \otimes_\Lambda C_*$ is also acyclic, by Nakayama's Lemma. Hence the maps $e_\pm$ induce isomorphisms on the absolute homology modules, and similarly on the relative homology. $\qquad\qquad\square$

The *Gassner representation* of E is $\gamma(E) = H_1(e_+)^{-1} H_1(e_-)$ in $Aut(H_1(D_0, *; \Lambda_{\mu S})) = GL(\mu, \Lambda_{\mu S})$. The *reduced* Gassner representation is the corresponding automorphism $\tilde{\gamma}(E)$ of $H_1(D_0; \Lambda_{\mu S})$. It is immediate from the definitions that these representations are homomorphisms. (Using cohomology instead of homology gives the dual representation, with respect to the Kronecker pairing.)

We may use the total linking number infinite cyclic cover instead to define the (reduced) *Burau* representation $\tau\gamma$, with values in $GL(\mu, \Lambda_S)$ (where $\Lambda_S = \mathbb{Z}[t, t^{-1}, (t - 1)^{-1}]$), for any μ-component string link. As automorphisms of $F(\mu)$ which conjugate and permute the basis elements preserve the homomorphism $\tau : F(\mu) \to \mathbb{Z}$ defined by $\tau(x_i) = 1$ for $1 \leq i \leq \mu$, they induce automorphisms of $\Lambda_S^\mu \cong \Lambda_S \otimes_{\check{\Gamma}} I(\check{F}(\mu))$. These representations are homomorphisms on the full μ-component string link monoid. (We may also construct representations of string links with fixed colourings, interpolating between the Burau and Gassner representations.)

The group of the closure $L = L(E)$ of a braid E is $\pi L \cong F(\mu)/\langle\langle g^{-1}Ar(E)((g))\rangle\rangle$. The Artin representation Ar induces the Gassner representation on $H_1(F(\mu); \Lambda_\mu)$, and it follows easily that $I - \gamma(E)$ is a presentation matrix for $A(L)$. There is a similar connection between the Burau representation on $H_1(F(\mu); \Lambda_1) = K/K'$ (where $K = \mathrm{Ker}(\tau)$) and the reduced Alexander module: $I - \tau\gamma(E)$ is then a presentation matrix for $A_{red}(L)$. (See Chapter 3 of [**Bir**].)

These results extend readily to string links. If $L = L(E)$ for E a μ-component string link $\pi L \cong \pi E/\langle\langle e_+(g) = e_-(g), \forall g \in F(\mu)\rangle\rangle$. If E is pure then $I - \gamma(E)$ is a presentation matrix for $A(L)_S$, by a Mayer-Vietoris argument. Similarly, if E is an arbitrary string link then $I - \tau\gamma(E)$ is a presentation matrix for $A_{red}(L)_S$.

LEMMA 14.5. *If $E(0)$ and $E(1)$ are I-equivalent $(1,\mu)$-disc links then $\gamma(E(0)) = \gamma(E(1))$.*

PROOF. Let $\mathcal{E}$ be an I-equivalence between $E(0)$ and $E(1)$. The inclusions of $X(E(0))$ and $X(E(1))$ into $X(\mathcal{E})$ induce isomorphisms on homology with coefficients $\Lambda_{\mu S}$, as in the previous lemma. The compositions of these isomorphisms with $H_1(e_+)$ are identical, as are the compositions with $H_1(e_-)$. The lemma follows easily. □

In particular, the Gassner representation factors through $C_{1,\mu}$ [**LD92**]. It is also trivial for boundary string links.

LEMMA 14.6. *Let E be a boundary string link. Then $\gamma(E) = 1$ and $\tilde\gamma(E) = 1$.*

PROOF. If E is boundary there is a map $f : X \to D_0$ such that $fe_+ = fe_- = id_{D_0}$ and $f(\{*\} \times D^1) = *$. Hence the homomorphisms $H_1(f)$ on relative and absolute homology modules are isomorphisms, so $H_1(e_+)^{-1}H_1(e_-) = H_1(f)H_1(f)^{-1} = 1$. □

Let $\langle\,,\,\rangle_S$ be the intersection pairing of $H_1(D_0; \Lambda_{\mu S})$ with itself into $\Lambda_{\mu S}$. This is skew-hermitean, and is perfect since $H_*(\partial D_0 : \Lambda_{\mu S}) = 0$. The following result shows that the reduced Gassner representation is unitary with respect to this pairing.

THEOREM 14.7. *Let E be a $(1, \mu)$-disc link. Then*

$$\langle \tilde{\gamma}(E)(\alpha), \tilde{\gamma}(E)(\beta) \rangle_S = \langle \alpha, \beta \rangle_S, \quad \text{for all } \alpha, \beta \in H_1(D_0; \Lambda_\mu S).$$

PROOF. Let $\partial_\pm X = e_\pm(D_0)$, and let $\langle \, , \, \rangle_{XS}$ be the intersection pairing of $H_2(X, \partial_+ X \cup \partial_- X; \Lambda_\mu S)$ with $H_1(X; \Lambda_\mu S)$. Given $\alpha \in H_1(D_0; \Lambda_\mu S)$ there is a class A_α in $H_2(X, \partial_+ X \cup \partial_- X; \Lambda_\mu S)$ whose image in $H_1(\partial_+ X; \Lambda_\mu S) \oplus H_1(\partial_- X; \Lambda_\mu S)$ under the connecting homomorphism is $\delta(A_a) = (\tilde{\gamma}(E)(\alpha), -\alpha)$. Then $\langle \tilde{\gamma}(E)(\alpha), \tilde{\gamma}(E)(\beta) \rangle_S = \langle A_\alpha, H_1(e_+)(\beta) \rangle_{XS} = \langle A_\alpha, H_1(e_-)(\beta) \rangle_{XS} = \langle \alpha, \beta \rangle_S$. $\square$

There is a similar result for the Burau representation and the localized intersection form on the homology of the infinite cyclic cover D_0^τ. This unitarity was first observed for braids by Squier, who showed that the images of the standard generators of B_n are unitary reflections with respect to an algebraically defined hermitean form on $(K/K') \otimes_\Lambda (\Lambda[s]/(s^2 - t))$ [**Sq84**]. Goldschmidt embedded D_0^τ in $\mathbb{R}^3$ and used linking numbers to give a geometric interpretation of this construction [**Go90**]. (The Squier-Goldschmidt pairing may be related to the intersection form via the formula $\ell(\alpha_+, \beta) - \ell(\alpha_-, \beta) = \langle \alpha, \beta \rangle$, for 1-cycles α, β on D_0^τ. It turns out to be $(s - \bar{s})^{-1} = s/(t-1)$ times this form, and thus is indeed hermitean rather than skew-hermitean.) He derived a sesquilinear pairing with values in $\mathbb{Q}(t)/\Lambda$ on $TA_{red}(L(E))$, for E a braid. (It is not clear how it is related to the Blanchfield pairing. The terminology in [**Go90**] differs from ours. In particular, his "Alexander module" is $H_1(X; \Lambda_1) = H_1(X^\tau; \mathbb{Z})$.)

Let Γ be a compact connected Lie group. Le Dimet uses Vogel's extension theorem (Theorem 13.9) to construct a representation of $Aut_0(\check{F}(\mu))$ in the group of germs of diffeomorphisms at the identity of Γ^μ, and shows that the representation obtained as the differential of the restriction to a maximal torus of Γ decomposes as the direct sum of the trivial representation and copies of the Gassner representation [**LD98**].

14.7. High dimensions

One difficulty in using homology surgery for general links is that there is no universal candidate for a target space or group.

Le Dimet overcame this problem for disc links (in principle), using the (Vogel) localization of $\vee^\mu S^1$ as a universal target for reference maps for (n, μ)-disc links; algebraically, the free group $F(\mu)$ is enlarged to its algebraic closure $\check{F}(\mu)$. Fix a homotopy equivalence $s : X_\mu^{n+1} = X(E_0^{n-1,\mu}) \to \vee^\mu S^1$, and let $\eta_\vee = \eta_{\vee^\mu S^1}$. Let G_{n+1} be the set of homotopy classes (rel ∂X_μ^{n+1}) of maps $h : X_\mu^{n+1} \to E(\vee^\mu S^1)$ such that $h|_{\partial X_\mu^{n+1}} = \eta_\vee s|_{\partial X_\mu^{n+1}}$. Since $X_\mu^{n+1} \cong X_\mu^n \times [0,1]$ we may add and invert such homotopy classes by stacking and flipping with respect to the $[0,1]$ factor, and so G_{n+1} has a natural group structure, which is abelian if $n \geq 2$. Let $\eta_{X(E)} : X(E) \to E(\vee^\mu S^1)$ be the Vogel localization of $X(E)$. The *Le Dimet homotopy invariant* of an (n, μ)-disc link E is the element $\partial([E]) = \eta_{X(E)}|_{X_\mu^{n+1} \times \{0\}} \in G_{n+1}$. This depends only on the concordance class of E, and ∂ is a homomorphism of abelian groups if $n \geq 2$.

The set G_{n+1} may also be described as the set of homotopy classes of maps from $\natural^\mu(S^n \times S^1)$ to $E(\vee^\mu S^1)$ which induce the inclusion $\phi : F(\mu) \to \check{F}(\mu)$ on fundamental groups. The bijection sends a map h as above to $h \cup s$. Since $\partial X(E) \cong \partial X_\mu^{n+2} \cong \natural^\mu(S^n \times S^1)$, we may then define $\partial([E])$ as the class of $f|_{\partial X(E)}$, where $f : X(E) \to E(\vee^\mu S^1)$ is a homology equivalence such that $\pi_1(f|_{\partial X(E)}) = \phi : F(\mu) \to \check{F}(\mu)$. (See [**LMO93**].)

Let Φ be the diagram

$$
\begin{array}{ccc}
\mathbb{Z}[F(\mu)] & \xrightarrow{\;1\;} & \mathbb{Z}[F(\mu)] \\
\phi_* \downarrow & & \varepsilon \downarrow \\
\mathbb{Z}[\check{F}(\mu)] & \xrightarrow{\;\varepsilon\;} & \mathbb{Z}.
\end{array}
$$

The set $\mathcal{S}_{n+2}$ of homology equivalences of bounded $(n+2)$-manifolds with X_μ^{n+2} (rel homeomorphisms of the boundary) may be identified with the homology surgery obstruction group $\Gamma_{n+3}^h(\Phi)$ (if $n \geq 4$), and there is a long exact sequence

$$
\ldots \Gamma_{n+3}^h(\Phi) \to C_{n,\mu} \to G_{n+1} \to \cdots \to G_4 \to \Gamma_5^h(\Phi)
$$

An (n, μ)-disc link (with $n \geq 4$) is concordant to a SHB link if and only if it has trivial Le Dimet homotopy invariant [**LMO93**]. This is always the case if $E(\vee^\mu S^1)$ is aspherical, for then the groups G_n are trivial.

The group $B(n, \mu)$ may be identified with a Γ-group if $n \geq 4$. In particular, $B(2q, \mu) = 0$ for all $q \geq 2$ and so boundary disc links of even dimension ≥ 4 are nullconcordant.

Closure determines a bijection from $C_{n,\mu}$ to the set of concordance classes of weakly based links $C'_{n,\mu}$, and forgetting basings induces a bijection from the set of left $\breve{\mathcal{H}}(\mu)$ orbits in $C_{n,\mu}$ to $C_n(\mu)$. Actual computation of these groups and actions remains a difficult problem. (See [**LeD**].)

A $2q$-link L is null concordant if there is a 2-connected, normally surjective homomorphism from πL to a finitely presentable group G such that $H_i(G; \mathbb{Z}) = 0$ for $1 < i \leq 2q + 2$. (Note that all string link groups have the latter property.) This was proven for $q \geq 2$ by De Meo, using homological surgery [**De81**], and for $q = 1$ by Cochran (under slightly weaker hypotheses on G), using Spin cobordism arguments [**Co84**]. In particular, even dimensional SHB links are slice. It is an open question whether this is so for all even dimensional links. More generally, one may ask whether every high dimensional link is concordant to a SHB link.

Bibliography

Books

[AM] Atiyah, M. and MacDonald, I.G. *Introduction to Commutative Algebra*,
Addison-Wesley, Reading - Menlo Park - London - Don Mills (1969).

[Bai] Bailey, J.L. *Alexander Invariants of Links*,
Ph.D. thesis, University of British Columbia (1977).

[BM] Bass, H. and Morgan, J.W. (editors) *The Smith Conjecture*,
Academic Press, New York - London (1984).

[Bir] Birman, J. *Braids, Links and Mapping Class Groups*,
Annals of Mathematics Study 82,
Princeton University Press, Princeton (1974).

[Bou] Bourbaki, N. *Commutative Algebra*,
Addison-Wesley/Hermann et Cie, Boston - Paris (1972).

[Bsf] Bousfield, A.K. *Homological Localizations for Groups and π-Modules*,
Memoir 186, American Mathematical Society, Providence (1977).

[Bre] Bredon, G. *Introduction to Compact Transformation Groups*,
Academic Press, New York - London (1972).

[BK] Brieskorn, E. and Knörrer, H. *Plane Algebraic Curves*,
Birkhäuser Verlag, Basel - Boston - Stuttgart (1986).

[BZ] Burde, G. and Zieschang, H. *Knots*,
Second edition. de Gruyter Studies in Mathematics 5,
W. de Gruyter, Berlin - New York (2003).

[Coc] Cochran, T.D. *Derivatives of Links*,
Memoir 427, American Mathematical Society, Providence (1990).

[Coh] Cohn, P.M. *Free Rings and their Relations*,
Academic Press, New York - London (1985).

[CF] Crowell, R.H. and Fox, R.H. *Introduction to Knot Theory*,
Ginn and Co., Boston (1963).
Second revised edition, Graduate Texts in Mathematics 57,
Springer-Verlag, Berlin - Heidelberg - New York (1977).

[EN] Eisenbud, D. and Neumann, W.D. *Three-dimensional Link Theory and Invariants of Plane Curve Singularities*,
Annals of Mathematics Study 110,

Princeton University Press, Princeton (1985).

[Fen] Fenn, R. (editor) *Low Dimensional Topology*
London Mathematical Society Lecture Note Series 95,
Cambridge University Press, Cambridge - New York - Melbourne (1985).

[For] Fort, M.K. (Jr) (editor) *Topology of 3-Manifolds and Related Topics*
Prentice-Hall, Englewood Cliffs, N.J. (1962).
Reprinted with new introduction by D.Silver,
Dover Publications Inc, Mineola, N.Y. (2010).

[FQ] Freedman, M.H. and Quinn, F. *Topology of 4-Manifolds,*
Princeton University Press, Princeton (1990).

[Hau] Hausmann, J.-C. (editor) *Knot Theory,*
Proceedings, Plans-sur-Bex, Switzerland 1977,
Lecture Notes in Mathematics 685,
Springer-Verlag, Berlin - Heidelberg - New York (1978).

[Hem] Hempel, J. *3-Manifolds,*
Annals of Mathematics Study 86,
Princeton University Press, Princeton (1976).

[Hil] Hillman, J.A. *2-Knots and theit Groups,*
Australian Mathematical Society Lecture Series 5,
Cambridge University Press, Cambridge - New York - Melbourne (1989).

[Jun] Jung, H.W.E. *Einführung in die Theorie der algebraischer Funktionen zwei Verändlicher,* Akademie-Verlag, Berlin (1951).

[Kai] Kaiser, U. *Link Theory in Manifolds,*
Lecture Notes in Mathematics 1669,
Springer-Verlag, Berlin - Heidelberg - New York (1997).

[Ka1] Kawauchi, A. (editor) *Knots 90,*
W. de Gruyter, Berlin (1992).

[Ka2] Kawauchi, A. *A Survey of Knot Theory,*
Birkhäuser Verlag, Basel - Boston - Berlin (1996).

[Kaz] Kazez, W.H. (editor) *Geometric Topology,*
American Mathematical Society/International Press (1997).

[KN] Koschorke, U. and Neumann, W.D. (editors) *Differential Topology, Siegen*
Lecture Notes in Mathematics 1350,
Springer-Verlag, Berlin - Heidelberg - New York (1988).

[Lam] Lam, T.Y. *Serre's Conjecture,*
Lecture Notes in Mathematics 635,
Springer-Verlag, Berlin - Heidelberg - New York (1978).

[Lan] Lang, S. *Algebra,*
Addison-Wesley, Reading - Menlo Park - London - Don Mills (1965).
Revised 3rd edition, Graduate Texts in Mathematics 211,
Springer-Verlag, Berlin - Heidelberg - New York (2002).

[LeD] Le Dimet, J.-Y. *Cobordisme d'Enlacements de Disques*,
       Mem. Soc. Math. France 32 (1988).
[Les] Lescop, C. *Global Surgery Formula for the Casson-Walker Invariant*,
       Annals of Mathematics Study 140,
       Princeton University Press, Princeton (1996).
[Lev] Levine, J. *Algebraic Structure of Knot Modules*,
       Lecture Notes in Mathematics 772,
       Springer-Verlag, Berlin - Heidelberg - New York (1980).
[Lom] Lomonaco, S.J., Jr, (editor) *Low Dimensional Topology*,
       Contemporary Mathematics 20,
       American Mathematical Society, Providence (1983).
[McC] McCleary, J. *User's Guide to Spectral Sequences*,
       Mathematics Lecture Series 12,
       Publish or Perish, Inc., Wilmington (1985).
[MKS] Magnus, W., Karras, A. and Solitar, D. *Combinatorial Group Theory*,
       Interscience Publishers, New York - London - Sydney (1966).
       Second revised edition, Dover Publications inc, New York (1976).
[Mat] Matlis, E. *Torsion-free Modules*,
       University of Chicago Press, Chicago - London (1972).
[Mil] Milnor, J.W. *Singularities of Complex Hypersurfaces*,
       Annals of Mathematics Study 61,
       Princeton University Press, Princeton (1968).
[Mor] Morisita, M. *Knots and Primes: An Introduction to Arithmetic Topology*,
       Unitext, Springer-Verlag, Berlin - Heidelberg - New York (2011).
[Neu] Neuwirth, L.P. (editor) *Knots, Groups and 3-Manifolds*,
       Annals of Mathematics Study 84,
       Princeton University Press, Princeton (1975).
[Qua] Quach thi Cam Van *Invariants des noeuds classiques fibrés*,
       doctoral thesis, Université de Genève (1981).
[Rob] Robinson, D.J.S. *A Course in the Theory of Groups*,
       Second edition. Graduate Texts in Mathematics 80,
       Springer-Verlag, Berlin - Heidelberg - New York (1996).
[Rol] Rolfsen, D. *Knots and Links*,
       Publish or Perish, Inc., Berkeley (1976).
[RS]   Rourke, C. and Sanderson, B.J. *Introduction to Piecewise-linear Topology*,
       Ergebnisse der Mathematik 69,
       Springer-Verlag, Berlin - Heidelberg - New York (1972).
[Ser] Serre, J.-P. *Corps Locaux*,
       Hermann et Cie, Paris (1962). Translated by M.J.Greenberg as
       Local Fields, Graduate Texts in Mathematics vol. 67,
       Springer-Verlag, Berlin - Heidelberg - New York (1979).

[She] Sheiham, D. Invariants of Boundary Link Cobordism,
Memoir 784, American Mathematical Society, Providence (2003).

[Sto] Stoltzfus, N. *Unravelling the Integral Knot Concordance Group*,
Memoir 192, American Mathematical Society, Providence (1977).

[Tur] Turaev, V.G. *Introduction to Combinatorial Torsions*
(Notes taken by F.Schlenk), Lectures in Mathematics ETH Zürich,
Birkhäuser Verlag, Basel (2001).

[Wal] Wall, C.T.C. *Surgery on Compact Manifolds*,
Academic Press, New York - London (1970).

The books [**BZ**] and [**Ka2**] have comprehensive bibliographies of papers on
knot theory published before 2003 and 1995, respectively.

Journal articles

[A'C73] A'Campo, N. Le nombre de Lefshetz d'une monodromie,
Indag. Math. 35 (1973), 113–118.

[A'C73'] A'Campo, N. Sur la monodromie des singularités isolées d'hypersurfaces complexes, Invent. Math. 20 (1973), 147–169.

[Ad75] Adams, J.F. Idempotent functors in homotopy theory,
in *Manifolds Tokyo 1973* (edited by A.Hattori),
Mathematical Society of Japan,
University of Tokyo Press, Tokyo (1975), 247–253.

[ASR94] Adams, S. and Sarnak, P. Betti numbers of congruence varieties, (with
an appendix by Z.Rudnick), Israel J. Math. 88 (1984), 31–72.

[AS88] Aitchison, I.R. and Silver, D.S. On certain fibred ribbon disc pairs,
Trans. Amer. Math. Soc. 306 (1988), 529–551.

[AFR97] Aravinda, C.S., Farrell, F.T. and Roushon, S.K. Surgery groups of knot
and link complements, Bull. London Math. Soc. 29 (1997), 400–406.

[AY81] Asano, K. and Yoshikawa, K. On polynomial invariants of fibered 2-knots,
Pacific J. Math. 97 (1981), 267–269.

[AB62] Auslander, M. and Buchsbaum, D.A. Invariant factors and two criteria
for projectivity of modules, Trans. Amer. Math. Soc. 104 (1962), 516–522.

[Ba85] Barlet, D. Forme hermitienne canonique sur la cohomologie de la fibre de
Milnor d'une hypersurface a singularité isolée,
Invent. Math. 81 (1985), 115–153.

[Ba05] Barlet, D. Modules de Brieskorn et formes hermitiennes pour une singularité isolée d'hypersurface, in *Singularités*,
Inst. Élie Cartan, 18, Univ. Nancy, Nancy (2005), 19–46.

[BT99] Bartels, A. and Teichner, P. All two dimensional links are null homotopic,
Geom. Topol. 3 (1999), 235–252.

[Ba01] Bartels, A. Higher dimensional links are singular slice,
Math. Ann. 320 (2001), 547–576.

[Ba64] Bass, H. K-theory and stable algebra,
Publ. Math. I.H.E.S. 22 (1964), 1–50.

[Ba64'] Bass, H. Projective modules over free groups are free,
J. Algebra 1 (1964), 367–373.

[Ba69] Baumslag, G. Groups with the same lower central series as a relatively
free group. II. Properties, Trans. Amer. Math. Soc. 142 (1969), 507–538.

[BDH80] Baumslag, G., Dyer, E. and Heller A. The topology of discrete groups,
J. Pure Appl. Math. 16 (1980), 1–47.

[Be98] Bellis, P. Realizing homology boundary links with arbitrary patterns,
Trans. Amer. Math. Soc. 350 (1998), 87–100.

[BS98] Benedetti, R. and Shiota, M. On real algebraic links in S^3,
Boll. Unione Mat. Italiana 8 (1998), 585–609.

[BP94] Berceanu, B. and Papadima, S. Cohomologically generic 2-complexes and
3-dimensional Poincaré complexes, Math. Ann. 298 (1994), 457–480.

[Bl57] Blanchfield, R.C. Intersection theory of manifolds with operators with
applications to knot theory, Ann. Math. 65 (1957), 340–356.

[BP08] Boileau, M. and Paoluzzi, L. On cyclic branched coverings of prime knots,
J. Topol. 1 (2008), 557–583.

[BW83] Boileau, M. and Weber, C. Le problème de Milnor sur le nombre gor-
dien des noeuds algébriques, in Noeuds, Tresses et Singularités (edited by
C.Weber), Monographie No 31 de L'Enseign. Math., Geneva (1983), 49-98.

[Br70] Brieskorn, E. Die Monodromie der isolierten Singularitäten von
Hyperflächen, Manus. Math. 2 (1970), 103-161.

[Br81] Brin, M. Torsion free actions on 1-acyclic manifolds and the loop theorem,
Topology 20 (1981), 353–363.

[BE77] Buchsbaum, D.A. and Eisenbud, D. What annihilates a module?,
J. Algebra 47 (1977), 231–243.

[Bu66] Burde, G. Alexander Polynome Neuwirtscher Knoten,
Topology 5 (1966), 321–330.

[Bu78] Burde, G. Uber periodische Knoten,
Arch. Math. (Basel) 30 (1978), 487–492.

[BM70] Burde, G. and Murasugi, K. Links and Seifert fibre spaces,
Duke Math. J. 37 (1970), 89–93.

[CCMP12] Cantarella, J., Cornish, J., Mastin, M. and Parsley, J. The 27 possible
intrinsic symmetry groups of two-component links,
arXiv[math.GT]:1201.2722

[CS80] Cappell, S.E. and Shaneson, J.L. Link cobordism,
Comment. Math. Helvetici 55 (1980), 20–49.

[CG86] Casson, A. and Gordon, C.McA. Cobordism of classical knots,
in A la Recherche de la Topologie Perdue, (edited by L.Guillou and A.Marin),
Progress in Mathematics, Vol. 62, Birkhäuser Verlag (1986), 181–197.

[CP84] Cenkl, B. and Porter, R. Lazard completion of a group and free graded
differential algebra models over subrings of the rationals,
Topology 23 (1984), 445–464.

[CF88] Cervantes, L. and Fenn, R. Boundary links are homotopy trivial,
Quarterly J. Math. (Oxford) 39 (1988), 151–158.

[CK99] Cha, J.C. and Ko, K.H. Signature invariants of links from irregular covers
and non-abelian covers, Math. Proc. Cambridge Phil. Soc. 127 (1999), 67–81.

[Ch04] Cha, J.C. A characterization of the Murasugi polynomial of an
equivariantly slice knot, arXiv:math.GT/0404403 v1 (22 April 2004).

[Ch51] Chen, K.T. Integration in free groups,
Ann. Math. 54 (1951), 147–162.

[CFL58] Chen, K.T., Fox, R.H. and Lyndon, R.C. Free differential calculus IV.
The quotient groups of the lower central series, Ann. Math. 68 (1958), 81–95.

[Co84] Cochran, T.D. Slice links in S^4,
Trans. Amer. Math. Soc. 285 (1984), 389–401.

[Co87] Cochran, T.D. Link concordance invariants and homotopy theory,
Invent. Math. 90 (1987), 635–645.

[Co90] Cochran, T.D. Links with trivial Alexander's module but non vanishing
Massey products, Topology 29 (1990), 189–204.

[Co91] Cochran, T.D. k-cobordism for links in S^3,
Trans. Amer. Math. Soc. 327 (1991), 641–654.

[Co96] Cochran, T.D. Non-trivial links and plats with trivial Gassner matrices,
Math. Proc. Cambridge Phil. Soc. 119 (1996), 43–53.

[CL91] Cochran, T.D. and Levine, J. Homology boundary links and the Andrews-
Curtis conjecture, Topology 30 (1991), 231–239.

[CO93] Cochran, T.D. and Orr, K.E. Not all links are concordant to boundary
links, Ann. Math. 138 (1993), 519–554.

[CO94] Cochran, T.D. and Orr, K.E. Homology boundary links and Blanchfield
forms: concordance classification and new tangle-theoretic constructions,
Topology 33 (1994), 397–427.

[CO98] Cochran, T.D. and Orr, K.E. Stability of lower central series of compact
3-manifold groups, Topology 37 (1998), 497–526.

[CO99] Cochran, T.D. and Orr, K.E. Homology cobordism and generalizations
of Milnor's invariants, J. Knot Theory Ramif. 8 (1999), 429–436.

[COT03] Cochran, T.D., Orr, K.E. and Teichner, P. Knot concordance, Whitney
towers and L^2 signatures, Ann. Math. 157 (2003), 433–519.

[CS99] Cohen, D.C. and Suciu, A.I. Alexander invariants of complex hyperplane
arrangements, Trans. Amer. Math. Soc. 351 (1999), 4043–4067.

[CN96] Coleman, R.F. and Nicolau, M. Periodicity and the monodromy theorem,
Duke Math. J. 82 (1996), 369–380.

[Cn59] Conner, P.E. Transformation groups on a $K(\pi, 1)$, II,

Michigan Math. J. 6 (1959), 413–417.

[Co70] Conway, J.H. An enumeration of knots and links, and some of their algebraic properties, in *Computational Problems in Abstract Algebra* (edited by J.C.Leech), Pergamon, Oxford (1970), 329–358.

[Cp82] Cooper, D. The universal abelian cover of a link, in *Low-Dimensional Topology* (edited by R.Brown and T.L.Thickstun), London Mathematical Society Lecture Note Series 48 (1982), 51–66.

[CM83] Coray, D. and Michel, F. Knot cobordism and amphicheirality, Comment. Math. Helvetici 58 (1983), 601–616.

[Cr61] Crowell, R.H. Corresponding group and module sequences, Nagoya Math. J. 19 (1961), 27–40.

[Cr64] Crowell, R.H. The annihilator of a link module, Proc. Amer. Math. Soc. 15 (1964), 696–700.

[Cr65] Crowell, R.H. Torsion in link modules, J. Math. Mech 14 (1965), 289–298.

[Cr71] Crowell, R.H. The derived module of a homomorphism, Adv. Math. 6 (1971), 210–238.

[Cr76] Crowell, R.H. (letter to N.F.Smythe, May 1976).

[CSt69] Crowell, R.H. and Strauss, D. On the elementary ideals of link modules, Trans. Amer. Math. Soc. 142 (1969), 93–109.

[CS00] Cutkosky, S.D. and Srinivasan, H. The algebraic fundamental group of a curve singularity, J. Algebra 230 (2000). 101–126.

[Da95] Davis, J.F. The homology of cyclic branched covers of S^3, Math. Ann. 301 (1995), 507–518.

[Da06] Davis, J.F. A two-component link with Alexander polynomial one is concordant to the Hopf link, Math. Proc. Camb. Phil. Soc. 140 (2006), 265–268.

[DL91] Davis, J.F. and Livingston, C. Alexander polynomials of periodic knots, Topology 30 (1991), 551–564.

[DL91'] Davis, J.F. and Livingston, C. Periodic knots, Smith theory and Murasugi's congruence, L'Enseign. Math. 37 (1991), 1–9.

[DN02] Davis, J.F. and Naik, S. Alexander polynomials of equivariant slice and ribbon knots in S^3, Trans. Amer. Math. Soc. 358 (2006), 2949–2964.

[deC99] de Campos, J.E.P.P. Boundary string links, J. Knot Theory Ramif. 8 (1999), 855–878.

[DW90] Del Val, P. and Weber, C. Plans' theorem for links, Top. Appl. 34 (1990), 247–255.

[De81] DeMeo, R. Cobordism of even dimensional links, Duke Math. J. 48 (1981), 23–33.

[deR67] de Rham, G. Introduction aux polynomes d'un noeud, L'Enseign. Math. 13 (1967), 187–194.

[DS78] Dicks, W. and Sontag, E.D. Sylvester domains,
   J. Pure Appl. Alg. 13 (1978), 243–275.

[DOS89] Dror Farjoun, E., Orr, K.E. and Shelah, S. Bousfield localization as an
   algebraic closure of groups, Israel J.M. 66 (1989), 143–153.

[Du74] Durfee, A. Fibred knots and algebraic singularities,
   Topology 13 (1974), 47–59.

[Du75] Durfee, A. The characteristic polynomial of the monodromy,
   Pacific J. Math. 59 (1975), 21–26.

[Du86] Duval, J. Forme de Blanchfield et cobordisme d'entrelacs bords,
   Comment. Math. Helvetici 61 (1986), 617–635.

[Dw75] Dwyer, W.G. Homology, Massey products and maps between
   groups, J. Pure Appl. Alg. 6 (1975), 177–190.

[DF87] Dwyer, W.G. and Fried, D. Homology of free abelian covers, I,
   Bull. London Math. Soc. 19 (1987), 350–352.

[Ed84] Edmonds, A. Least area Seifert surfaces and periodic knots,
   Top. Appl. 18 (1984), 109–113.

[EL83] Edmonds, A. and Livingston, C. Group actions on fibered 3-manifolds,
   Comment. Math. Helvetici 58 (1983), 529–542.

[Er69] Erle, D. Die quadratische Form eines Knotens und ein Satz über Knoten-
   mannigfaltigkeiten, J. Reine Angew. Math. 236 (1969), 174–218.

[Fa77] Farber, M.S. Duality in an infinite cyclic covering and even-dimensional
   knots, Izv. A. N. Ser. Mat. 41 (1977), 794–828 (Russian);
   English translation: Math. USSR Izvestija 11 (1977), 749–781.

[Fa83] Farber, M.S. The classification of simple knots,
   Uspekhi Mat. Nauk 38:5 (1983), 59–106 (Russian);
   English translation: Russian Math. Surveys 38:5 (1983), 61–117.

[Fa91] Farber, M.S. Hermitean forms on link modules,
   Comment. Math. Helvetici 66 (1991), 189–236.

[Fa92] Farber, M.S. Stable homotopy and homology invariants of boundary links,
   Trans. Amer. Math. Soc. 334 (1992), 455–477.

[Fa92'] Farber, M.S. Noncommutative rational functions and boundary links,
   Math. Ann. 293 (1992), 543–568.

[FV92] Farber, M.S. and Vogel, P. The Cohn localization of the free group ring,
   Math. Proc. Cambridge Phil. Soc. 111 (1992), 433–443.

[Fl85] Flapan, E. Infinitely periodic knots,
   Canad. J. Math. 37 (1985), 17–28.

[Fo53] Fox, R.H. Free differential calculus,
   Ann. Math. 57 (1953), 547–560.

[Fo62] Fox, R.H. A quick trip through knot theory,
   in [For], 120–167.

[Fo62'] Fox, R.H. Knots and periodic transformations,

in [**For**], 177–182.

[FS64] Fox, R.H. and Smythe, N.F. An ideal class invariant of knots,
        Proc. Amer. Math. Soc. 15 (1964), 707–709.

[FT54] Fox, R.H. and Torres, G. Dual presentations of the group of a knot,
        Ann. Math. 59 (1954), 211–218.

[Fr88] Freedman, M.H. Whitehead₃ is a "slice" link,
        Invent. Math. 94 (1988), 175–182.

[Fr93] Freedman, M.H. Link compositions and the topological slice problem,
        Topology 32 (1993), 145–156.

[FT95] Freedman, M.H. and Teichner, P. 4-Manifold topology
        I: subexponential groups, Invent. Math. 122 (1995), 509–529;
        II: Dwyer's filtration and surgery kernels, *ibid.* 122 (1995), 531–557.

[Fr87] Fried, D. Homology of free abelian covers, II,
        Bull. London Math. Soc. 19 (1987), 353–358.

[Fr06] Friedl, S. Algorithm for finding boundary link Seifert matrices,
        J. Knot Theory Ramif. **15** (2006), 601–612.

[FV11] Friedl, S. and Vidussi, S. A survey of twisted Alexander polynomials,
        in *The mathematics of knots*, Contrib. Math. Comput. Sci., 1,
        Springer-Verlag, Berlin - Heidelberg - New York (2011), 45–94.

[FV11'] Friedl, S. and Vidussi, S. Twisted Alexander polynomials detect fibered
        3-manifolds, Ann. Math. 173 (2011), 1587–1643.

[Ga87] Gabai, D. Foliations and the topology of 3-manifolds, III,
        J. Diff. Geom. 26 (1987), 479–536.

[GL01] Garoufalidis, S. and Levine J. Homology surgery and invariants of
        3-manifolds, Geom. Topol. 5 (2001), 551–578.

[GL02] Garoufalidis, S. and Levine J. Analytic invariants of boundary links,
        J. Knot Theory Ramif. 11 (2002), 283–293.

[Gi67] Giffen, C.H. On transformations of the 3-sphere fixing a knot,
        Bull. Amer. Math. Soc. 73 (1967), 913–914.

[Gi76] Giffen, C.H. F-isotopy and I-equivalence
        unpublished manuscript, The University of Virginia (1976).

[Gi79] Giffen, C.H. Link concordance implies link homotopy,
        Math. Scandinavica 45 (1979), 243–254.

[Gi93] Gilmer, P.M. Link cobordism in rational homology 3-spheres,
        J. Knot Theory Ramif. 2 (1993), 285–320.

[GL92] Gilmer, P. and Livingston, C. The Casson-Gordon invariant and link
        concordance, Topology 31 (1992), 475–492.

[Go90] Goldschmidt, D.M. Classical link invariants and the Burau
        representation, Pacific J. Math. 144 (1990), 277–292.

[Go74] Goldsmith, D. Homotopy of braids - in answer to a question of E. Artin,
        in *Topology Conference, VPISU (1973),*

(edited by Raymond F. Dickman, Jr, and Peter Fletcher),
Lecture Notes in Mathematics 375,
Springer-Verlag, Berlin - Heidelberg - New York (1974), 91–96.

[Go75] Goldsmith, D. Symmetric fibred links,
in [**Neu**], 3–23.

[Go78] Goldsmith, D. A linking invariant of classical link concordance,
in [**Hau**], 135–170.

[Go79] Goldsmith, D. Concordance implies homotopy for classical links in M^3,
Comment. Math. Helvetici 54 (1979), 347–355.

[GAR99] González-Acuña, F. and Ramírez, A. A composition formula in the
rank two free group, Proc. Amer. Math. Soc. 127 (1999), 2779–2782.

[Go72] Gordon, C.McA. Knots whose branched covers have periodic
homology, Trans. Amer. Math. Soc. 168 (1972), 357–370.

[Go81] Gordon, C. McA. Ribbon concordance of knots in the 3-sphere,
Math. Ann. 257 (1981), 157–170.

[GL78] Gordon, C. McA. and Litherland, R.A. On the signature of a link,
Invent. Math. 47 (1978), 53–70.

[GL79] Gordon, C. McA. and Litherland, R.A. On a theorem of Murasugi,
Pacific J. Math. 82 (1979), 69–74.

[GL84] Gordon, C. McA. and Litherland, R.A. Incompressible surfaces in
branched coverings, in [**BM**], 139–152.

[GLM81] Gordon, C. McA., Litherland, R.A. and Murasugi, K. Signatures of
covering links, Canad. J. Math. 33 (1981), 381–394.

[Gr74] Grima, M.-C. La monodromie rationelle ne determine pas la topologie
d'une hypersurface complexe,
in *Fonctions de Plusieurs Variables Complexes*,
Lecture Notes in Mathematics 409,
Springer-Verlag, Berlin - Heidelberg - New York (1974), 580–602.

[GM86] Groves, J.R.J. and Miller, C.F., III Recognizing free metabelian groups,
Illinois J. Math. 30 (1986), 246–254.

[Gr80] Grünenfelder, L. Lower central series, augmentation quotients and
homology of groups, Comment. Math. Helvetici 55 (1980), 159–177.

[GDC99] Gusein-Zade, S.M., Delgado, F. and Campillo, A. The Alexander
polynomial of plane curve singularities and rings of functions on curves,
Uspekhi Mat. Nauk 54 (1999), 157–8 (Russian);
English translation: Russian Math. Surveys 54 (1999), 634–5.

[Gu72] Gutiérrez, M.A. Boundary links and an unlinking theorem,
Trans. Amer. Math. Soc. 171 (1972), 491–499.

[Gu79] Gutiérrez, M.A. Concordance and homotopy, I: fundamental group,
Pacific J. Math. 82 (1979), 75–91.

[Ha92] Habegger, N. Applications of Morse theory to link theory, in [**Ka1**], 389–394.

[HL90] Habegger, N. and Lin, X.-S. The classification of links up to homotopy, J. Amer. Math. Soc. 2 (1990), 389–419.

[HL98] Habegger, N. and Lin, X.-S. On link concordance and Milnor's $\bar{\mu}$-invariants, Bull. London Math. Soc. 30 (1998), 419–428.

[HO99] Habegger, N. and Orr, K. Milnor link invariants and quantum 3-manifold invariants, Comment. Math. Helvetici 74 (1999), 322–344.

[Ha85] Hain, R. Iterated integrals, intersection theory and link groups, Topology 24 (1985), 45–66.

[Ha79] Hartley, R. Metabelian representations of knot groups, Pacific J. Math. 82 (1979), 93–104.

[Ha80] Hartley, R. Knots and involutions, Math. Z. 171 (1980), 175–185.

[Ha80'] Hartley, R. Invertible amphicheiral knots, Math. Ann. 252 (1980), 103–109.

[Ha81] Hartley, R. Knots with free period, Canad. J. Math. 33 (1981), 91–102.

[Ha83] Hartley, R. The Conway potential function for links, Comment. Math. Helvetici 58 (1983), 365–378.

[Ha83'] Hartley, R. Lifting group homomorphisms, Pacific J. Math. 105(1983), 311–320.

[HM77] Hartley, R. and Murasugi, K. Covering linkage invariants, Canad. J. Math. 29 (1977), 1312–1319.

[HM78] Hartley, R. and Murasugi, K. Homology invariants, Canad. J. Math. 30 (1978), 655–670.

[HK78] Hausmann, J.-C. and Kervaire, M. Sur le centre des groupes de noeuds multidimensionelles, C.R. Acad. Sci. Paris 287 (1978), 699–702.

[Hi78] Hillman, J.A. Alexander ideals and Chen groups, Bull. London Math. Soc. 10 (1978), 105–110.

[Hi81] Hillman, J.A. Blanchfield pairings with squarefree Alexander polynomial, Math. Z. 176 (1981), 551–563.

[Hi81'] Hillman, J.A. New proofs of two theorems on periodic knots, Arch. Math. (Basel) 37 (1981), 457–461.

[Hi81"] Hillman, J.A. Finite knot modules and the factorization of certain simple knots, Math. Ann. 257 (1981), 261–274.

[Hi85] Hillman, J.A. Topological concordance and F-isotopy, Math. Proc. Cambridge Phil. Soc. 98 (1985), 107–110.

[Hi85'] Hillman, J.A. The kernel of the cup product, Bull. Austral. Math. Soc. 32 (1985), 261–274.

[Hi86] Hillman, J.A. Finite simple even-dimensional knots, J. London Math. Soc. 34 (1986), 369–374.

[Hi87] Hillman, J.A. The kernel of integral cup product,
J. Austral. Math. Soc. 43 (1987), 10–15.

[Hi05] Hillman, J.A. Singularities of plane algebraic curves,
Expos. Math. 23 (2005), 233–254.

[HLN06] Hillman, J.A., Livingston, C. and Naik, S. Twisted Alexander
polynomials of plane curves, Alg. Geom. Topol. 6 (2006), 143–167.

[HS97] Hillman, J.A. and Sakuma, M. On the homology of finite abelian
coverings of links, Canad. Bull. Math. 40 (1997), 309–315.

[HSW10] Hillman, J.A., Silver, D.S. and Williams, S.G. On reciprocality of
twisted Alexander invariants, Alg. Geom. Topol. 10 (2010), 1017–1026.

[Hr97] Hironaka, E. Alexander stratifications of character varieties,
Ann. Institut Fourier 47 (1997), 555–583.

[Hr80] Hirschhorn, P.S. Link complements and coherent group rings,
Illinois J. Math. 24 (1980), 159–163.

[Ht79] Hitt, L.R. Examples of higher-dimensional slice knots which are not
ribbon knots, Proc. Amer. Math. Soc. 77 (1979), 291–297.

[HS66] Holmes, R. and Smythe, N.F. Algebraic invariants of isotopy of links,
Amer. J. Math. 88 (1966), 646–654.

[Hs58] Hosokawa, F. On ∇-polynomials of links,
Osaka Math. J. 10 (1958), 273–282.

[Ho84] Hoste, J. The Arf invariant of a totally proper link,
Top. Appl. 18 (1984), 163–177.

[Hu93] Hughes, J. Structured groups and link-homotopy,
J. Knot Theory Ramif. 2 (1993), 37–63.

[IO01] Igusa, K. and Orr, K. Links, pictures and the homology of nilpotent
groups, Topology 40 (2001), 1125–1166.

[JW93] Jiang, B.J. and Wang, S.C. Twisted topological invariants associated
with representations, in *Topics in knot theory (Erzurum, 1992)*, NATO Adv.
Sci. Inst. Ser. C, 399, Kluwer Acad. Publ., Dordrecht (1993), 211–227.

[Ji87] Jin, G.T. On Kojima's η-function of links,
in [**KN**], 14–30.

[KY00] Kadokami, T. and Yasuhara, A. Proper links, algebraically split links
and Arf invariant, J. Math. Soc. Japan 52 (2000), 591–608.

[Ka92] Kaiser, U. Homology boundary links and fusion constructions,
Osaka J. Math. 29 (1992), 573–593.

[Ka92'] Kaiser, U. Geometric properties of strong fusion,
J. Knot Theory Ramif. 1 (1992), 297–302.

[Ka93] Kaiser, U. Homology boundary links and strong fusion,
Kobe J. Math. 10 (1993), 179–188.

[KT76] Kauffman, L.R. and Taylor, L. Signature of links,
Trans. Amer. Math. Soc. 216 (1976), 351–365.

[Kw77] Kawauchi, A. On quadratic forms of 3-manifolds,
Invent. Math. 43 (1977), 177–198.

[Kw78] Kawauchi, A. On the Alexander polynomials of cobordant links,
Osaka J. Math. 15 (1978), 151–159.

[Kw80] Kawauchi, A. On links not cobordant to split links,
Topology 19 (1980), 321–334.

[Kw84] Kawauchi, A. On the Robertello invariants of proper links,
Osaka J. Math. 21 (1984), 81–90.

[Kw86] Kawauchi, A. Three dualities on the integral homology of infinite cyclic
coverings of manifolds, Osaka J. Math. 23 (1986), 633–651.

[Kw02] Kawauchi, A. An intrinsic Arf invariant on a link and its surface-link
analogue, Top. Appl. 121 (2002), 255–274.

[Ke75] Kearton, C. Blanchfield duality and simple knots,
Trans. Amer. Math. Soc. 202 (1975), 141–160.

[Ke75'] Kearton, C. Cobordism of knots and Blanchfield duality,
J. London Math. Soc. 10 (1975), 406–408.

[Ke79] Kearton, C. Signatures of knots and the free differential calculus,
Quart. J. Math. Oxford 30 (1979), 157–182.

[Ke81] Kearton, C. Hermitian signatures and double-null-cobordism of knots,
J. London Math. Soc. 23 (1981), 563–576.

[KW92] Kearton, C. and Wilson, S.M.J. Cyclic group actions and knots,
Proc. Roy. Soc. London A 436 (1992), 527–535.

[Ke65] Kervaire, M.A. Les noeuds de dimensions supérieures,
Bull. Soc. Math. France 93 (1965), 225–271.

[Ke65'] Kervaire, M.A. On higher dimensional knots,
in *Differential and Combinatorial Topology*
(A Symposium in Honor of Marston Morse) (edited by S.S.Cairns),
Princeton University Press, Princeton (1965), 105–109.

[KW78] Kervaire, M.A. and Weber, C. A survey of higher dimensional knots,
in [**Hau**], 61–134.

[Ki78] Kidwell, M.A. On the Alexander polynomials of certain three-component
links, Proc. Amer. Math. Soc. 71 (1978), 351–354.

[Ki78'] Kidwell, M.A. Alexander polynomials of links of small order,
Illinois J. Math. 22 (1978), 459–475.

[Ki79] Kidwell, M.A. On the Alexander polynomials of alternating 2-component
links, Internat. J. Math. Math. Sci. 2 (1979), 229–237.

[Ki58] Kinoshita, S. On knots and periodic transformations,
Osaka Math. J. 10 (1958), 43–52.

[Ki61] Kinoshita, S. On the Alexander polynomials of 2-spheres in a 4-sphere,
Ann. Math. 74 (1961), 518–531.

[KT57] Kinoshita, S. and Terasaka, H. On unions of knots,

Osaka Math. J. 9 (1957), 131–153.

[Ki97] Kirby, R.C. Problems in low-dimensional topology,
in [**Kaz**],Part 2, 35–473.

[KL99] Kirk, P. and Livingston, C. Twisted Alexander invariants, Reidemeister torsion and Casson-Gordon invariants, Topology 38 (1999), 635–662.

[KL99'] Kirk, P. and Livingston, C. Twisted Alexander invariants: inversion, mutation and concordance, Topology 38 (1999), 663–672.

[KLW01] Kirk, P., Livingston, C. and Wang, Z. The Gassner representation for string links, Comm. Contemp. Math. 3 (2001), 87–135.

[Ki96] Kitano, T. Twisted Alexander polynomial and Reidemeister torsion, Pacific J. Math. 174 (1996), 431–442.

[Ki07] Kitayama, T. Normalization of twisted Alexander invariants, arXiv[math.GT]:0705.2371.v4.

[Kl93] Klyachko, A. Funny property of sphere and equations over groups, Comm. Alg. 21 (1993), 2555–2575.

[Ko87] Ko, K.H. Seifert matrices and boundary link concordance, Trans. Amer. Math. Soc. 229 (1987), 657–681.

[KS07] Ko, K.H. and Song, W.T. Seifert matrices of periodic knots, J. Knot Theory Ramif. 16 (2007), no. 1, 45–57.

[Ko82] Kobelskii, V.L. Modules of multidimensional links of codimension two, Zap. LOMI 122 (1982), 109–116 (Russian).

[Ko82'] Kobelskii, V.L. Isotopy classification of odd-dimensional links in codimension two, Izv. A.N. 46 (1982), 983–993 (Russian); English translation: Math. USSR Izvestiya 21 (1983), 281–290.

[Ko89] Kobelskii, V.L. Modules of two-component codimension two links, Algebra i Analiz 1 (1989), 160–171 (Russian); English translation: Leningrad Math J. 1 (1990), 727–739.

[Ko83] Kojima, S. Milnor's $\bar{\mu}$-invariants, Massey products and Whitney's trick in 4 dimensions, Top. Appl. 16 (1983), 43–60.

[KY79] Kojima, S. and Yamasaki, M. Some new invariants of links, Invent. Math. 54 (1979), 213–228.

[KM94] Kronheimer, P.B. and Mrowka, T.S. The genus of embedded surfaces in the projective plane, Math. Res. Letters 1 (1994), 797–808.

[KT98] Krushkal, V.S. Additivity properties of Milnor's $\bar{\mu}$-invariants, J. Knot Theory Ramif. 7 (1998), 625–637.

[KT97] Krushkal, V.S. and Teichner, P. Alexander duality, gropes and link homotopy, Geom. Topol. 1 (1997), 51–69.

[Ku93] Kulikov, V.S. Alexander polynomials of plane algebraic curves, Izv. A.N. 57 (1993), 77–101 (Russian); English translation: Russ. Acad. Sci. Izv. Math. 42 (1994), 67–89.

[Lf71] Laufer, H. Some numerical link invariants,

Topology 10 (1971), 119–131.

[LD92] Le Dimet, J.Y. Enlacements d'intervalles et représentation de Gassner, Comment. Math. Helvetici 67 (1992), 306–315.

[LD98] Le Dimet, J.Y. Représentations du groupe des tresses généralisées dans des groupes de Lie, Manus. Math. 96 (1998), 507–515.

[LD01] Le Dimet, J.Y. Enlacements d'intervalles et torsion de Whitehead, Bull. Soc. Math. France. 129 (2001), 215–235.

[Le72] Lê Dũng Trang, Sur les noeuds algébriques, Compositio Math. 25 (1972), 281–321.

[Le10] Lê Thang Ty Quoc, Homology torsion growth and Mahler measure, arXiv[math.GT]:1010.4199

[Le33] Lehmer, D.H. Factorization of certain cyclotomic functions, Ann. Math. 34 (1933), 461–479.

[Le00] Letsche, C.F. An obstruction to slicing knots using the eta invariant, Math. Proc. Camb. Phil. Soc. 128 (2000), 301–319.

[LV68] Levin, G. and Vasconcelos, W.V. Homological dimensions and Macaulay rings, Pacific J. Math. 25 (1968), 315–323.

[Le65] Levine, J. A characterization of knot polynomials, Topology 4 (1965), 135–141.

[Le66] Levine, J. Polynomial invariants of knots of codimension two, Ann. Math. 84 (1966), 537–554.

[Le67] Levine, J. A method for generating link polynomials, Amer. J. Math. 89 (1967), 69–84.

[Le69] Levine, J. Knot cobordism groups in codimension two, Comment. Math. Helvetici 44 (1969), 229–244.

[Le70] Levine, J. An algebraic classification of some knots of codimension two, Comment. Math. Helvetici 45 (1970), 185–198.

[Le77] Levine, J. Knot modules. I, Trans. Amer. Math. Soc. 229 (1977), 1–50.

[Le82] Levine, J. The module of a 2-component link, Comment. Math. Helvetici 57 (1982), 377–399.

[Le83] Levine, J. Localization of link modules, in [**Lom**], 213–229.

[Le87] Levine, J. Links with Alexander polynomial zero, Indiana Univ. Math. J. 36 (1987), 91–108.

[Le87'] Levine, J. Surgery on links and the $\bar{\mu}$-invariants, Topology 26 (1987), 45–61.

[Le88] Levine, J. Symmetric presentations of link modules, Top. Appl. 30 (1988), 183–198.

[Le88'] Levine, J. An approach to homotopy classification of classical links, Trans. Amer. Math. Soc 306 (1988), 361–387.

[Le88"] Levine, J. The $\bar{\mu}$-invariants of based links,
in [**KN**], 87–103.

[Le89] Levine, J. Link concordance and algebraic closure of groups,
Comment. Math. Helvetici 64 (1989), 236–255;

[Le89'] Levine, J. Link concordance and algebraic closure of groups, II,
Invent. Math. 96 (1989), 571–592.

[Le94] Levine, J. Link invariants via the η-invariant,
Comment. Math. Helvetici 69 (1994), 82–119.

[LMO93] Levine, J., Mio, W. and Orr, K.E. Links with vanishing homotopy
invariant, Comm. Pure Appl. Math. 46 (1993), 213–220.

[Li77] Liang, C.-C. An algebraic classification of some links of codimension two,
Proc. Amer. Math. Soc. 67 (1977), 147–151.

[Li02] Libgober, A. Hodge decomposition of Alexander invariants,
Manus. Math. 107 (2002), 251–269.

[Li97] Lin, X.-S. Power series expansions and invariants of links,
in [**Kaz**], Part 1, 184–202.

[Li01] Lin, X.-S. Representations of knot groups and twisted Alexander
polynomials, Acta Math. Sinica (English series) 17 (2001), 361–380.

[Li81] Lipman, J. and Teissier, B. Pseudorational local rings and a theorem of
Briançon-Skoda about integral closures of ideals,
Michigan Math. J. 28 (1981), 97-116.

[Li90] Lipson, A.S. Link signatures, Goeritz matrices and polynomial invariants,
L'Enseign. Math. 36 (1990), 93–114.

[Li79] Litherland, R.A. Signatures of iterated torus knots,
in *Topology of Low-Dimensional Manifolds* (edited by R.Fenn)
Lecture Notes in Mathematics 722,
Springer-Verlag, Berlin - Heidelberg - New York (1979), 71–84.

[Li84] Litherland, R.A. Cobordism of satellite knots,
in *Four-Manifold Theory* (edited by C.McA.Gordon and R.C.Kirby),
CONM 35, American Mathematical Society,
Providence, R.I. (1984), 327–362.

[Li95] Livingston, C. Lifting representations of knot groups,
J. Knot Theory Ramif. 4 (1995), 225–234.

[Lo84] Long, D. Strongly +amphicheiral knots are algebraically slice,
Math. Proc. Cambridge Phil. Soc. 95 (1984), 309–312.

[Lu80] Lüdicke, U. Darstellungen der Verkettungsgruppe,
Abh. Math. Sem. Hamburg 50 (1980), 232–237.

[Ly50] Lyndon, R.C. Cohomology theory of groups with a single defining relation.
Ann. of Math. 52 (1950), 650–665.

[Mc06] McClure, J.E. On the chain-level intersection pairing for PL manifolds,
Geom. Topol. 10 (2006), 1391–1424.

Correction, Geom. Topol. 13 (2009), 1775–1777.

[Mc02] McMullen, C.T. The Alexander polynomial of a 3-manifold and the Thurston norm on cohomology, Ann. E.N.S. 35 (2002), 153–171.

[MM83] Maeda, T. and Murasugi, K. Covering linkage invariants and Fox's Problem 13, in [**Lom**], 271–283.

[Ma49] Mal'cev, A.I. Nilpotent torsion free groups, Izv. A.N. Ser. Mat. 13 (1949), 201–212 (Russian).

[MP92] Markl, M. and Papadima, S. Homotopy Lie algebras and fundamental groups via deformation theory, Ann. Inst. Fourier 42 (1992), 905–935.

[MP93] Markl, M. and Papadima, S. Moduli spaces for fundamental groups and link invariants derived from the lower central series, Manus Math. 81 (1993), 225–242.

[Ma68] Massey, W.S. Higher order linking numbers, preprint, University of Chicago (1968); published in J. Knot Theory Ramif. 7 (1998), 393–419.

[Ma80] Massey, W.S. Completion of link modules, Duke Math. J. 47 (1980), 399–420.

[Ma77] Matumoto, T. On the signature invariants of a nonsingular complex sesquilinear form, J. Math. Soc. Japan 29 (1977), 67–71.

[MM82] Mayberry, J.P. and Murasugi, K. Torsion-groups of abelian coverings of links, Trans. Amer. Math. soc. 271 (1982), 143–173.

[MS86] Meeks, W.H.,III and Scott, P. Finite group actions on 3-manifolds, Invent. Math. 86 (1986), 287–346.

[MY81] Meeks, W.H., III and Yau, S.-T. Equivariant Dehn's lemma and loop theorem, Comment. Math. Helvetici 56 (1981), 225–239.

[MT96] Meng, G. and Taubes, C.H. $SW = $ Milnor torsion, Math. Research Letters 3 (1996), 661–674.

[Mi54] Milnor, J.W. Link groups, Ann. Math. 59 (1954), 177–195.

[Mi57] Milnor, J.W. Isotopy of links, in *Algebraic Geometry and Topology*, a Symposium in Honour of S.Lefshetz (edited by R.H.Fox, D.C.Spencer and W.Tucker), Princeton University Press, Princeton (1957), 280–306.

[Mi62] Milnor, J.W. A duality theorem for Reidemeister torsion, Ann. Math. 76 (1962), 137–147.

[Mi66] Milnor, J.W. Whitehead torsion, Bull. Amer. Math. Soc. 72 (1966), 358–426.

[Mi68] Milnor, J.W. Infinite cyclic coverings, in *Conference on the Topology of Manifolds* (edited by J.G.Hocking), Prindle, Weber and Schmidt, Boston - London - Sydney (1968), 115–133.

[Mi69] Milnor, J.W. On isometries of inner product spaces,
Invent. Math. 8 (1969), 83–94.

[Mi87] Mio, W. On boundary link cobordism,
Math. Proc. Cambridge Phil. Soc. 101 (1987), 259–266.

[Mo75] Montesinos, J.M. Surgery on links and double branched covers of S^3,
in [**Neu**], 227–259.

[Mo07] Morifuji, T. A Torres condition for twisted Alexander polynomials,
Publ. Res. Inst. Math. Sci. 43 (2007), 143–153.

[Mu65] Murasugi, K. On the center of the group of a link,
Proc. Amer. Math. Soc. 16 (1965), 1052–1057.
Corrigendum *ibid.* 18 (1967), 1142.

[Mu65'] Murasugi, K. On a certain numerical invariant of link types,
Trans. Amer. Math. Soc. 117 (1965), 387–422.

[Mu66] Murasugi, K. On Milnor's invariant for links,
Trans. Amer. Math. Soc. 124 (1966), 94–110.

[Mu70] Murasugi, K. On Milnor's invariant for links. II. The Chen group,
Trans. Amer. Math. Soc. 148 (1970), 41–61.

[Mu70'] Murasugi, K. On the signature of links,
Topology 9 (1970), 283–298.

[Mu71] Murasugi, K. On periodic knots,
Comment. Math. Helvetici 46 (1971), 162–174.

[Mu80] Murasugi, K. On symmetry of knots,
Tsukuba J. Math. 4 (1980), 331–347.

[Mu80'] Murasugi, K. (letter to J.A.Hillman, January 1980)

[Mu85] Murasugi, K. Nilpotent coverings of links and Milnor's invariant,
in [**Fen**], 106–142.

[Mu85'] Murasugi, K. On the height of 2-component links,
Top. Appl 19 (1985), 227–243.

[Mu85"] Murasugi, K. Polynomial invariants of 2-component links,
Rev. Mat. Iberoamericana 1 (1985), 121–144.

[Mu86] Murasugi, K. Milnor's $\bar{\mu}$-invariant and 2-height of reducible plane curves,
Arch. Math. (Basel) 46 (1986), 466–472.

[Na78] Nakagawa, Y. On the Alexander polynomials of slice links,
Osaka J. Math. 15 (1978), 161–182.

[Na94] Naik, S. Periodicity, genera and Alexander polynomials of knots,
Pacific J. Math. 166 (1994), 357–371.

[Ni92] Nicolau, M. Semifree actions on spheres,
Pacific J. Math. 156 (1992), 337–358.

[Ne88] Neto, O.M. Total linking number modules,
Trans. Amer. Math. Soc. 307 (1988), 503–533.

[Og99] Ogusa, E. Link cobordism and the intersection of slice discs,

Bull. London Math. Soc. 31 (1999), 729–736.

[Oh82] Ohkawa, T. The pure braid group and the Milnor $\bar{\mu}$-invariants,
Horishima Math. J. 12 (1982), 485–489.

[O'N79] O'Neill, E.J. Higher order Massey products and links,
Trans. Amer. Math. Soc. 248 (1979), 37–46.

[Or87] Orr, K.E. New link invariants and applications,
Comment. Math. Helvetici 62 (1987), 542–560.

[Or89] Orr, K.E. Homotopy invariants of links,
Invent. Math. 95 (1989), 379–394.

[Or91] Orr, K.E. Link concordance invariants and Massey products,
Topology 30 (1991), 699–710.

[OZ81] Osborne, R.P. and Zieschang, H. Primitives in the free group on two
generators, Invent. Math. 63 (1981), 17–24.

[Pa97] Papadima, S. Campbell-Hausdorff invariants of links,
Proc. London Math. Soc. 75 (1997), 641–670.

[Pa01] Papadima, S. Braid commutators and homogeneous Campbell-Hausdorff
tests, Pacific J. Math. 197 (2001), 383–416.

[Pa02] Papadima, S. Generalized $\bar{\mu}$-invariants for links and hyperplane
arrangements, Proc. London Math. Soc. 84 (2002), 492–512.

[Pl53] Plans, A. Apportación al estudio de los grupos de homología de los
recubrimientos cíclicos ramificados correspondientes a un nudo,
Rev. real. Acad. Cienc. exact. Fis. Natur. Madrid 47 (1953), 161–193.

[Pl86] Platt, M.L. Insufficiency of Torres' conditions for two-component links,
Trans. Amer. Math. Soc. 296 (1986), 125–136.

[Pl88] Platt, M.L. Alexander ideals of links with all linking numbers 0,
Trans. Amer. Math. Soc. 306 (1988), 597–605.

[Po71] Poenaru, V. A note on the generators for the fundamental group of the
complement of a submanifold of codimension 2, Topology 10 (1971), 47–52.

[Po80] Porter, R. Milnor's $\bar{\mu}$-invariants and Massey products,
Trans. Amer. Math. Soc. 257 (1980), 39–71.

[Po04] Porti, J. Mayberry-Murasugi's formula for links in homology 3-spheres,
Proc. Amer. Math. Soc. 132 (2004), 3423–3431.

[RRV99] Retakh, V., Reutenauer, C. and Vaintrob, A. Noncommutative rational
functions and Farber's invariants of boundary links, in *Differential topology,
infinite-dimensional Lie algebras, and applications*, Amer. Math. Soc. Transl.
Ser. 2, 194, Amer. Math. Soc., Providence, RI (1999), 237–246.

[Ri72] R. Riley, Parabolic representations of knot groups, I,
Proc. London Math. Soc. 24 (1972), 217–242.

[Ri71] Risler, J.-J. Sur l'idéal jacobien d'une courbe plane,
Bull. Soc. Math. France 99 (1971), 305–311.

[Ro65] Robertello, R. An invariant of knot cobordism,

Comm. Pure Appl. Math. 18 (1965), 543–555.

[Ro72] Rolfsen, D. Isotopy of links in codimension two,
J. Indian Math. Soc. 36 (1972), 263–278.

[Ro74] Rolfsen, D. Some counterexamples in link theory,
Canad. J. Math. 26 (1974), 978–984.

[Ro75] Rolfsen, D. Localized Alexander invariants and isotopy of links,
Ann. Math. 101 (1975), 1–19.

[Ro85] Rolfsen, D. Piecewise linear I-equivalence of links,
in [**Fen**], 161–178.

[Ru80] Rubinstein, J.H. Dehn's lemma and handle decompositions of some
4-manifolds, Pacific J. Math. 86 (1980), 565–569.

[Sa90] Sabbah, C. Modules d'Alexander et $\mathcal{D}$-modules,
Duke Math. J. 60 (1990), 729–814.

[Sa71] Saito, K. Quasihomogene isolierte Singularitäten von Hyperflächen,
Invent. Math. 14 (1971), 123-142.

[Sa77] Sakai, T. A remark on the Alexander polynomials of knots,
Math. Sem. Notes Kobe University 5 (1977), 451–456.

[Sa83] Sakai, T. Polynomials of invertible knots,
Math. Ann. 266 (1983), 229–232.

[Sk79] Sakuma, M. The homology groups of abelian coverings of links,
Math. Sem. Notes Kobe University 7 (1979), 515–530.

[Sk81] Sakuma, M. Periods of composite links,
Math. Sem. Notes Kobe University 9 (1981), 445–452.

[Sk81'] Sakuma, M. On the polynomials of periodic links,
Math. Ann. 257 (1981), 487–494.

[Sa56] Samuel, P. Algébricité de certains points singuliers algébroides,
J. Math. Pures Appl. 35 (1956), 1-6.

[Sa81] Sato, N. Alexander modules of sublinks and an invariant of classical link
concordance, Illinois J. Math. 25 (1981), 508–519.

[Sa81'] Sato, N. Free coverings and modules of boundary links,
Trans. Amer. Math. Soc. 264 (1981), 499–505.

[Sa81"] Sato, N. Algebraic invariants of boundary links,
Trans. Amer. Math. Soc. 265 (1981), 359–374.

[Sa84] Sato, N. Alexander modules,
Proc. Amer. Math. Soc. 91 (1984), 159–162.

[Sa86] Sato, N. Boundary link modules I: below the middle dimension,
preprint, Texas A. and M. University (1986).

[Sc85] Scharlemann, M.G. Unknotting number 1 knots are prime,
Invent. Math. 82 (1985), 37–55.

[Sc80] Scherk, J. On the monodromy theorem for isolated hypersurface
singularities, Invent. Math. 58 (1980), 289-301.

[Se70] Sebastiani, M. Preuve d'une conjecture de Brieskorn,
Manus. Math. 2 (1970), 301-308.

[Se34] Seifert, H. Uber das Geschlecht von Knoten,
Math. Ann. 110 (1934), 571–592.

[SW99] Silver, D.S. and Williams, S.G. Knot invariants from symbolic dynamical
systems, Trans. Amer. Math. Soc. 351 (1999), 3243–3265.

[SW99'] Silver, D.S. and Williams, S.G. Periodic links and augmented groups,
Math. Proc. Cambridge Phil. Soc. 127 (1999), 217–236.

[SW02] Silver, D.S. and Williams, S.G. Mahler measure, links and homology
growth, Topology 41 (20020, 979–991.

[SW02'] Silver, D.S. and Williams, S.G. Torsion numbers of augmented groups,
with applications to knots and links, L'Enseign. Math 48 (2002), 317–343.

[SW06] Silver, D.S. and Williams, S.G. Twisted Alexander polynomials detect
the unknot, Alg. Geom. Topol. 6 (2006) 1893–1901.

[Si80] Simon, J. Wirtinger approximations and the knot groups of F^n in S^{n+2},
Pacific J. Math. 90 (1980), 177–190.

[Sm89] Smolinsky, L. A generalization of the Levine-Tristram invariant,
Trans. Amer. Math. Soc. 315 (1989), 205–217.

[Sm89'] Smolinsky, L. Invariants of link cobordism,
Top. Appl. 32 (1989), 161–168.

[Sm66] Smythe, N.F. Boundary links,
in *Topology Seminar, Wisconsin 1965* (edited by R.H.Bing),
Annals of Math. Study 60, Princeton University Press,
Princeton (1966), 69–72.

[Sm67] Smythe, N.F. Isotopy invariants of links and the Alexander matrix,
Amer. J. Math. 89 (1967), 693–704.

[Sm70] Smythe, N.F. n-linking and n-splitting,
Amer. J. Math. 92 (1970), 272–282.

[Sq84] Squier, C. The Burau representation is unitary,
Proc. Amer. Math. Soc. 90 (1984), 199–202.

[St62] Stallings, J. On fibering certain 3-manifolds,
in [**For**], 95–100.

[St65] Stallings, J. Homology and central series of groups,
J. Algebra 2 (1965), 170–181.

[St89] Stein, D. Computing Massey products of links,
Top. Appl. 32 (1989), 169–181.

[St90] Stein, D. Massey products in the cohomology of groups with applications
to link theory, Trans. Amer. Math. Soc. 318 (1990), 301–325.

[St12] Steinitz, E. Rechteckige Systeme und moduln in algebraische Zahlkörpern.
I, Math. Ann. 71 (1912), 328–354.

[St00] Stevens, W. Recursion formulas for some abelian knot invariants,

J. Knot Theory Ramif. 9 (2000), 413–422.

[St79] Stolzfus, N.W. Equivariant concordance of invariant knots,
Trans. Amer. Math. Soc. 254 (1979), 1–45.

[Su77] Sullivan, D. Infinitesimal computations in topology,
Publ. Math. IHES 47 (1977), 269–332.

[Su74] Sumners, D.W. On the homology of finite cyclic coverings of higher-dimensional links, Proc. Amer. Math. Soc. 46 (1974), 143–149.

[Su75] Sumners, D.W. Smooth Z_p-actions on spheres which leave knots pointwise fixed, Trans. Amer. Math. Soc. 205 (1975), 193–203.

[SW77] Sumners, D.W. and Woods, J. The monodromy of reducible plane curves,
Invent. Math. 40 (1977), 107–141.

[Sw77] Swarup, G.A. Relative version of a theorem of Stallings,
J. Pure Appl. Alg. 11 (1977), 75–82.

[Te59] Terasaka, H. On null-equivalent knots,
Osaka J. Math. 11 (1959), 95–113.

[To53] Torres, G. On the Alexander polynomial,
Ann. Math. 57 (1953), 57–89.

[Tr82] Traldi, L. A generalization of Torres' second relation,
Trans. Amer. Math. Soc. 269 (1982), 593–610.

[Tr82'] Traldi, L. The determinantal ideals of link modules. I,
Pacific J. Math. 101 (1982), 215–222. II, *ibid.* 109 (1983), 237–245.

[Tr83] Traldi, L. Linking numbers and the elementary ideals of links,
Trans. Amer. Math. Soc. 275 (1983), 309–318.

[Tr84] Traldi, L. Milnor's invariants and completions of link modules,
Trans. Amer. Math. Soc. 284 (1984), 401–424.

[Tr85] Traldi, L. On the Goeritz matrix of a link,
Math. Z. 188 (1985), 203–213.

[Tr69] Tristram, A.G. Some cobordism invariants for links,
Math. Proc. Cambridge Phil. Soc. 66 (1969), 251–264.

[Tr61] Trotter, H.F. Periodic automorphisms of groups and knots,
Duke Math. J. 28 (1961), 553–557.

[Tr62] Trotter, H.F. Homology of group systems with applications to knot theory,
Ann. Math. 76 (1962), 464–498.

[Tr73] Trotter, H.F. On S-equivalence of Seifert matrices,
Invent. Math. 20 (1973), 173–207.

[Tr77] Trotter, H.F. Knot modules and Seifert matrices,
in [**Hau**], 291–299.

[Tu76] Turaev, V.G. The Milnor invariants and Massey products,
Zap. Nauchn. Sem. Leningrad Otdel. Mat. Inst. Steklov. (LOMI)
66 (1976), 189–203 (Russian);
English translation: J. Sov. Math. 66 (1976), 189–203.

[Tu86] Turaev, V.G. Reidemeister torsion in knot theory,
Uspekhi Mat. Nauk 41 (1986), 97–147 (Russian);
English translation: Russian Math. Surveys 41 (1986), 119–182.

[Tu88] Turaev, V.G. On Torres-type relations for the Alexander polynomials of
links, L'Enseign. Math. 34 (1988), 69–82.

[Tu89] Turaev, V.G. Elementary ideals of links and manifolds: symmetry and
asymmetry, Algebra i Analiz 1 (1989), 223–232 (Russian);
English translation: Leningrad Math. J. 1 (1990), 1279–1287.

[Tu98] Turaev, V.G. A combinatorial formulation for the Seiberg-Witten
invariants of 3-manifolds, Math. Res. Letters 5 (1998), 583–598.

[Va79] Van Buskirk, J. The converse of Kawauchi's amphicheiral knot conjecture,
Notices Amer. Math. Soc. (1979), abstract 768-55-2.

[Vi73] Viro, O.G. Branched coverings of manifolds with boundary, and invariants
of links, Izv. A.N. Ser. Mat. 37 (1973), 1241-1258 (Russian);
English translation: Math. USSR Izvestija 7 (1973), 1239–1256.

[Vo88] Vogel, P. 2 × 2-Matrices and application to link theory,
in *Algebraic Topology and Transformation Groups*,
Proceedings, Göttingen 1987, Lecture Notes in Mathematics 1361,
Springer-Verlag, Berlin - Heidelberg - New York (1988), 269–298.

[Vo92] Vogel, P. Representations of link groups,
in [**Ka1**], 381–387.

[Wa94] Wada, M. Twisted Alexander polynomial for finitely presentable groups,
Topology 33 (1994), 241–256.

[Wa72] Waldi, R. *Wertehalbgruppe und Singularitäte einer ebene algebraischen
Kurve*, Dissertation, Regensburg (1972).

[Wa80] Wanna, S. A spectral sequence for group presentations with applications
to links, Trans. Amer. Math. Soc. 261 (1980), 271–285.

[We80] Weber, C. Sur une formule de R.H.Fox concernant l'homologie d'une
revêtement ramifié, L'Enseign. Math. 25 (1980), 261–272.

[Wh69] Whitten, W. Symmetries of links,
Trans. Amer. Math. Soc. 135 (1969), 213–222.

[Wh50] Whitehead, J.H.C. Simple homotopy types,
Amer. J. Math. 72 (1950), 1–57.

[Yj64] Yajima, T. On simply knotted spheres,
Osaka J. Math. 1 (1964), 133–152.

[Yj69] Yajima, T. On a characterization of knot groups of some knots in $\mathbb{R}^4$,
Osaka Math. J. 6 (1969), 435–446.

[Ym84] Yamamoto, M. Classification of isolated algebraic singularities by their
Alexander polynomials, Topology 23 (1984), 277–287.

[Yn69] Yanagawa, T. On ribbon 2-knots. The 3-manifold bounded by the 2-knots,
Osaka J. Math. 6 (1969), 447–464.

[Yn70] Yanagawa, T. On ribbon 2-knots III. On the unknotting ribbon 2-knots
in S^4, Osaka J. Math. 7 (1970), 165–172.

[Yo81] Yoshikawa, K. On fibering a class of n-knots,
Math. Sem. Notes Kobe Univ. 9 (1981), 241–245.

Index

SERIES ON KNOTS AND EVERYTHING

Editor-in-charge: Louis H. Kauffman *(Univ. of Illinois, Chicago)*

The Series on Knots and Everything: is a book series polarized around the theory of knots. Volume 1 in the series is Louis H Kauffman's Knots and Physics.

One purpose of this series is to continue the exploration of many of the themes indicated in Volume 1. These themes reach out beyond knot theory into physics, mathematics, logic, linguistics, philosophy, biology and practical experience. All of these outreaches have relations with knot theory when knot theory is regarded as a pivot or meeting place for apparently separate ideas. Knots act as such a pivotal place. We do not fully understand why this is so. The series represents stages in the exploration of this nexus.

Details of the titles in this series to date give a picture of the enterprise.

*The complete list of the published volumes in the series can also be found at
http://www.worldscibooks.com/series/skae_series.shtml